# Praise for *The Permaculture City*



"Half the world's people now live in cities, and as Toby Hemenway convincingly demonstrates, they can be at the very forefront of the revolution in how we live. This book will thrill you!"

—**Bill McKibben**, author of *Deep Economy*

"Many people who are searching for a more fulfilling life, wanting to reduce their ecological footprint and build resilience for uncertain futures, grasp that permaculture might be part of the solution but are often unsure how it applies to their particular situation. For residents of towns and cities in the modern affluent world, *The Permaculture City* shows how permaculture design makes common sense."

—**David Holmgren**, co-originator of the permaculture concept

"I'm someone with a strong bias toward country living and I've always thought that the phrase 'urban permaculture' is oxymoronic. I've often thought master planners should be working to revive small towns, not build more cities. Toby Hemenway has shown me the error of my ways. The function of a well-conceived city, he says, is to inspire. His book inspires."

—**Albert Bates**, president, Global Village Institute for Appropriate Technology

"What a great, accessible, and timely book! *The Permaculture City* is a must-read for anyone who loves where they live, wishes to deepen their relationship with and pleasure in it, and realizes that our food, water, and community resilience may depend upon it. Toby Hemenway offers great guidance for applying the lens of intentional design to increasing food self-reliance (and pleasure!), improving water efficiency and usage, and growing community, three elements that promise to improve the quality of relatedness to place, as well as resilience in the face of weather and other uncertainties.

"Whether the topic is gardening in a window box or creating a community garden, catching and channeling rainwater, or redesigning an edible landscape around a suburban home, this timely book offers everyone a window into the joy and long-term fulfillment of permaculture."

—**Nina Simons**, cofounder of Bioneers and founder of Everywoman's Leadership

"Toby Hemenway is among the true visionaries who can turn vision into practical action. *The Permaculture City* is a landmark book that will be used for decades as a compass and field guide to regenerate our world and communities. Toby depicts the virtuous circle people are already creating across the country and world, from small acts an individual can take, to larger systemic changes that only communities and societies can make. This is the gospel of building resilience from the ground up, and Toby is a true hero of our age—he shows us we're all invited to the party."

—**Kenny Ausubel**, cofounder of Bioneers

"*The Permaculture City* is a triumph in bringing the wisdom of permaculture practices to city dwellers. This book is a 'bridge book' for greening our urban landscapes. Rich in practical knowledge, Toby Hemenway is a trailblazer in demystifying the art of living sustainably within ecosystems: teaching how YOU can be a collaborative partner in a healthy urban biosphere. This book's impact will be increasingly significant as we inevitably march toward living in built environments. For urban planners, architects, green builders, and simply citizens who want to enjoy a higher quality of life, *The Permaculture City* is The Book to lead the way."

—**Paul Stamets**, author of *Mycelium Running*

"Toby Hemenway combines the skill of a storyteller with the vigor of experience and insight. He shows us an urban landscape with gardens, food, energy systems, and architecture that can ensure genuine sustainability. Beyond these vital elements, he also creates a template for a new kind of city: a human-scale collection of village communities where quality of life is valued above quantity of output. With the majority of the human race becoming city dwellers, this is vital information for a more collaborative, intelligent, and resilient urban landscape, one that will enable us to face serious challenges now, and in the future."

—**Maddy Harland**, cofounder and editor of *Permaculture* magazine

"We stand at a crossroads where long-held societal beliefs are shattered and assumptions fall before knowledge and reality. Human civilization is in the balance. Few people have grasped the significance of the moment in the way that Toby Hemenway has. His voice is particularly important at this time. *The Permaculture City* is his attempt to understand what we are confronted with and to skew the discussion toward sustainability. There is no doubt that his message is timely and relevant. Read this book like the life of your children depended on it . . . because it does."

—**John D. Liu**, director of the Environmental Education Media Project and visiting fellow at the Netherlands Institute of Ecology

"Toby Hemenway's *Gaia's Garden* is the go-to book I always recommend for those interested in permaculture. His new book, *The Permaculture City*, is the much-needed urban version, a great introduction and full of important information on adapting permaculture to an urban environment."

—**Starhawk**, permaculture designer and author of *The Empowerment Manual*

"Permaculture is applied ecology, and its practice is evolving as society becomes more urbanized. As Toby Hemenway puts it, 'We're not just gardening plants but people, neighborhoods, and even cultures.' Whether you're new to permaculture or a seasoned 'permie,' *The Permaculture City* is essential: It captures the explorative state of the art in readable, often delightful, prose. And, like all good permaculture books, it is eminently helpful at solving a myriad of practical problems in the home and garden."

—**Richard Heinberg**, senior fellow, Post Carbon Institute

# The Permaculture City

# The Permaculture City

## Regenerative Design for Urban, Suburban, and Town Resilience

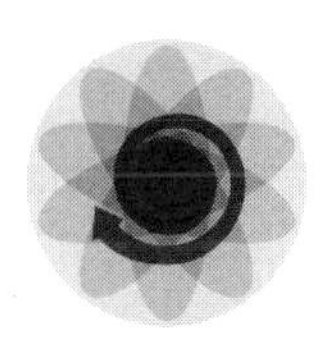

**TOBY HEMENWAY**

Chelsea Green Publishing
White River Junction, Vermont

Project Manager: Patricia Stone
Developmental Editor: Makenna Goodman
Copy Editor: Alice Colwell
Proofreader: Eileen M. Clawson
Indexer: Linda Hallinger
Designer: Melissa Jacobson
Page Composition: Abrah Griggs

Printed in the United States of America.
First printing July, 2015.
10 9 8 7 6 5 4 3 2 1    15 16 17 18

Chelsea Green Publishing is committed to preserving ancient forests and natural resources. We elected to print this title on 100-percent postconsumer recycled paper, processed chlorine-free. As a result, for this printing, we have saved:

**125 Trees (40' tall and 6-8" diameter)**
**56 Million BTUs of Total Energy**
**10,766 Pounds of Greenhouse Gases**
**58,392 Gallons of Wastewater**
**3,909 Pounds of Solid Waste**

Chelsea Green Publishing made this paper choice because we and our printer, Thomson-Shore, Inc., are members of the Green Press Initiative, a nonprofit program dedicated to supporting authors, publishers, and suppliers in their efforts to reduce their use of fiber obtained from endangered forests. For more information, visit: www.greenpressinitiative.org.

Environmental impact estimates were made using the Environmental Defense Paper Calculator. For more information visit: www.papercalculator.org.

**Our Commitment to Green Publishing**
Chelsea Green sees publishing as a tool for cultural change and ecological stewardship. We strive to align our book manufacturing practices with our editorial mission and to reduce the impact of our business enterprise in the environment. We print our books and catalogs on chlorine-free recycled paper, using vegetable-based inks whenever possible. This book may cost slightly more because it was printed on paper that contains recycled fiber, and we hope you'll agree that it's worth it. Chelsea Green is a member of the Green Press Initiative (www.greenpressinitiative.org), a nonprofit coalition of publishers, manufacturers, and authors working to protect the world's endangered forests and conserve natural resources. *The Permaculture City* was printed on paper supplied by Thomson-Shore that contains 100% postconsumer recycled fiber.

**Library of Congress Cataloging-in-Publication Data**
Hemenway, Toby, 1952- author.
The permaculture city : regenerative design for urban, suburban, and town resilience / Toby Hemenway.
pages cm
Includes bibliographical references and index.
ISBN 978-1-60358-526-2 (pbk.)—ISBN 978-1-60358-527-9 (ebook)
1. Permaculture. 2. Urban gardening. I. Title.

S494.5.P47H46 2015
631.5'8—dc23

2015009981

Chelsea Green Publishing
85 North Main Street, Suite 120
White River Junction, VT 05001
(802) 295-6300
www.chelseagreen.com

To the Wednesday Back Alley Group
and for Kiel

# Contents

# INTRODUCTION

# Looking at Cities through a Permaculture Lens

The word "permaculture" evokes images of verdant vegetable polycultures erupting beneath laden fruit trees, of herb spirals and keyhole garden beds, of swales snaking along contour lines to moisten soil and build humus, and of mulch, deep layers of mulch everywhere. Thus it's tempting to think that urban permaculture is simply ecological gardening that's been scaled down small and tight, a specialized subset of high-intensity growing methods designed to coax food from the few unpaved niches missed by relentless urban infilling and building. Indeed, cities do pose special challenges to food growing and habitat preservation for which permaculture offers solutions. But as Jane Jacobs observes in her classic manifesto on the livable metropolis, *The Death and Life of Great American Cities*, thinking of urban ecology as an impoverished shadow of wild nature, where pigeons battle rats for dominance, misses the vibrant ecosystem that suffuses all cities: the human one.[1] Along the same lines, urban permaculture is only slightly about gardening, and mostly about people. The human ecosystem that is the city is rich, and it includes much more than food. To understand, work within, and enhance that ecosystem, we need to understand not just how we feed ourselves in cities and towns but how we meet all our needs: How do we build, move about, use water and energy, feel secure, make decisions, solve problems, sustain ourselves, develop policies, live together?

This book documents the rise of a new sophistication, depth, and diversity in the approaches and thinking of permaculture designers and practitioners. Within it are stories and examples of hope and opportunity for our cities, towns, and suburbs; the people who live in them; and the wild and tame nature that is so heavily impacted by them. Urban permaculture takes what we have learned in the garden and applies it to a much broader range of human experience. We're not just gardening plants but people, neighborhoods, and even cultures.

Permaculture is applied ecology; that is, it is a design approach based on finding and applying to our own creations some of the guiding axioms at work in natural ecosystems. We search for the principles that generate life's resilience, immense productivity, diversity, interconnectednesss, and elegance. Permaculturists are also learning how nature does that so well while powering itself with sunlight; working at ambient, comfortable temperatures; assembling nontoxic materials via life-friendly processes into an ever-evolving and deepening webwork; and recycling all of it in ways that spur yet more diversity and opportunities. Life's wisdom can teach us to build sorely needed replacements for industrial production's murderous conditions, nonrenewable fuels, and toxic, landfill-clogging products. But nature can do more than improve how we make things. It can also teach us how to cooperate, make decisions, and arrive at good solutions.

This means that permaculturists can come from all walks of life. In the early days of the discipline, permaculture attracted gardeners, farmers, landscape designers, and nascent permaculture teachers. But because permaculture's concepts can be applied in so many ways, permaculturists today include software designers; water, waste, and energy engineers; social justice activists; educators and school administrators from the kindergarten level to graduate studies; community organizers and government officials; restoration ecologists; teachers of yoga, bodywork, and spiritual practice; in short, almost anyone. As permaculture teacher Larry Santoyo says, it's not that we "do" permaculture, but rather that we use permaculture in what we do—whether that is farming, law, medicine, science, or accounting. Permaculture has something to offer everyone. All humans, because we plan, dream, and manipulate the environment with our hands and brains, are designers of some sort. Permaculture is a universal design tool; thus each of us can benefit from using it. Nature has tricks to teach everyone.

The most straightforward, easily comprehended way to learn nature's approach is by mimicking the form and processes of the natural environment. Thus permaculturists took first to the garden because it offered the most obvious nursery in which to learn natural-systems design. Want to create a human-designed yet still healthy ecosystem? Start with the stuff of natural ones: plants, soil, water, and sunlight. When we use the ingredients that life has worked wonders with for four billion years, following what we know of nature's design rules, we will be aided by beneficial creatures and processes that show up spontaneously to help stitch together our crude constructions into functioning microecosystems. Our efforts in the garden and farm are the baby steps that permaculturists have been taking to learn how nature operates: plant guilds and polycultures, rotational and sequential grazing, tree crops on contour, water harvesting, food forests, and building soil from the top down.

After thirty years we're starting to know what we're doing. Thousands of nature-mimicking, food-filled yards and hundreds of farms have been designed and built using permaculture's toolkit. Permaculture has been used to design tiny container gardens on apartment balconies, modest (but inspiring) plots in community gardens, microfood forests in narrow urban backyards, food-producing wildlife gardens in suburbia, and productive farms out in the country.

And now we're realizing that the same rules and strategies that apply to understanding and designing living ecosystems for the home and farm hold true for many other human endeavors. Permaculture design applies broadly to many forms of what we have come to call complex adaptive systems. These types of systems include not just ecosystems but brains and nervous systems, businesses and economic networks, communities and neighborhoods, legal and social systems, and a host of other multicomponent, interconnected, flexible, and responsive systems that all follow a similar set of governing principles and have many general properties in common. What permaculturists have learned in the garden is precisely what systems theorists, ecologists, neuroscientists, economists, and many others have been learning in the lab, field, and office: When many parts are assembled so they can interact and influence each other, new properties emerge, such as self-regulation, feedback loops, self-organization, and resilience. We are beginning to understand how that happens and how we can apply what we know about these complex adaptive systems to our own designs.

The relevance of these systems to this book is this: Urban permaculture is vastly more than gardening in the city. What permaculturists have learned by applying whole-systems thinking to food production and habitat restoration has profound and immediate application to the human ecology of the urban environment. So this book cannot only be about gardening; in fact, only about one-quarter of it is. That's because the technical problems of food growing are just a fraction of the challenges confronting us in our essential and increasingly urgent task of learning how to coexist with a finite planet and with each other. Currently we are failing at that task. But I'm renewed and inspired by the growing realization, both my own and that of countless other ecologically minded people, that the principles, strategies, and methods that have been organized under the heading of permaculture and refined in gardens and farms apply directly to designing and working not only with energy, water, and waste systems and other parts of the built environment but also with what we call the invisible structures: businesses, currencies, and economies; communities, families, and other human groups; legal, justice, and decision-making processes; and many other intangible aspects of our culture.

This book is themed loosely around my own modification of David Holmgren's permaculture flower, in which each petal of the flower represents a basic human need that must be met in a sustainable—or, better yet, regenerative—way if we are to build an equitable, ecologically sound culture.[2] This flower pattern of Holmgren's brilliantly encapsulates many of permaculture's core concepts in a single graphic, and it's become a permaculture icon. The flower expresses three central ideas. The first is that permaculture's ethics, principles, and methods can be applied not just to gardens but to all essential needs. The second is that in order to create a sustainable culture, we need to address all of these needs, and the flower is an iconic way to keep them all in front of us. The third is that we must meet these needs not just at the personal level—we don't just stock up food in our bunker—but at the local and regional levels as well. Those needs include the physical ones, such as food, energy, and water, but equally important the nonphysical ones, such as community and livelihood.

In a sense, each of the physical petals of the permaculture flower tells us *what* to do: grow food, harvest water, rely on the sun. We call these the visible structures. The nonphysical petals—the invisible structures of our culture—tell us *how* to do it: share the work, include stakeholders, help people feel secure, build social capital. It is probably occurring to you now that the what is the much easier part. Planting a seed

## What Is Permaculture?

The first glimmerings of permaculture arose in the mind of Bill Mollison, a charismatic and iconoclastic Australian. Mollison grew up in the wildlands of Tasmania and worked over the years as a forester, trapper, and guide and later as a field biologist, researcher, and college instructor. While doing fieldwork in 1959, inspired by the complex interactions of marsupials browsing in the Tasmanian forest, Mollison scratched a seminal note into his journal: "I believe that we could build systems that would function as well as this one does." Over the next twelve years, he fleshed out that idea. Based on his observations of natural ecosystems as well as indigenous cultures, he identified key principles and patterns of sustainable design. In 1972, while a tutor at the University of Australia, he met David Holmgren, a student at a nearby college who grasped the importance of what Mollison was working on. Holmgren wrote an undergraduate thesis that eventually, under Mollison's guidance, became *Permaculture One*, the revolutionary first enunciation of permaculture's concepts.

The word "permaculture" was originally a portmanteau of "permanent" and "agriculture," signaling its focus on sustainable food production. Mollison, Holmgren, and many others quickly realized that a set of design principles able to tranform row-crop farm fields into functional ecosystems had vast scope. By now, these principles have been applied to design in many other arenas: water and energy use, waste treatment, and even community design and economic systems. Thus "permaculture" today means "permanent culture."

What does permaculture look like? Here is one way to express it. If we think of practices such as organic gardening, graywater reuse, natural building, renewable energy—and even less tangible activities such as more equitable decision-making and social-justice methods—as tools for sustainability, then permaculture is the toolbox that helps us organize and decide when and how to use those tools. Permaculture is not a discipline in itself or a set of techniques but rather a design approach that connects different disciplines and makes use of a wealth of strategies and techniques. It, like nature, uses and combines the best features of whatever is available to it. Although there are certain methods that are used often because they illustrate permaculture principles beautifully, such as keyhole beds and a decision-making process called dynamic governance, there are few if any techniques that belong only to permaculture. The art of good permaculture design is in deciding which techniques and strategies will solve a particular challenge in the most ecologically and socially sound manner. And that challenge, or design problem, can be of almost any sort: agricultural, technical, social, or financial. In fact, my own short-form definition of "permaculture" is "a set of decision-making tools, based on natural systems, for arriving at regenerative solutions to design challenges of all kinds." This book will give you examples of how astonishingly broad the scope of permaculture is for solving, in sustainable ways, the problems that human beings face.

or installing a rain barrel is simple compared to finding affordable land on which to plant that seed or agreeing on rainwater use policies. Human needs cover a number of whats as well as hows, and in the same way each chapter of this book explores permaculture's approach to meeting one of these needs in an urban environment.

The book begins with an overview of cities from a permacultural point of view: one of whole-systems thinking. A chapter on how

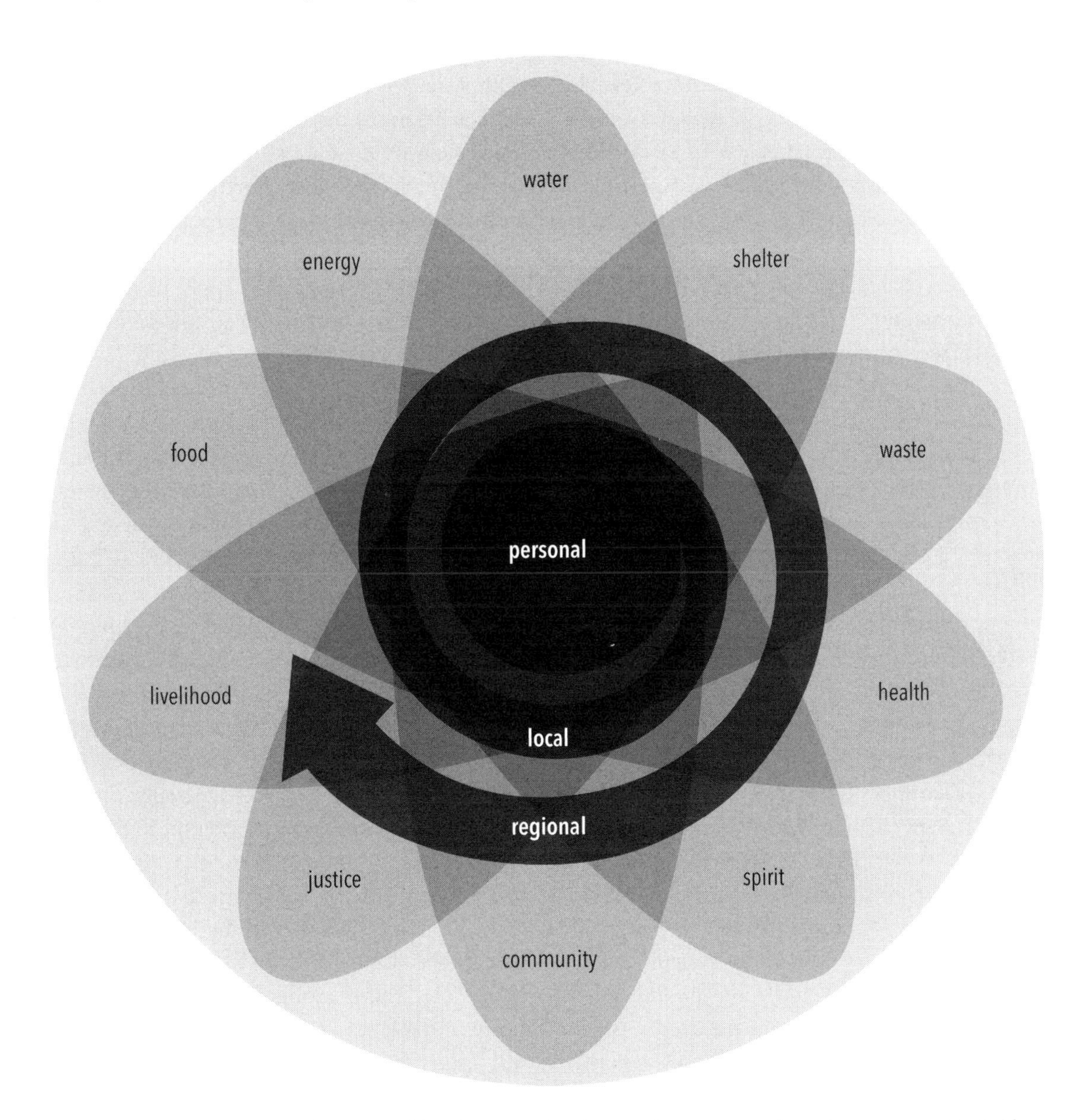


**FIGURE I-1.** The permaculture flower, modified from David Holmgren. Each petal is a basic human need. Adapted from artwork by Jonathan Woolson of www.ThinkPlayDesign.com.

permaculture's design tools apply to cities, towns, and suburbs follows. My aim here is to focus on the special circumstances of permaculture design in our paved-over places, so this chapter dives deep into aspects of permaculture that most other references rarely mention. Although I have attempted to make the design chapter, as well as the entire book, accessible to the permaculture novice, if readers want a more basic or general approach to elementary permaculture design principles and methods, some entire books, many book chapters, and hundreds of web resources focus solely on this topic. Some of these sources are listed in the text and bibliography.

After the introductory and design sections come three chapters on gardening in cities, the last of which focuses on food growing in community, since shared garden spaces are more common in cities than in places where everyone has a yard, and using them, or simply finding a space for a garden in town, brings unique challenges. In the garden-related chapters, I try not to repeat information on urban gardening that is readily and extensively available elsewhere, such as high-intensity growing methods and container gardening, but instead I offer the reader tools for thinking like a permaculturist about urban growing: What functions do we want our yards to fill? How do we take advantage of rather than fight the special conditions of the urban environment? These chapters explore examples of highly productive and diverse small landscapes and show how whole-systems thinking can help us choose and build those features that are most relevant to our own lives in town. For more basic information about permaculture's design and gardening methods, I will direct the reader to my own book, *Gaia's Garden: A Guide to Home-Scale Permaculture*. Besides providing a detailed introduction to permaculture garden design and implementation, *Gaia's Garden* also contains a chapter on urban permaculture that covers different material from that in the present book. Other permaculture books cover similar ground in different ways.

Following the gardening section, this book takes us out of the garden and into a chapter on the special problems and opportunities of using and saving water in cities and towns. A look at using, conserving, and thinking about energy in the urban home from a permacultural point of view is next. That completes the tour of the major visible elements that apply to town life (waste reduction and recycling are folded into each of those chapters).

The next chapter is on livelihood. It explores ways of making an ethical living in the city, as well as giving an overview of equitable currency systems and economics. The penultimate chapter covers community, working in groups, developing policies for more sustainable individual and collective lives in cities, and decision-making. Three of the other linchpins of invisible structures, however—health, spirituality, and justice—are not addressed in this book. A permaculture approach to health needs a book of its own. Spirituality is an intensely personal, sprawlingly diverse, and often deeply divisive topic, and I am not going to attempt to describe or prescribe a course of spiritual action for anyone. And as a white American middle-class male, I am acutely aware of the justice system's strong bias in my favor. Thus, although the theme of a more equitable society undergirds many of the ideas and examples in this book, I will leave a formal permacultural perspective on justice and the law to others who have broader experience and a view less likely to be colored by my own unearned privilege.

The final chapter is my attempt to tie all the pieces together to show what emerges when the elements of the permaculture flower—basic human needs—are evaluated and met using the toolkit and mindset that stem from permaculture's whole-systems view.

## STRATEGIES FOR THE PARADIGM SHIFT

The approach of this book is much like that of the classic seventy-two-hour permaculture design course, the formal two-week class that is many people's first introduction to, and life-changing immersion in, permaculture's whole-systems way of looking at the world. Like that course, this book is an overview that spans many topics, mentioning much, whetting appetites, focusing on the relationships and interconnections among the many subjects more than on their details, and thus can only rarely burrow into any single topic too deeply. I'm offering guidelines for assembling a diverse and flexible toolkit for regenerative urban living rather than compiling a detailed instruction manual on how to use each tool. You'll see that I present a swarm of methods in this book, on gardening, using water, saving energy, creating community, making a secure living, and all the book's other topics. In the interests of not writing a 900-page tome costing hundreds of dollars (and preferring to return someday to an active life rather than remain chained to a keyboard), I only occasionally give enough instructions to fully implement the techniques that I describe. Readers can easily find full instructions for using any of the mentioned methods, especially in this marvelous age in which we have not only books and magazines galore that offer how-to guides on every imaginable technique but also free offerings at websites, in forums, and in how-to videos. I also offer a trail through the information jungle by listing some of my favorite references in each chapter.

Learning a technique is the easy part. It mostly takes a clear set of instructions and some practice. Much harder tasks are figuring out which technique is appropriate for the conditions and designing a strategy for using the techniques in the proper place and order. Those—strategy, planning, and decision-making—are permaculture's strong suits. As a permaculture book, this volume's aim is to show how to develop plans for smart urban living and how to best choose the techniques, out of the multitudes available that will serve those plans.

This focus, on strategies and tool selection rather than tool use, may be disorienting at first to readers who are used to the typical Western-culture focus on technique. It may be jarring when I outline one promising method for, say, reducing noise pollution around your home, and instead of giving details on exactly how to do it, I move on to an alternative method. Technique-hungry readers may be equally impatient when I give a potted history of city water systems or explain a seemingly arcane energy concept before moving on to what to do. I'm telling these stories to place both good and bad design solutions in a context that helps us make better decisions about how to arrive at the good ones.

The goal of this book is to teach readers how to think like permaculturists, to become adept at a whole-systems approach to living in and finding solutions in cities, towns, and suburbs. Concentrating on techniques and the latest hot design idea is precisely how *not* to do that; that's why we can easily track beginning permaculturists by the number of abandoned herb spirals and needless swales they leave in their wake—and I count myself among the guilty. Our culture is enamored with things and how to make them and spends little time exploring which things we truly need, where (and when) they belong, and what their effects will be when we make and use them. Thus this book focuses on tools for thinking, designing, and creating from a whole-systems perspective, which is the key to arriving at resilient, regenerative solutions. I've also included many examples of people and groups that have successfully implemented these tools. If I've been successful, after reading this book you'll have a mental toolkit for developing strategies and approaches to solving challenges in multiple arenas, as well as a long list of techniques for doing this, in the physical environment as well as in the personal and interpersonal realms.

In design, knowing how to set goals, plan, and develop strategies holistically is the first step. Once you create those, you'll then be able to figure out what tools and methods to use to get to your goals. Those steps, of planning and choosing techniques, are the hard parts. Building what you've designed is easy in comparison.

Given how many tools we use, if tool use were the principal part of living sustainably we would have gotten there ages ago. Choosing goals and planning the strategies to arrive at them without trashing our environment or our relationships is a much more difficult task. For most of human history, our focus has simply been on meeting our needs: How do we get food? How do we make shelter? How do we stay healthy? We have developed vast warehouses of tools for dealing with each of those problems. Because we live on a planet that for millennia was immense compared to the human population and its needs and impact, our species could focus on meeting its needs without paying much attention to the ecological—and often even the social—consequences. We could just move on. But our industrial civilization has chewed up ecosystems and cultures relentlessly. With seven billion of us and counting, there's really no place left to go.

We are learning that without healthy eco- and social systems, humans—and everything else—suffer, potentially to the point of extinction. So we can no longer employ just any method at hand that answers the question, "How do we meet human needs?" Some ways of doing that are too destructive. We need evaluation criteria and decision-making tools to tell us which tools we need to stop using or use more often. To the exhortation to "meet human needs" we now must add a second clause, "while preserving ecosystem and social health." That addition represents an enormous paradigm shift from piecemeal to whole-systems thinking. It also explains why gaining a deep grasp of permaculture—seeing that it's not just keyhole beds and guild-planting—takes some work. Permaculture is applied whole-systems thinking, and in a culture that teaches us to focus on individual things and rarely on the relationships among them or on the consequences of making and using them, whole-systems thought—the idea that we need to look not just at the parts but at connections and relations between parts, and how the whole is more than just the sum of the parts—is alien. Thus the approach of this book is to explore what it means to meet human needs in a whole-systems manner. For those of you unfamiliar with whole-systems thinking, I'll describe it in more detail shortly.

Cities and urban life are ideal platforms for tackling all this, as they are where we most vigorously work at meeting our needs, and they are the places from which our effects on this planet radiate most powerfully.

So let's begin.

# CHAPTER ONE

# The Surprisingly Green City

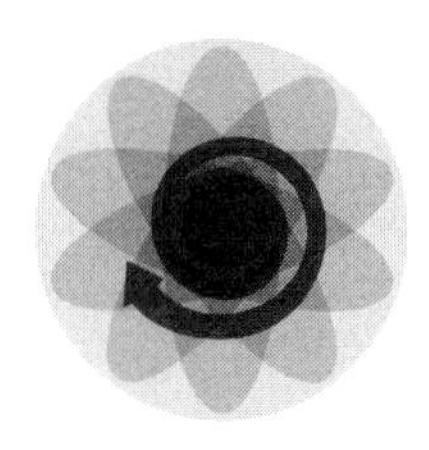

In the 1990s my wife, Kiel, and I moved from an urban existence in Seattle to a deeply rural one in southern Oregon. One of the many reasons for this seismic life shift was our yearning to trim our ecological footprint. We were going to grow much of our food, make more of the things we used, and simplify our lives. The move to country life turned out to be easy for us, and we reveled in the feeling of slowing down after the bustle and grit of the city.

But after a year or two, I noticed a few persistent glitches in our dream. Even though we worked at home, our gasoline use skyrocketed, because we were miles from any supplies, friends, and culture. The nearest grocery store was a twenty-minute drive, and organic food was a two-hour round trip. In sparsely populated and enormous Douglas County (as large as Connecticut), some of our friends lived ninety minutes away, and we burned a lot of gas to fuel our social life.

I kept scraping against other troubling downsides of modern rural life. Winter rains hammered our gravel road each year, and along the quarter mile that we maintained, every spring's repairs swallowed 40 to 50 cubic yards of trucked-in rock. When the cable company strung a line to my neighbor's home, the wire to that single house stretched over 1,500 feet. The pipe from our well meandered over a half mile, since the recalcitrant geology didn't conveniently place water or a clear trench run. The well also needed its own half-mile of power line; an insulated, electrified pump house; and a separate meter. Then there was the septic tank, distribution box, and a sizable grid of perforated drainpipe set in gravel-lined trenches. All this *stuff* for one house, and every other house around us. We lived out in the wild, but our long umbilical cord to civilization used a lot of metal, plastic, rock, and fuel to build and sustain.

We spent a wonderful ten years in Douglas County, but over time we felt a growing sense that our much-wanted seclusion and self-reliance were becoming more like isolation.

We had also accomplished much of what we had set out to do. All this took us back to the city, up I-5 to the burgeoning green scene in Portland. There I quickly noticed that our house tied us to services via a few feet of pipe and wire. Friends, stores, and cafés were within walking distance or a couple of minutes of biking or driving, and our car sat for days in the driveway—a driveway that was 20 feet long and never needed resurfacing. Downtown was a six-minute bus ride.

When we moved back to the city, our energy use plummeted. Of course, it's possible to live in the country and consume fewer resources than in the city, but virtually no one in Western society does that. Rural self-sufficiency rarely exists. Almost the same proportion of rural people commute to jobs as city folk, and they drive longer distances to do it. As modern life in the United States is set up, rural people use more resources, especially energy, than city dwellers.

I wasn't alone in noticing this. At about the same time as our return to the city, an article appeared in the *New Yorker* by David Owen called "Green Manhattan."[1] Owen had moved from Manhattan to rural Connecticut and found that his modest country house used nearly eight times the electricity as his apartment. He and his wife worked at home, but they still drove 30,000 miles each year, mostly for errands, while they had not even owned a car in the city. Owen reported that 82 percent of Manhattanites travel to work via public transit, bike, or on foot, ten times the national average. Granted, cities have plenty of problems, Owen acknowledged, but in resource and energy consumption, dense urban centers such as New York City, Chicago, San Francisco, and Washington, DC, are paragons of conservation compared to American suburban and rural settlements.

I'm not saying that cities are gleaming exemplars of sustainability. Of course they aren't, any more than virtually any other aspect of contemporary life is sustainable. And that's why it is so important to bring the whole-systems thinking, methods, and practices of permaculture into cities: Cities are a leverage point. Over 50 percent of the world's population lives in urban areas, and that number is rising.[2] Much of humanity's production and consumption occurs in cities, and the vast majority of all goods move through cities. What's more, most ideas and cultural trends come from cities. If we can't create regenerative urban cultures, what happens elsewhere hardly matters. Unsustainable cities will drag the rest of society down with them. We need to make the transition to regenerative, resilient, life-supporting cities and towns. To get a better idea of where we want to go on that journey, we ought to take a brief look at how we got here—how cities arose and what they do. Then we can put this history in a whole-systems context.

## THE RISE OF THE CITY

The first important question is, where did cities come from? Though the earliest cities most likely arose in Mesopotamia, societies the world over have independently arrived at cities as a dynamic, effective resolution of the social and political forces that pulse through all civilizations, and cities in radically different cultures share essential features and functions. Whether it is Herodotus in the fifth century BCE writing of exotic Babylon in his *History* or Bernal Díaz del Castillo, a soldier with Cortés in the sixteenth century, penning in his diary awestruck accounts of the temples, bridges, and plazas that gleamed in Aztec Tenochtitlán's tropical sun, chroniclers of metropolises recognized in all cities many activities, features, and physical layouts that were familiar even when built by a society jarringly alien to their own. They noted that in every city and town, marketplaces teem with crowds buzzing in the age-old banter of buying and selling. At every urban core sits a central park, a gleaming complex of temples, or some other embodiment of dedicated, even sacred public space. The city itself lies in a strategic and easily defended site—Tenochtitlán stood on an island linked to the mainland by causeways and ringed by food-producing *chinampas*, while wall-enclosed Babylon overlooked and controlled the Euphrates River, and every other long-lived city has been built where the landforms offered easy transport and mingling of people and goods, places to gather, and defensibility.

These three essential functions—commerce, community, and security—are common to every city. It is through this lens of function—in permaculture parlance, of matching needs with yields—that this book will explore how life in towns, cities, suburbs can be enhanced, redesigned, or simply viewed differently to help us live more engaged and resilient lives with each other and with the natural world.

But what exactly is a city? The US census defines a city as having more than 25,000 people. I will fudge that a bit, because sheer numbers are not the only quality needed to create a municipality that holds many features and patterns in common with an official city. Much smaller settlements have multistory buildings, miles of pavement, enough traffic to cause gridlock, an absence of wildlife, and other hallmarks of a metropolis. Plenty of communities that might be more properly called towns, suburbs, or even villages have imposing edifices, heavy traffic, hip spots to hang out, a music or performance scene, and an ecology that is recognizably not that of a rural area—if natural ecosystems remain at all. These are some of the primary patterns of urban life, and pattern rather than number—quality rather than quantity—is central to permaculture design. Main Street in a small town of 1,500 throbs with patterns more akin to those pulsing through Chicago's Michigan Avenue or Manhattan's Broadway than with the patterns of a forest or prairie. Thus this book will use the term "urban permaculture" broadly, to mean permaculture that is practiced wherever the technological and social functions of the built environment outweigh the biological processes of nature, or, to put it more succinctly, wherever we live amid pavement and people more than with plants.

What brought humans into these unnatural environments? The earliest known site that displays the lofty architecture typical of cities is Göbekli Tepe, a 25-acre expanse of multiton stone monuments in Turkey that was built about 12,000 years ago (see the color insert, page 1). This was before farming was invented, which means that this complex site and its titanic, richly carved monuments were built by hunter-gatherers, a feat they were supposedly incapable of. Archaeologists think Göbekli Tepe was a religious center, which raises the possibility that humans gathered in large numbers for spiritual reasons before farming arose. This tips the conventional theories of urban origins upside down, because the dominant hypotheses all posit that people needed to first develop agriculture and its storable surpluses in order to grow populations large enough to spare a nonfarming labor force that could build imposing monuments. It may have taken an estimated 500 people to carve and drag the site's many 40-ton monuments into place. Feeding those crowds of worshippers and builders, the scenario goes, overwhelmed the natural carrying capacity of the surrounding land. The novel stresses from packing many hungry mouths into one place, some revisionist archaeologists now believe, drove humans to contrive the high-yielding synthesis of techniques called farming. The site's principal excavator, Klaus Schmidt, uses the slogan "First the temple, then the city" to describe this inversion of the conventional thinking. This also acknowledges that the deep link between food and the sacred that can be felt by anyone who has put hands into the soil is almost certainly far older than civilization.

While religious sites such as Göbekli Tepe may have been the first places that people gathered in large numbers, true cities—places where people dwelt rather than simply worshipped—came later. No one is sure when or where that first city arose. The honor usually bounces, depending on the latest findings and local boosterism, among the Middle Eastern sites of Byblos, Jericho, Damascus, Aleppo, and several others, all dating from roughly 5000 to 3000 BCE. However, a recently discovered settlement, now submerged off the west coast of India, dates from 7500 BCE, which could push the origin of cities back by several millennia. Wherever they began, by roughly 2000 BCE about ninety 90 percent of the population of Mesopotamia was living in cities.[3] Cities, judging by how quickly they attracted the people around them, must have filled some fundamental needs.

## ➡ CITIES EMERGE ⬅

What were those needs? They are revealed in the special functions that they serve, ones that aren't well filled by rural and village settlements. Although rural villages allow people to gather, worship, and trade, they are hampered by the limits of scale. Before the petroleum age blasted us through the limits of a solar budget, most people in agricultural societies were farmers. That meant that a village of a few hundred or thousand could support only a small number of specialists to produce nonfarm goods or more esoteric services. It may have taken 1,000 farm families to support 50 or 100 nonfarmers, and most of the latter also produced basic products and services: harnesses, clothing, bookkeeping, and so forth. In a farm-village economy, there simply wasn't enough free time and labor to move beyond the basics. At some point in population growth, an expanding cadre of specialists allowed the emergence of the unique functions and special payoffs that are the emblems of urban life. The ancient shift of human groups from farms, forests, and savannas into cities also created a new social order in which tribal and clan relationships were transcended and replaced by larger, more stratified social, commercial, and spiritual communities that met human needs in more formal, centralized ways.

Economist Edward Glaeser contends that the benefits of cities flow from their lowered cost of moving a critical mass of goods, people, and ideas over much shorter distances; that is, from commerce of all sorts.[4] That's not a surprising pronouncement given Glaeser's profession. Practitioners of other disciplines agree, but they list some other benefits that emerge from assembling an urban critical mass.

Beyond the assets that build under an accelerated flow of goods and ideas, sociologically oriented architects such as Lewis Mumford, Kevin Lynch, and Joel Kotkin cite security and the attraction of monumental public spaces as critical functions of cities. In hunter-gatherer societies, where people rarely accrued the surpluses of food and other goods that would attract raiders or traders, there was little need for the security and enhanced commercial opportunities of the city. But once agriculture's storable (and therefore stealable) surpluses were concentrated in large granaries, and specialized workers produced piles of trade goods, safety and markets became prime concerns. We can think of security and ready markets as by-products of urban surpluses. They are the novel functions that emerged from the dense concentration and grand scale of urban abundance and population. Through all but the last few years of human history, markets have been physical gathering places. These, if Göbekli Tepe is typical, already existed in the form of temples and plazas, and the necessary crowd of people was also present. Hence the physical environment of the city plus its ability to allow a critical mass of people to gather and be supported there created the conditions for a set of unique properties to develop—rapid dissemination of goods and ideas, inspirational gathering places for diverse groups to mingle, and security in the numbers and order of the city. The location of these three functions in the same place simultaneously allowed links and synergies among them to arise; for example, a secure marketplace meant valuable goods could be created and displayed.

## WHAT FUNCTIONS TELL US

When a permaculturist sees words such as "function" and "synergy," it sets off lightbulbs in his or her head. Function, for example, indicates a relationship, a connection between two or more elements. A road functions to move traffic, thus the road has a relationship with vehicles, and it mediates the movement—that is, it makes connections—between the traffic, its origin, and its destination. Knowing a function, in turn, leads us to identify the items and processes necessary to fill that function and also points to the yields created when that function is filled. Thinking in terms of functions, then, is a powerful leverage point, because it identifies needs, yields, relationships, and goals, and it helps us spot blockages, missing elements, buildup of waste, and inefficiencies in the various flows and linkages that are part of that function's workings.

### Functions of Cities

1. Gathering places
    - Celebration and worship
    - Social, commercial, and leisure gathering
    - Inspiration from art and culture
    - Projection of power via public and government monuments
2. Security
    - Protection against outsiders
    - Security through local rule of law
    - Projection of government power to citizens, region, and foreign lands
3. Trade
    - Markets for produced goods
    - Markets for labor, services, and skills
    - Markets for rare materials and services
    - Economies of scale
    - Collection and distribution of goods for region
    - Reduced transport costs

## Is Food Growing an Essential Function of Cities?

One note that may tweak some foodies: You may have noticed that food growing is missing from the list of critical urban functions. Obviously, people in cities need access to healthy food, but food *production* has never been a fundamental role of cities. The world's cities, towns, and villages have always relied on their surrounding regions for much of their food and other raw materials. Even early towns were simply too densely settled to fit in much growing space. Ancient villages and preindustrial cities were nearly always close-knit clusters of adjoining houses, temples, and shops, separated by paved streets and enclosed by walls. Daily, villagers walked from this built-up, densely populated hub to the farm fields that ringed it.[5] Substantial urban gardens were rarities afforded only by the rich. While some cities have grown significant tonnages of high-value, low-calorie vegetables and larger urban yards sport a few chickens or pigs (fed on country-grown, imported fodder), the extensive acreage needed for growing the grains, dairy products, and meat that make up the caloric bulk of most diets is available only outside the town walls. A look at a map of nearly any preindustrial village or city will confirm this. Ancient Jericho, for example, covered less than 10 acres and held between 500 and 3,000 people.[6] Not much room for urban farming there!

Urban food production today is highest in the sprawling, fast-growing metropolises of China and other developing nations that are filled with newly arrived farmers drawn from the countryside. The data support the notion that as these recent immigrants give up their country ways, urban farming declines. In Beijing, for example, which is past its principal population explosion of the 1960s and

This means that when we look at cities, their residents, and the other components of urban life in terms of their functions, we can spot the factors that influence how well they are able to perform those functions. Then we can study, understand, and direct those factors and influences in ways that will create and enhance the functions and properties of cities that are beneficial, such as community-building public plazas, parks, and structures; open and supportive marketplaces; and habitat-creating green space; as well as human elements such as responsive policy processes. We can also spot and damp down the negative factors. Once we've done this, the next step is to evaluate, to see how well our changes have moved us toward a more livable, and life-filled, environment. That is the heart of design.

The importance of the three primary functions of cities—inspirational gathering space, security, and trade—is also visible in the negative. When cities grow ugly or inhumanly scaled, when they are crime-ridden or prone to raids, or when their industries fail, urbanites retreat if they can to the suburbs, the hinterlands, or another more functional city. Those who can't leave often crowd—or are forced—into ghettos and enclaves. The movement of people in and out of a city is useful feedback about how well that city functions and what needs to be redesigned.

1970s, urban farmland declined by over half from 1991 to 2001,[7] while in fast-growing Shenzhen, urban farming is still increasing. The surge in urban farming of the last few decades also may be a side effect of the automobile. Most urban agriculture is done not in the dense urban core but in the much more open peri-urban regions that were recently rural but are now spattered with the houses and yards of car-driving commuters.

I am not arguing that we should abandon urban food growing, as there are many other reasons to practice it, only that it is a bonus feature—not the essence—of a vibrant city. Urban produce is an added-on quality of cities, practiced by farm immigrants, occasionally done out of economic necessity by those few urban poor who are lucky enough to have land for gardens or seen as a way to make cities more livable by activists and by the affluent with yards. Cities, until cheap oil burst the ancient pattern, have always been fed by the surrounding land; this is why New Jersey's license plates display the baffling moniker "Garden State." The vanished truck gardens of rural New Jersey, New York State, and Connecticut once fed New York City. This book's structure reflects urban food-growing's role among the other critical functions of cities, which is why the gardening chapters are balanced by thicker sections on permaculture's approach to meeting the other important needs that city life must fill.

Where cities and food intersect most forcefully is in the political and economic power that city dwellers can wield to improve virtually every aspect of the food system by demanding high nutritional and ecological standards for the vast quantities of food that is grown to feed them.

## THE BENEFITS OF CITY SCALE

But don't villages and other small settlements fill the functions of trade, security, and gathering spaces well enough? After all, each of us doesn't need more than a few friends, customers, and merchants to fill our social and economic needs. Big cities don't have a monopoly on public squares and churches. Small towns have police and a legal system, and during unstable times a walled village of a few hundred people would be large enough to intimidate all but the largest bands of marauders. So why are cities, especially the largest ones, such as New York, Los Angeles, Mexico City, and Mumbai, such powerful magnets? Do unique benefits emerge at the larger urban scale?

Indeed, something special does happen as population centers grow. The theoretical physicist Geoffrey West, working at the Santa Fe Institute, collaborated with a high-powered team of economists and technology experts to sift through reams of data from dozens of cities, scanning for patterns in everything from inventions filed to bank deposits, R&D startups, and wealth creation.[8] They found that indicators of innovation and creativity didn't scale up at a simple linear rate but at one that mathematicians call superlinear. A city that was ten times larger than another didn't produce just ten times more patents or new startups but seventeen times more. And the multiples of exponential growth increased as cities got bigger. A city fifty times larger than another generated 150 times more ideas.

Something in cities stimulates us. On average, a person in a city of five million souls, the report suggests, will be three times more creative than a person in a town of a hundred thousand, in terms of generating original works, ideas, patents, publications, performances, and other innovations. Edward Glaeser reports that 96 percent of all product innovation occurs in cities, and he credits proximity to people and goods as the key element. It's as if proximity drives an exponential relationship similar to the inverse square law: Cutting the distance in half between people, ideas, or goods doesn't just double the number of their interactions and exchanges; it squares or even cubes the creativity that emerges. Long ago, urban activist Jane Jacobs observed that "great cities are not like towns only larger"; the statistics show that, in terms of how creative we are in them, she was right.

Big cities are different from towns in some important respects, such as how they spur innovation. But as I wrote earlier, towns and cities of all sizes also share many qualities, and it's on those similarities—the functions and properties that are found in all cities and that make town life so different from rural existence—that I want to focus now. Early in this chapter I referred to cities as complex adaptive systems. In those systems, while simply having certain elements, functions, and properties is important, their dynamism and adaptability stems from how richly their parts connect, combine, overlap, and affect one another. Although the parts that make up these systems help define their qualities—a collection of cells and a group of people aren't the same—their interactions and relationships are what gives these systems their character. The ability of large numbers of parts to interact dynamically is what gives complex systems their responsiveness and ability to behave in unpredictable, novel ways. This is why cities spur creativity: simply interacting with other people and being influenced by their ideas stimulates the emergence of creativity. We've all had the experience of being around someone brilliant and feeling a little more brilliant ourselves. Cities increase the odds that we'll all run into those inspirational people and circumstances.

Researchers who study complex systems have learned that certain critical functions, such as adaptability or creativity, aren't carried out well, or at all, in systems until a certain threshold of complexity is reached. That tipping point holds for cities, too: Individual farms and households not only lack the organizational horsepower to perform some of the functions that towns and cities do until they cluster into larger settlements, but they don't have the diversity of functions that, by their sheer ability to combine in immensely flexible ways, can generate a rich set of novel possibilities. A few farm households near each other don't need—and can't spare the labor for—repair shops where tinkerers can play, accountants to track their resources, lawyers, colleges, or the host of other livelihood possibilities that emerge and offer new benefits as settlements grow.

By gathering in large numbers, people and groups can probe the huge space of novel, unexplored opportunities that emerge from the combinatorial explosion erupting from many autonomous parts that are able to interact in diverse ways. That creative, novelty-exploring stew is a big chunk of what makes urban life special. Besides, the elements we find in cities—people, knowledge, customs, ideas, skills—are not static but dynamic, learning and evolving themselves, so not only can they combine in many configurations because of sheer numbers, but their malleability and responsiveness means they can combine in particularly rich ways that can adapt and change. Their ability to learn and grow generates even more novelty, even more possibilities.

## CITIES AS COMPLEX SYSTEMS

The sciences of complexity studies arose in the 1960s and 1970s and spread, because they were so widely applicable, from the arid realms of theoretical physics and mathematics to other disciplines. A subdiscipline of urban planning, sometimes called complexity theory of cities, emerged in the 1980s and has since generated a blizzard of publications and experiments in urban design. I will give an overview of the origins and tenets of complexity theory of cities as it relates to permaculture. For those interested in exploring the intersection of urban design with complexity theory in more detail than I can offer here, a good place to start is an anthology of articles collected under the title *Complexity Theories of Cities Have Come of Age*, edited by Juval Portugali and others.[9]

Understanding that cities are a form of complex adaptive system has helped urbanists restore some vibrancy to moribund metropolises, so it's worth understanding a little about these systems. The general "messiness" of cities has been irritating urban theorists and planners for centuries, but it wasn't until recently that urbanists truly understood that it is just that messiness that gives cities their life.

The urge to rationalize and give order to cities—which, incidentally, culminated in the dehumanizing urban-renewal projects of the 1960s—has its seeds back in the Enlightenment era. Philosophers and scientists of that day, inspired by the successes of Newton, Galileo, and Kepler at finding simple laws that explained and predicted mechanical action, began thinking of nature and the universe as a machine that could be dissected, rebuilt, and controlled. Once they saw that planets and falling bodies operated by simple rules, some of them began extending the machine metaphor to the living world. Soon farming and forestry were remade in the image of the machine, and this mechanical worldview spread to human systems as well. The standardized, abstract measurements of the metric system supplanted local and traditional units that once kept their uses connected to natural objects and activities. An acre, for example, was the area of flat land that a pair of oxen could plow in a day; an inch was the length of three grains of barley laid end to end. A meter is just, well, a meter—and since the 1983 General Conference on Weights and Measures, defined as, "the length of the path travelled by light in vacuum during a time interval of 1/299,792,458 of a second." How's that for abstract?

Tested land-use customs that had been culture- and site-specific were swept aside by nationwide property laws, official languages taught in state schools extinguished dialects and indigenous speech, and major cities such as Paris and Washington, DC, were rebuilt on rigid geometric patterns.

This attempt to impose a clockwork order on the confusing welter of urban life, while making cities more comprehensible to travelers and tax officials, reached its peak in the neighborhood-razing visions of New York's Robert Moses, the sterile facades and inhuman whole-city plans of Le Corbusier, and the crime-ridden high-rise projects of south Chicago and countless other cities. As the failures of what has been called high modernism became obvious in the 1970s and 1980s, architects, planners, officials, and urban dwellers began to see that a machine city is a dead city.[10]

Right at that time, though, several countering forces were emerging. One was an activist revolt against large-scale urban planning. As so often happens in the simultaneous emergence of parallel ideas whose time has come, this grassroots movement was also gaining academic legitimacy in work by theorists in the developing new complexity sciences. Mathematicians, ecologists, economists, and planners alike began to spot the consonance between complex systems such as weather, forests, neural networks, markets, and cities. Some of these complex systems could adapt and learn, while others, like the weather, could not. The former came to be called complex adaptive systems, or CAS. Researchers soon determined that to be able to learn, adapt, and evolve, CAS needed to possess certain features:

1. They are composed of *autonomous agents*; that is, their parts work according to their own internal operating rules, whether they are nerve cells, trees, or people.
2. These agents *interact* with each other according to certain (often simple) rules. A rule for a bird in a flock may be, "Keep the bird ahead of you at a 45-degree angle and 3 feet away." These simple rules can result in stunningly complex behaviors, as anyone can attest who has watched a shimmering flock of birds spin patterns against the sky.
3. Those new behaviors are an example of *emergence*, which is the appearance of novel properties that can't be predicted by studying the parts in isolation. Watching a single bird in flight would never let you predict the intricate, captivating dance of a swooping flock of birds. Studying one cell of a slime mold would never suggest that as a group they can merge to fashion a bizarre mushroomlike colonial structure for reproduction.
4. The agents respond to changes in their environment via *feedback*. They sense some of the effects of their actions, which allows them to adapt and learn.
5. CAS usually exhibit *homeostasis*; that is, they self-regulate and "tune" their behavior to certain states that are preferred over other, less stable states, and they can return to these states after a disturbance. These states are usually far from equilibrium. A mammal, for example, maintains its body temperature independent of both the air temperature and how hard it is exercising. If it were at equilibrium, it would be at air temperature—and it would be dead.
6. These systems maintain themselves in a rich, possibility-filled region between perfect order and total randomness that complexity thinkers call *the edge of chaos*. An organism, for example, contains proteins that are made to a specific pattern but are constantly moving in and out of that pattern as they are built up and broken down in metabolism. But metabolism isn't chaotic. It follows specific pathways and rules. We can see this also in our genes. They generally are built to a set DNA sequence and pattern, but occasional mutation and regular recombination permit new possibilities to emerge. Perfect order

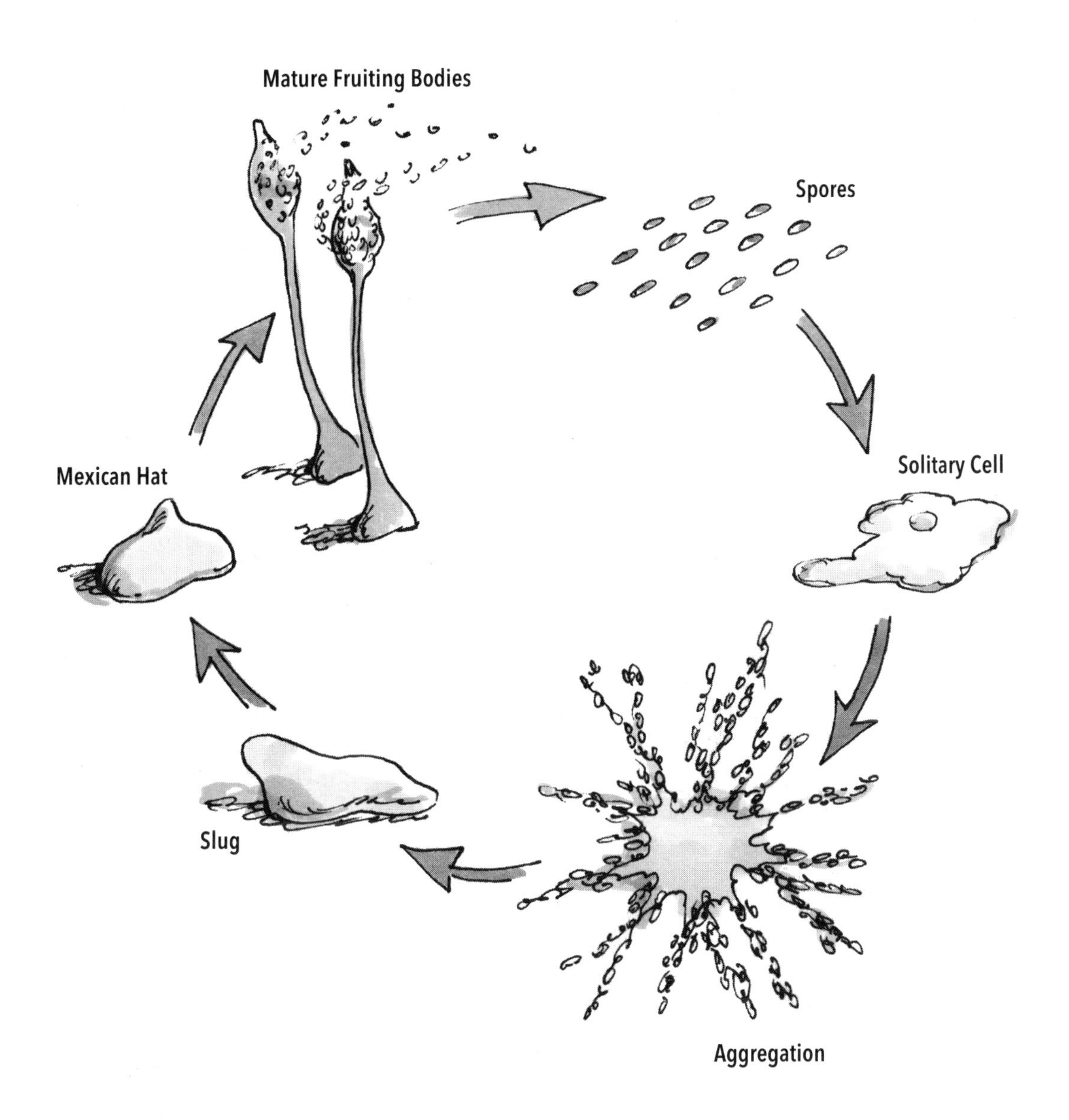


**FIGURE 1-1.** Emergence in action. The slime mold *Dictyostelium* germinates from spores as individual cells that remain independent until food becomes scarce. At that point the cells aggregate and can move as a multicellular organism in the pseudoplasmodium or slug stage. This "slug" slithers to a well-lighted, open place and transforms into a mushroomlike fruiting body that then releases spores. The slug and the collective fruiting body possess properties not present in the individual cells, such as the ability to form complex shapes, solve mazes (in the slug phase), and release spores (the fruiting body). Illustration by Elara Tanguy

is dead, while complete chaos allows no structure. Life and other complex adaptive systems attune themselves to the fecund, creative place between frozen order and seething randomness, to the edge of chaos, and thrive there. Healthy cities do the same.

In summary, CAS contain many autonomous parts, they respond to changes via feedback, and they form self-organizing, self-maintaining assemblages that display emergent properties. So how do the principles of CAS apply to urban permaculture?

Those principles suggest that rigid planning that leaves no room, or even not enough room, for spontaneous self-organization will create sterile cities. Strict top-down planning is anathema to CAS, including cities; it imposes a rigidity that eliminates adaptability and spontaneity. On the other hand, pure bottom-up accretion of elements with no rules or pattern at all approaches chaos and can result in grossly unequal distribution of resources, incoherent layout, gentrification, food deserts, and the other ills that plague many cities. Thus urban design methods that provide enough organization in the form of simple rules but create the conditions for spontaneity to occur can take advantage of the ways that cities behave as CAS. What does that look like?

One of the first to grasp the importance of urban life's lack of tidiness was Patrick Geddes, a biologist who later turned to sociology and urban planning. Geddes was a student of Thomas Huxley, the man known as "Darwin's bulldog" for his fierce defense of the theory of natural selection, and Geddes brought his own appreciation for evolution and life's spontaneity to urban design. During the late nineteenth century, when Geddes was practicing, the common view was that cities were simply "architecture writ large," mechanical elements assembled on a large scale. Geddes taught that every city evolves in both a historical context and a unique geographical setting, and any planning that ignores or attempts to remake these will harm those who live there. But Geddes was nearly a lone voice against the rising influence of those who saw the city as a machine, and their views dominated the first six decades of the twentieth century.

## RAGING AGAINST THE MACHINE CITY

The first significant counter to high modernism was the work of urban activist Jane Jacobs, who battled entrenched bureaucrats in New York City whose power stemmed from mammoth, top-down urban renewal projects. Her classic book, *The Death and Life of Great American Cities*, was an impassioned and influential attack on the idea that the city could be made rational and orderly and still suit human beings. She argued that healthy city life depended on urban buildings and neighborhoods that had diverse uses, ages, appearances, layouts, and income levels. She believed that whatever small amount of planning was necessary should be done locally and transparently, and not via a top-down process. Jacobs was an advocate of complexity before there was such a discipline, and her book, arriving at the cresting of the high-modernist tide, became a rallying point for those angered or victimized by inhuman schemes of urban renewal.

We can see a perfect illustration of the defects of central planning versus the invigorating effects of organic, human-scale organization in cities by comparing Jacobs's work with that of Le Corbusier. Le Corbusier, born as Charles-Édouard Jeanneret-Gris, was more than an architect; he was an influential visionary with plans large enough to remake whole cities, nations, and societies. His work epitomizes the megalomania of high modernism. At various times Le Corbusier proposed citywide plans for rebuilding Buenos Aires, Paris, Stockholm, Geneva, Barcelona, and several other comparable metropolises. His plan for Paris, typical of all his work, was to bulldoze the core of the city and replace it with rigidly geometric buildings and streets. Most of his designs at this scale were, thankfully, never built.

Le Corbusier found the seeming disorder and confusion of cities offensive. "We must refuse to afford even the slightest concession to what is: to the mess we are in now," and "We claim, in the name of the steamship, the airplane, and the automobile, the right to health, logic, daring, harmony, and perfection" are statements typical of his certainty and bombast. His ideal form was "an unbroken straight line."[11] The principal feature of his city-scale plans were their geometrical perfection when seen from far outside. His drawings for cities are usually from the perspective of an airplane, where their precision and order can be seen, and his artwork for a new Buenos Aires is as if viewed from a far-off ocean liner "after a two-week crossing." From those distances, the people are invisible. Nothing about the inhabitants or their quality of life mars his perfect vision.

One of the few cities built to his plans, Chandigarh, the capital of the Indian state of Punjab, was designed by its zoning ordinances and by the huge scale of the roads specifically to prevent street life from occurring. Urban activist Madhu Sarin, writing about her experience of that city, said, "The scale is so large and the width between meeting streets so great that one sees nothing but vast stretches of concrete paving with a few lone figures here and there. The small-scale street traders, the hawker, and the *rehris* [vendor's wheelbarrow] have been banned."[12] Le Corbusier's attempts to cause, as he himself put it, "the death of the street," were in vain even there. An unplanned city has grown up around the planned one, and it is there that most of the city's activity occurs.

Along with geometric precision, Le Corbusier's answer to the messiness of cities was a radical separation of functions: workplaces were to be split away from houses, factories from other commerce, and every other activity—entertainment, athletics, gathering, dining—was to be cloistered in its own zone. Seduced by the theoretical elegance of Le Corbusier's ideas and by his certainty and charisma, urban planners of the day enthusiastically began cleaving cities into fragmented functional areas. This move quickly proved deadening and destructive, leaving downtowns empty and dangerous at night and neighborhoods silent and sterile by day.

In contrast, Jane Jacobs did not view cities from planes or distant cruise ships but from strolls around her neighborhood, chats with shopkeepers, and people-watching in parks. *The Death and Life of Great American Cities* is her attack on the soulessness of the urban planning of her day, and it galvanized a movement to return cities to their inhabitants. Jacobs claimed that a crucial error made by planners is to confuse visual with functional disorder. "The leaves dropping from the trees in autumn, the interior of an airplane engine, the entrails of a rabbit, the city desk of a newspaper, all appear to be chaos if they are seen without comprehension," she wrote. The failure of the high modernists was to view order only as aesthetic and not functional. By reducing the city to geometry and isolated areas of activity, they killed it. A living thing is never built of straight lines. In a phrase that anticipated the formal science by twenty years, Jacobs referred to the city as "organized complexity."

Her analysis went deeper than the physical layout of the city to examine the patterns of human interaction. She argued that a healthy social order does not follow the same rules as architectural order and cannot be imposed by a plan. A social fabric is not created or maintained by laws, police, and officials. The public, daily life of cities, she writes, "is kept by an intricate, almost unconscious network of voluntary controls and standards among the people themselves, and enforced by the people themselves." She tells the story of watching from her window as a man seemed to be quietly accosting a little girl. She was gathering the nerve to intervene when she saw that the butcher's wife, a fruit vendor, the deli owner, and two bar patrons were approaching the man, and several other people made it clear that they, too, were watching. The man retreated. In living communities, police are rarely needed, but if that neighborhood, as planned, had been rebuilt along a proposed and later rejected high-modernist scheme, that would have eliminated those watchful, caring eyes.

Jacobs's book became a fulcrum on which urban planning pivoted away from rigid, deadening order toward spontaneity, organic development, and a human scale. Another advocate for organic design as an essential element of urban life was Christopher Alexander, whose Ph.D. thesis, *Notes on the Synthesis of Form*, later published as a book, advocated that the best models for vibrant cities came from the preindustrial world and that bottom-up design was essential. A subsequent essay, "A City Is Not a Tree," argued that planned towns and cities failed because they were designed in a simple organization-chart hierarchy, where subassemblies on one level connected only to one point in a higher level, and elements at that level then connected only to a single point in the next higher level, and so on.[13] He proposed a more network- or latticelike structure in which many elements had mutual influences upon other elements in multiple levels.

Alexander is best known for *A Pattern Language*, one of a trilogy of related books on human-scale design.[14] *A Pattern Language* describes 253 "patterns," each one, in Alexander's usage, a successful, life-enhancing solution to a commonly encountered need in design, ranging in scale from ring roads that encircle entire cities to clusters of village nodes, from crenellated house fronts that engage our eyes to window seats that invite us to nap in the sun. Choosing the right patterns for the places and things we build creates successful, usable designs and results in what the author calls "places that live." These patterns are not isolated building blocks but are combined in supportive relationships in which each pattern needs other patterns that are larger, smaller, and the same size in order to function well. Alexander was among the first to conceive of patterns as solutions to regularly encountered problems, needs, or functions in design. This idea—a pattern as a design solution—is a core concept in permaculture as well.

What allows patterns in this sense to guide us toward creating "places that live" is the seeming disorder within their orderliness. A pattern in nature—a ripple, a spiral, the branching of a tree—is formed according to a few rules or a simple algorithm that generates the features that each example of that pattern has in common;

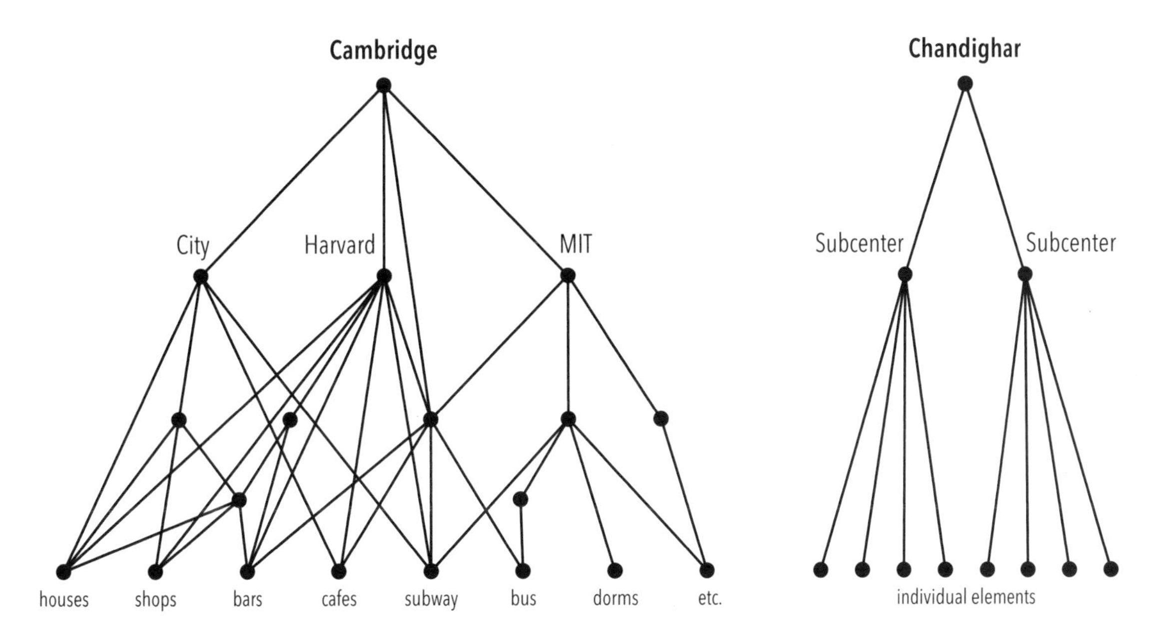


**FIGURE 1-2.** The features of an organically developed city such as Cambridge, Massachusetts, *left*, are connected in a semilattice, in which elements are linked to each other in multiple ways, allowing overlap and interconnections among layers. In a planned city such as Chandighar, India, the top-down structure means that elements connect only at the next level above. This allows little feedback, interaction, and other qualities that bring a city (and any complex system) to life. Illustration inspired by *A City Is Not a Tree* by Christopher Alexander.

that is, every ripple on a beach, in a cloudy sky, or in a piece of rumpled fabric is unique, but all have common qualities that we are able to group together as forms of the pattern "ripple." This is the genius and elegance of natural systems design. It offers just enough order to create a functional framework but plenty of room for variation, spontaneity, and adaptation to the context. In a sense, this is design without design.

## DESIGN WITHOUT DESIGN

No ecosystem or city or any other form of complex adaptive system is designed in the conventional sense of the word, but each follows patterns shaped by and specific to its elements and context. The possibilities are bounded by the laws of thermodynamics and the rules of chemistry and physics and constrained further by the properties of each system's elements and their interactions. Over time, patterns develop, and a good observer can begin to spot some of the operating principles and rules of thumb at work. In an ecosystem, patterns of interactions take the form of food webs, succession over time, and development of niches, to name just a few. In city ecosystems, we see patterns such as gentrification, neighborhood development and decay, and shifting concentrations of political and economic power. All complex adaptive systems develop characteristic patterns of activity and behavior, and as observers, designers, and participants in some of these systems, we can learn to see these patterns, categorize and understand them, and interact with them. We can even begin to gauge their health and nudge them in ways that we think will be beneficial. But we can't "design" them the same way we can design a car or a chair. They are far too dynamic and complex.

With CAS, design resembles real-world garden design: we come up with a plan, we build it, then nature takes over and carries it in directions we didn't foresee. Every gardener has been surprised by what happens to a careful scheme to design a yard: A tree gets far bigger than you thought, some of the plants die and leave holes, one or two species get exuberant and occupy huge swaths that you'd planned for some other purpose, the dog digs up the dahlia bulbs, and for years you're moving plants and paths around in response to all the serendipity and unexpectedness. That's the nature of design in complex systems. We can't impose rigid order and dictate all the conditions; the system will always develop its own order. As natural farming pioneer Masanobu Fukuoka says, if we shove Mother Nature out the door, she will come back in through the window with a pitchfork.

This is where a deep understanding of patterns of systems is crucial. When we know that systems will evolve, will undergo succession in their own specific ways, and will self-organize in patterns characteristic of that system, we can then develop a "pattern language" for the particular patterns of the system we're working with. We can try to create the conditions that help some of the desired patterns emerge and discourage undesirable ones. But we can't do too much more than set up those conditions. Once we've done that, self-organization, emergence, feedback loops, and all the other qualities of CAS will develop and give the system its own life. At that point, we can tweak and nudge and redirect resources toward the events we want to occur, but we're not in control. The failure of the high-modernist program and of almost every top-down planning scheme in any large system, from cities to ecosystem restoration to nation-building, is proof of that.

The take-home here is that when we're working with complex adaptive systems, we can set up some conditions and guidelines in a general way that, in the best cases, will allow possibilities to emerge, creativity to thrive, and healthy and desirable outcomes to occur. Cities, as complex adaptive systems, seem best to be "designed" this way. In a sense, it is design without design. We set up a minimal number of conditions and guidelines, then let the autonomous agents within them—the people, the neighborhoods, the social groups—explore the possibilities that emerge. They will, if all goes well, self-organize to evolve creative and vibrant solutions that are shaped by the unique conditions that abide in that specific culture, the geography, and their needs and resources.

## ➡ APPLYING COMPLEXITY IN URBAN DESIGN ⬅

All this sounds great—if a little scary—in theory, but what does it look like in the real world? There are many examples, some as formal applications of CAS theory and others as grassroots innovations arrived at by observation in real complex systems, such as communities and towns. As an example of a large-scale project, a few cities are experimenting with self-organizing traffic lights; that is, "smart" traffic lights that operate under a few simple rules and adapt their sequencing to current road conditions. We've all waited at red lights where there was no cross traffic or otherwise been delayed and frustrated by not-very-smart traffic lights. Researchers in Europe have come up with a networked traffic light that counts the number of cars at an intersection, consults with nearby lights, and, using a few simple rules, decides when to flip from red to green. In models based on jam-prone neighborhoods such as one in Dresden that contains thirteen intersections, sixty-eight pedestrian crossings, and multiple tram lines, the smart signals reduced waiting times for cars by 10 percent, buses and trams by 56 percent, and pedestrians by 36 percent.[15]

In Oakland, California, entrepreneurs Alfonso Dominguez and Sarah Filley have created Popuphood, a business incubator for reinvigorating depressed neighborhoods that attempts to set up the conditions for success, then let it unfold. Teaming with the Oakland Redevelopment Agency and a local landlord, the pair was able to offer six months of rent-free space for six new businesses on one block in the city's Old Oakland neighborhood. A bicycle shop, a clothing store, and four other businesses quickly signed leases, did well, and stayed on after the free-rent period expired. The project jump-started retail business in the neighborhood and is now being repeated in other areas.

On a smaller scale, urban activists have created installations in cities that are designed to create spontaneous gatherings and interactions and to offer models of how cities could be more livable. One of these, called Play Me, I'm Yours by British artist Luke Jerram, scatters a few dozen pianos in public places around a city to encourage impromptu songfests and concerts by passersby. Over twelve hundred pianos have been installed in forty-three cities as of this writing. Many of the pianos have been decorated by local artists, and videos on the web attest to the popularity of the project and its ability to create spontaneous community.

Another "design for spontaneity" piece was Pop Rocks, which recycled plastic and fabric waste into giant beanbag chairs placed on a street in downtown Vancouver, British Columbia, for passersby to plop into. One of the most famous spontaneous design projects is Park(ing) Day, begun in 2005 by San Francisco art and design firm Rebar and now an annual event in dozens of cities. On the first Park(ing) Day, Rebar artists transformed a metered parking space into a temporary public park using sod, a bench, and a single potted tree. Park(ing) Day has since moved into dozens of cities and expanded far beyond the original tree-bench-lawn concept. Parking spaces have now been turned into temporary urban farms, bike repair shops, seminar rooms for courses, and free health clinics. A how-to manual at the website www.parkingday.org describes how to do all this legally.

In each of these examples, the designers have simply created the conditions for good outcomes to occur and let the "autonomous agents" in the system—usually people—self-organize into creative patterns that explore new possibilities for healthy urban living and more connected, enjoyable communities. The results, on large and small scale, suggest that we're starting to learn some of the principles behind urban design that can regenerate our cities.

The examples I've given here, while not specifically labeled "permaculture," are all in alignment with permaculture's design principles and methods. Permaculture design, as an attempt to grasp and articulate the strategies that nature uses to create evolving, self-renewing systems, is turning out to be beautifully suited to urban contexts. The following chapters of this book offer strategies, techniques, and examples for applying permaculture to life in city, town, and suburb, in the built environment as well as the social and economic realms. Before we get there, though, we need to spend a few pages exploring the principles and methods that permaculture uses to create whole-systems designs modeled on nature's wisdom.

# CHAPTER 2

# Permaculture Design with an Urban Twist

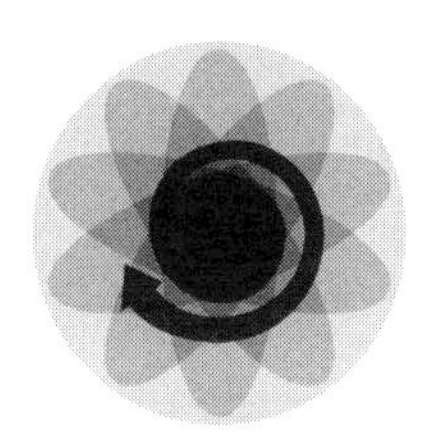

In the previous chapter I called permaculture a type of whole-systems design. It's time to explain that, back it up with some evidence, and show how to design in whole systems. First, though, let's look at what whole-systems design *isn't*. Automobile design offers a good example of the difference. Until recently, car design was done by isolated pods of specialists, each with a specific domain, each group striving to make their own piece a little better each year. One group tinkered with the engine, and its goals often varied: add more power one year, then, when fuel prices rose, increase efficiency; the next year, make it quieter. When fuel prices dropped, sheer horsepower again seized the reins. While all this was going on, another team was plotting incremental tweaks to the transmission, and others to the wheels, the ignition, brakes, steering, body, seats, the various electric and electronic packages, and so forth. At the end of each design round, another team puzzled out how to link these altered pieces into a functional whole, and yet another group piled on safety features to minimize the inherently dangerous behavior of this multiton collection of bolted-together, fast-moving parts.

The car itself had effects that rippled through society and the environment: Roads punched through countryside and were graveled, then paved, then widened; buses, trains, and trolleys languished and disappeared; isolated hamlets were pulled into easy reach; growing suburbs and lengthening commutes eviscerated urban downtowns at night; and ever-growing clouds of carbon dioxide billowed into the atmosphere.

The piecemeal approach to design had been adequate, more or less, for simple machines. But it didn't work well for complex systems. It seemed as if at some point in its refinement, every technology sailed toward its own apex of unsuitability, its own Peter Principle where simple, useful devices were plastered with complicated features until they failed. Think of the "unsafe at any speed" Corvair car made infamous by consumer advocate Ralph Nader, the many design-related crashes of the early DC-10 airliner, or the Mars climate orbiter, doomed by a missing metric-to-English conversion. Piecemeal design of complex technologies has often had harmful, or at best unexpected, effects on the physical, social, and ecological systems that they are embedded within.

In whole-systems design, engineers and designers learn that they are working with multiple, nested sets of systems that tie together to create larger systems that in turn link into higher-level systems, and so on up—as well as down—the chain. A change in one system, such as the engine or the brakes, might prompt shifts in the systems on the same level (steering or suspension), and some changes spin off effects that bounce up and down the levels, too.

From the beginning, whole-systems design looks at parts and assemblies of parts in relation to each other, not separately. A wheel isn't just a wheel. From this new point of view, a wheel comprises an interconnected system of parts: tire, rim, hubcap, bolts, and so on. Those parts are at a systems level "below" that of the wheel. None of them in isolation does what a wheel does, and they must be combined in a specific way to make this new thing that has wheelish properties.

Other subassemblies connect to the wheel: the brakes, axle, steering assembly, and suspension. We can think of each of those subassemblies as being on the same level, since they all connect to each other to create the higher-level structure of the chassis. Each subassembly in turn is made up of smaller parts that don't do much on their own, but in combination those parts form a functional component with more complex behavior. A brake pad, rotor, caliper, or other brake part is relatively inert and uninteresting alone, but assembled, the brake system moves and squeezes and has a range of behaviors that are more dynamic than any part alone.

When several subassemblies—wheels, brakes, steering, axle, and suspension—are combined, they yield a chassis, a pretty active rig that can roll down a hill in a steerable, brakeable way. None of the parts can do that. This emergent behavior at the "higher" whole-chassis level isn't present in any of the subassemblies. It arises from the parts being linked in a specific relationship. Hopping up one more level, the completed car has a complex set of properties that emerge when engine, body, and the rest of the subassemblies are linked together. Automobiles, in turn, are one piece in the transportation network along with roads, public transit, the fuel supply system, traffic laws, trucking, and other similar components. Moving up yet one more level, the transportation network links to other large-level components that make up society, such as commerce, legal systems, and others, each with its own many levels of components. In one sense these levels combine to form a hierarchy, but this term, based on the Greek word for a priest or sacred ruler, is a bad fit for describing a system in which each level is important, has its own properties, and must work with the others. The writer Arthur Koestler coined the term "holarchy" to describe this webwork of interdependent tiers, and each level he called a "holon," to reflect that each component is both a whole and a part. Figure 2-1 illustrates how the automobile can be thought of as a holon among the other holons in the transportation network.

Whole-systems thinkers recognize that any holon that they are working on—whether wheel, car, or highway network; seedling, raised bed, or market garden; patient, nursing staff, or hospital; local ordinance or town government—is a system made of connected holons and fits

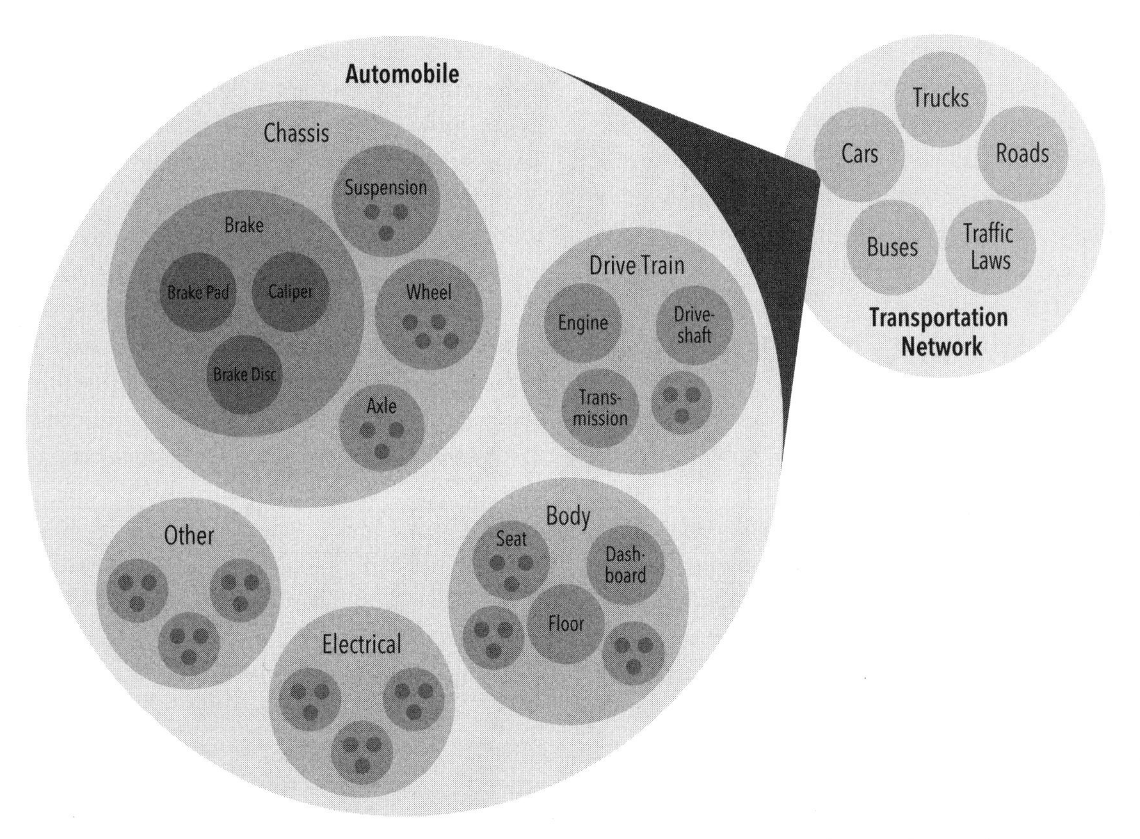


**FIGURE 2-1.** Holarchies and holons. Here, the automobile is a holon, made up of levels of other holons, in the transportation network holarchy. At each level, properties emerge that are not found in the level below. The smallest holons shown are, of course, made up of even smaller components, and the transportation network itself is a holon in larger social, economic, and political systems. We draw circles around specific holons as thinking tools, while remembering that much that we may not be taking into account lies outside and deep within the circles and would need to be considered in other contexts.

into a tier of different-level holons. To keep this potentially mind-boggling stack of tiers and nested wheels-within-wheels straight, and to keep the size of each piece of it workable, designers draw arbitrary boxes around various portions. But they're aware that those boxes are just mental conveniences to help us think and work coherently. The boxes do not have fixed boundaries that cleave reality into "real" separate chunks, any more than the divisions of the sciences into biology, chemistry, and so on reflect any true divisions in nature. We draw boxes around different things based on what aspects we are working with. The key elements to remember are that we draw a box as a tool and not as reality, and that the box's boundaries may need to be adjusted as we learn more. Designers are also aware that those boxes connect with each other and to other holons, and changes in one can propagate through many other boxes at many scales.

Permaculture and other methods of whole-systems design, then, are attempts to develop a set of tools that help us work constructively within this new framework of complex systems. Piecemeal tools aren't up to the job. To work with complex systems, we need design tools that focus on the connections and functions of the holons—the parts-that-are-wholes—that generate the various emergent features, rather than solely on the parts themselves.

## WHY FUNCTIONS MATTER

You'll often hear permaculture designers talk—and talk—about function. We'll exhort newbies, "Make each piece in a design serve multiple functions," or, "Support important functions in several ways." Why the focus on function? It turns out that function is a powerful leverage point for applying whole-systems design. Good design is not just having a bunch of cool stuff strewn around randomly. At its heart, design is about placing the right parts in useful relationships so that the desired processes can happen. If we want a rainwater tank to fill, it's got to be located below the height of the roof gutters that feed it, and of course we also want it to be connected somehow to whatever it is meant to water. The relationships are just as important as the parts—sometimes even more so. Often we can swap one similar part for another—a plastic tank for a pond—but the relationship must stay the same for the system to work.

Function, it turns out, is an indicator of, is created by, and tells us about relationships, and relationships are what unify a design and make it work. Function means that *this* does something to *that*. In other words, *this* and *that* are connected in some way. A whole system—an organism, a community, a business, or an ecosystem—exists only because of the connections and relationships among its parts. Remove the connections, and there's no system left.

Thinking in terms of functions—drawing the conceptual box that way—lets us more easily create useful relationships. For example, in rainwater harvesting, the roof functions to deliver rainwater to the tank. That's one relationship: roof to tank. The tank, in turn, functions to store water and send it to the soil and other users. Now we have more relationships: tank to soil, to plants, to faucets, to the animals (including humans) that need the water, to the pipes or delivery ditches, and to the slope or pump that moves the water. Thinking about the tank in terms of some static quality such as its shape or color or weight is only part of getting its location right. But if we think about its potential functions (and usually more than one), we know a lot about what it needs to be tied to and thus where an optimal location is for it.

It's important to realize that everything in a design has functions, properties, or features beyond the ones we've picked it for. That rainwater tank, yes, it holds water. It also is a big object, so perhaps it can act as a windbreak or privacy screen. It also casts shade. Full, it has thermal mass, so it can keep an area a bit warmer or cooler. It's got vertical surfaces, so we could trellis plants on it. Now we're really starting to dial in its location and how it is tied to other parts of our design. This is one reason an early step in permaculture design is to list the functions and properties of all the important elements—so we can spot potential connections among them.

## FOUNDATIONS OF THE PERMACULTURE DESIGN PROCESS

Let's start moving from whole-systems theory to how we actually design and build with it.

How do we design something that creates and preserves all those connections, the relationships that result in the resilience and abundance we see in living nature? This is the beauty of permaculture. It offers a set of design principles for creating useful relationships that guide us in formulating our plans, and a host of connection-building design methods that help us decide which techniques to use to implement those plans. Permaculture design is also ruled by and starts with a set of ethics, unlike many other design approaches, to increase the odds that our efforts are beneficial in the world rather than destructive. Ethics, in this case, are the things we do in the world that express our moral sense, our inner compass of right and wrong. Permaculture ethics, as set out by Bill Mollison in his book *Introduction to Permaculture*, are these:

- Care for the earth
- Care for people
- Return of surplus time, money, and material toward those ends

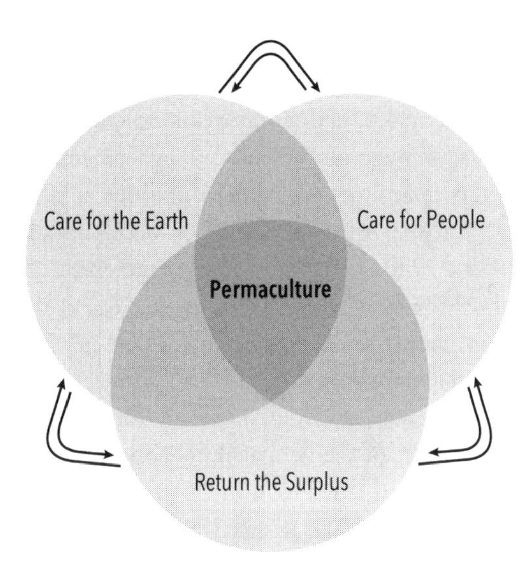


**FIGURE 2-2.** The three ethics of permaculture. Caring for the earth often turns into caring for people as well, and vice versa. Earth care and people care generate abundance, and the surplus from these is reinvested in earth and people, which preserves and supports the systems and their yields. The presence of all three ethics sets the stage for permaculture design.

Before and during every design project, permaculture designers ask, "Is what we are doing caring for the earth? Does it care for people? Are we reinvesting a portion of the design's products to support the earth and the people that make it all possible?" This is how permaculture keeps us in tune with the new paradigm of meeting human needs while preserving ecological and social health. Our tool-loving monkey brains are easily distracted by shiny toys and a "because it's there" mentality that can tempt us into actions that have tragic consequences. The physicist Robert Oppenheimer, who read sacred Hindu texts in the original Sanskrit and in many ways had a highly developed moral sense, couldn't help but design the atomic bomb that incinerated hundreds of thousands at Hiroshima and Nagasaki, because, he said, "When you see something that is technically sweet, you go ahead and do it."[1] The permaculture ethics rule out that kind of work. They steer us away from using good methods to do destructive, vengeful, and selfish things. In other words, no matter what techniques you are using, if you're designing a bomb, a broad-spectrum herbicide, or a factory that makes them, you aren't using permaculture design. While permaculturists favor some techniques because they nicely illustrate ecological principles—hence the propensity for mulch or hugelkultur—permaculture is not a set of techniques. It's a program of ethical, ecologically, and socially sound design approaches.

## Core Principles for Ecological Design

The other piece that permaculture designers hold and use throughout the entire design process is a set of permaculture principles. These are operating rules for design that are based on observing and working with ecosystems. They were meant to answer the question: What is nature doing that lets it create self-renewing, abundant, resilient ecosystems and processes? Some of permaculture's principles were laid out in rough form by the cooriginators of the permaculture concept, David Holmgren and Bill Mollison, in the 1970s. They and many others have refined them over the years. Holmgren offers his list of twelve principles in his book *Permaculture: Principles and Pathways beyond Sustainability*. Other designers and teachers use slightly different versions, but all are just different ways of articulating the same concepts. I use a set of fourteen principles that various designers and teachers in the US West have worked out.

### 1. Observe

Use protracted and thoughtful observation rather than prolonged and thoughtless action. Observe the site and its elements in all seasons. Design for specific sites, clients, and cultures.

### 2. Connect

Use *relative location*; that is, place the elements of your design in ways that create useful relationships and time-saving connections among all parts. The number of connections among elements, not the number of elements, creates a healthy, diverse ecosystem.

### 3. Catch and Store Energy and Materials

Identify, collect, and hold useful flows. Every gradient (in slope, charge, concentration, temperature, and so forth) can produce energy; every cycle is an opportunity for yield. Reinvesting resources builds the capacity to capture yet more resource flows.

### 4. Make Each Element Perform Multiple Functions

Choose and place each element in a design to perform as many functions as possible. Beneficial connections between diverse components creates a stable whole. Stack elements in both space and time.

### 5. Have Each Function Supported by Multiple Methods

Use multiple techniques to achieve important functions and to create synergies. Redundancy protects when one or more elements fail.

### 6. Make the Least Change for the Greatest Effect

Understand the system you are working with well enough to find its "leverage points" and intervene there, where the least work accomplishes the most change.

### 7. Start with Small-Scale, Intensive Systems

Begin at your doorstep with the smallest systems that will do the job, and build on your successes. Grow by chunking, that is, by developing a small system or arrangement that works well, and repeat it, with variations.

### 8. Optimize Edge

The edge—the intersection of two environments—is the most diverse place in a system, and it is where energy and materials accumulate or are transformed. Increase or decrease edge as appropriate.

### 9. Collaborate with Succession

Systems evolve over time, often toward more diversity and productivity. Work with this tendency, and use design to jump-start succession when needed.

### 10. Use Biological Resources before Technological Ones

Biological resources, including human networks, reproduce and increase over time, store energy, renew themselves, and interact with other elements. Technological resources—manufactured goods—often don't.

## Principles Based on Attitudes

The principles just listed, based in ecological science and indigenous wisdom, tell us what to do. Equally important is our mindset and how we think about design. The remaining four principles help us in our mental approach to permaculture.

### 11. Turn Problems into Solutions

Constraints can inspire creative design, and most problems usually carry not just the seeds of their own solution within them but the inspiration for simultaneously solving other problems. As Walt Kelly's Pogo has it, "We are confronted with insurmountable opportunities."

### 12. Get a Yield

Design for both immediate and long-term returns from your efforts: You can't work on an empty stomach. Set up positive feedback loops to build the system and repay your investment.

### 13. Recognize that Lack of Creativity Is the Greatest Limit

The designer's imagination and skill usually limit productivity and diversity before any physical limits are reached.

### 14. Learn from Mistakes

Evaluate your trials. Making mistakes is a sign you're trying to do things better. There is usually little penalty for mistakes if you learn from them.

## Using the Permaculture Principles

These principles inform all our work. I think of them as filters for sustainability. When you run a set of possible design ideas through the principles, the not-so-sustainable solutions are screened out, and the regenerative ones pass through. If what you are doing or designing follows the principles, the result has a resilience, an efficiency, even a gracefulness to it, and it tends to improve the systems around it instead of degrading them. The principles also act as an alarm system, in that an action that violates the principles—say, a piece of a design that has only one function—stands out clearly. After I worked with the principles for a while, they became like a set of mantras, always cycling in the back of my mind, guiding me toward more harmonious actions. I could spot them at work in a landscape, in any elegant solution to a problem, in seeing how people behave with each other, in my relationships.

The principles, then, are touchstones for regenerative design. One hint that they embody whole-systems design is that when you use or design with one or two of the principles, several more of them appear in the design almost by magic. If a part of a design works well with edge—say, a pond that has a lobed, noncircular shape—it very often also has multiple functions, connects well to other parts of the design, and provides yields. The principles are a package deal. This "use one and the others come along" aspect shows that they are deeply linked to each other and make up a whole system in themselves.

# THE FOUR LEVELS OF THE DESIGN PROCESS

With the ethics and principles in our mind, we have begun the process of design. The point of any design is to move toward some desired outcome—a productive garden, a rewarding business—with as much certainty as possible, some sureness that we're taking the right steps. Put simply, a design is a plan or a set of strategies toward a purpose. The design process, then, is a program for articulating that purpose and for giving us a sure set of procedures for choosing the steps toward it. To set us on the path, we can think of the design process as a holarchy of four major levels, or holons, of activities. As with any holarchy, each level encompasses and incorporates the levels below it.

Level 1: Mission—the overall purpose of the design
Level 2: Goals—projects done to achieve the mission
Level 3: Strategies—the series of actions for achieving each goal
Level 4: Techniques—the methods and tasks needed to enact a strategy

Each holon at an upper level can contain multiple holons from the level below it. In other words, a mission can cover several goals, and each goal may need several strategies to attain it. Here's an example to make the distinctions clear.

## Mission

To thrive during and after the transition to a post-fossil-fuel world.

## Goals

The list of goals needs to be comprehensive enough to leave no major gaps, no big pieces missing that block the success of the mission. Such as:

- Have secure energy sources for the home and basic needs
- Have dependable, long-term food and water sources
- Be part of a supportive community for mutual help and security
- Generate an ethical income to get out of debt and meet financial needs
- And so on . . .

## Strategies

That is, plans for reaching the first goal (secure energy sources), which include:

- Find ways to reduce energy consumption
- Develop multiple sources of energy so that if one fails, others function
- Take advantage of renewable energy subsidies while they last
- Explore neighborhood-scale power generation

At this step, we list strategies for the other goals, too, and keep an eye out for overlap and synergies among them all. Those overlaps often indicate potent leverage points that stack functions. Strategy itself is a deep subject that we will cover in more detail below.

## Techniques

Here is where we list out techniques and tasks for implementing all the strategies. Look for techniques that have multiple functions, meet several goals, or further several strategies. Those will be high-priority, powerful intervention points.Techniques for implementing the first strategy (to reduce energy consumption) could be:

- Move to a smaller house
- Insulate the house
- Moderate the house microclimate with plantings
- Install a solar water heater
- Use a clothesline instead of the dryer
- And so on . . .

This approach breaks down a seemingly daunting task—surviving peak everything—into a less overwhelming set of strategies to follow, and those strategies in turn comprise lists of very manageable tasks. This is what good design does. It keeps our eye on a large, holistic mission while delivering a set of interconnected, self-reinforcing, and doable steps to achieve it.

## THE CENTRAL ROLE OF STRATEGY

Permaculture focuses primarily at the level of strategy and goals. Strategy—planning—is the linchpin of good design. It also turns out to be the toughest, most slippery part of design. It's where many designs and designers go wrong. It's also the part that responds well to being guided by ethics and principles, and when we get our strategies right, they in turn lead us to choose the appropriate techniques. Strategy is much harder to do well than the other three levels of design. Devising the mission of a design comes from the heart, the psyche, and the life experiences of the people involved. It is what triggered the design project in the first place and is often fairly set by the time a designer comes on the scene: We need a new kitchen; we want to replace our lawn with something more sustainable; we want to start a business that makes solar water heaters. There's not a lot of give there, other than for the designer to assess whether the mission is doable given the resources or advisable to do at all. Similarly for goals: For a mission to be successful, most or all of the bases have to be covered. Basic needs must be met. If we look at the goals we listed for our "surviving peak everything" mission, we see that they each have to do with meeting a fundamental need such as food, energy, security, and livelihood.

At the other end, a technique, once chosen, is also well defined. A technique, whether it is building a swale, framing a wall, conducting a ritual, or voting via majority rule, is carried out following formulae and standard procedures that are varied only a bit for local conditions. Again, there's only a modest amount of give to adapt a technique to the circumstances. Applying a technique well depends mostly on having clear instructions and plenty of practice.

Strategy is where the flexibility, the creativity, and the greatest chance to go wildly, excruciatingly wrong in design all lie. I want to emphasize the importance of learning to develop effective strategies. All the other factors of design—a well-chosen mission, appropriate goals, having ample resources, knowing the right techniques, everything else—don't matter if the strategies aren't well chosen. Lavishly financed ventures have failed, overwhelmingly superior armies have been routed, brilliant ideas abandoned, vast fortunes squandered because of poor strategy. So repeat after me: Strategy is at the heart of design.

What is strategy? Simply put, it's a plan. It's a set of ordered steps to achieve a solution. One of my favorite definitions of strategy is that of management consultant Henry Mintzberg: "A strategy is a pattern in a stream of decisions."[2]

This highlights, in just a few words, several important aspects of strategy. First, a strategy is built of decisions, choices to be made and carried out, and nearly always not just one but several. The decisions are in a stream; that is, a flow or sequence, in which one choice leads to and opens the door for another. The decision and the sequence combine to make each other's successful outcomes more likely. And finally, the decisions are patterned: There is an order, a detectable rhythm, relationship, and harmony in the decision chain. Together, the decisions create a unified whole.

When strategy is properly developed, actions, techniques, and forces come together to drive inexorably toward the goal. Set up right, good strategy creates the same feeling as jumping multiple pieces in a checkers game—click, click, click, click, and there's no stopping it. That's what we're hoping for in good design, to create the conditions for our desired outcome to be inevitable. When it rains, the gutters harvest rain, the swales stop runoff and nourish the soil, and the cisterns and tanks simply fill. It has to happen. It's just physics. (Of course, this is an idealization. In the real world, uncertainties loom, bumping us away from perfection, but good design tremendously increases the odds of the desired outcome.)

How do we craft good strategies? Management consultant and professor of business at UCLA Richard Rumelt says that a good strategy has this structure:[3]

1. **A diagnosis.** We have the mission and goals in mind. Now, what is keeping us from reaching the goal? What are the challenges and obstacles? Also, what are the opportunities to build momentum?
2. **A guiding policy.** What is our overall approach to resolving the challenges that we have diagnosed? Here, the permaculture principles give us the approach. We use them to create an integrated, cohesive program (I'll give an example below).
3. **Coherent actions.** How do we create a coordinated series of steps to carry out the approach? Permaculture's design methods, which I will detail shortly, are ways of arriving at a well-patterned series of steps.

That format dovetails nicely with the permaculture design approach, and it's what I use to develop strategies. Here's an example of how to develop a smart strategy using it.

One of the goals listed above was to have reliable sources of healthy food during and after energy descent. First, we need to remember that the goal does not dictate any specific technique. There are potentially many ways to achieve it. Being members of a technique-focused culture, we're liable to think of the goal in method-specific, narrow terms, such as "I must grow all my own food." The goal doesn't limit us to that. It doesn't suggest how we do it. It just says, "Get reliable food sources." One way to reach the goal could indeed be growing some of our food. But that's subsistence farming, and it's a full-time job. Also, it relies on a single source—you—which violates permaculture principle number 5: each function is supported by multiple methods. What if you get hurt or sick? What if deer eat your garden? You need multiple pathways for meeting your food needs. Remember that a strategy is not a single decision but a coherent pattern in a stream of decisions. We want to develop a strategy that has multiple streams of decisions that reinforce one another. The strategy structure helps us develop multiple options via its first step of diagnosis, so let's do that.

## Strategy Structure Step 1: Diagnose the Challenges

Instead of swooping toward a single-method solution, as we're often tempted to do, the strategy structure says we should first ask, "What are the obstacles keeping us from our goal of reliable, ethical food sources?" In permaculture design, we call this the observation step: What are we working with? What are the features of the landscape, including the parts that may get in the way of our project? So we list them. Some potential obstacles to gaining secure food sources include:

- a short growing season, limited land, a busy schedule, and other challenges to growing it all ourselves;
- the high cost of buying food;
- the possible unreliability of food from far away during energy descent;
- supporting practices we don't agree with by buying food at chain stores.

After listing the obstacles and challenges, we move to the next step.

## Strategy Structure Step 2: Create a Guiding Policy

How do we build a guiding policy for solving these obstacles? The permaculture principles give us some clear avenues to that, so this is a good place to employ them. We've already begun using principle 1, observation, to list the challenges. As part of observation, we should also list the available resources and solutions, such as all the possible places we could obtain food.

Other principles that apply to food include:

- Principle 3, catching and collecting resources that are flowing by already, such as friends who garden or are chefs or who work in some aspect of the food system. We can also use food-buying clubs and even Dumpster diving at grocery stores.
- Principle 5, using multiple methods of meeting the challenges, such as growing food yourself, finding neighbors who garden, community gardens, community-supported agriculture (CSA: local farms that members subscribe to for a regularly supplied box of fresh food), farmers' markets, and shopping at stores that support local farms.
- Principle 6, finding the leverage points. If you're going to garden, this principle suggests that you get the most bang for your buck. What food is the most expensive? What is easiest to grow in your conditions? What do you eat the most of? Your answers to those three questions would be good candidates for high-priority crops to grow. For foods that you buy, what can you buy in bulk, in season, in quantities you can share with friends to bring down the price?
- Principle 7, starting at your doorstep locally. This also ties into principles 3 and 5. What makes sense to grow in your yard? Now what is left? Can neighbors grow that for you? If not, can you get it through a CSA or buying club? The closer you are to a food source, the more control and influence over it you have.

Notice that at this point we're making lists and finding possibilities, not making choices or specific plans. Much of the effort in good design is spent observing and brainstorming to make sure we're capturing as many ideas and options as possible.

Permaculture has other, powerful tools for narrowing down the choices and organizing them, which we will get to shortly. For now, it's important to fight our innate tendency to grab the quickest answer—"Hey, this method sounds like it would work!"—and make sure we're considering the widest range of possibilities. We don't want to collapse our field of action too soon.

It's also important to understand that a guiding policy is made up of specific, concrete activities. We can easily mistake a restatement of the goal or challenge as a policy, such as: "Work really hard to make sure I always have food." Simply being motivated is not a strategy.

This approach to developing strategies can apply to any activity that involves design or planning, whether it is building food security, planning a landscape, writing a business plan, creating a budget, or running a meeting. I've used it for all of these. Permaculture design really is a universal toolkit. It's a strategy for developing strategies, an überstrategy.

## Strategy Structure Step 3: Organize Coherent Actions

Now it's time to gather all our possibilities and solutions and organize them into sets of steps to take. This is when we begin the process of arranging how, when, and where to implement our solutions. It's the part that most people think of as the design process; that is, it answers the question, Where does everything go? But where stuff goes is only a small part of good design, a point I can't emphasize enough. Arranging the pieces of a design feels like the meaningful—and fun—part of design. It's satisfying, and it makes us feel that we're getting something done. But by being thorough in doing all the other steps and carefully applying the various design methods permaculture uses will give far more effective, ecologically and socially sound, lasting results than rushing to get to Where does everything go?

Permaculture uses a wide array of design methods to organize solutions and design elements, depending on the context and the designer's training. Here's a list of many of those methods, with a brief description of each:

1. **Observation:** noting what is present at the site (this can be nonselective, based on certain themes, using instrumentation or sensory)
2. **Mimicking nature:** imitating the structures and processes of natural systems; working with nature, not against it
3. **Data overlay:** making a base map and overlaying it with selected aspects of the site
4. **Flow diagrams:** brainstorming via bubble diagrams, organizing a sequence of steps by using process flowcharts
5. **Following traditional and indigenous cultures:** adopting the time-tested ways of those who have a long-term relationship with a place
6. **Random assembly:** linking random elements by using random connecting words (e.g., greenhouse *over* pond *next to* house)
7. **Highest use:** prioritizing the multiple uses or functions of a design element so that the maximum number of uses and functions remains after each step
8. **Needs-and-resources analysis:** connecting the yields or by-products of design elements so that they provide for the needs of other elements, and connecting the needs to the yields
9. **Zone analysis:** placement of elements by frequency of use or care
10. **Sector analysis:** placement of elements so they work with the influences arriving from off the site

The first five methods are used in many design fields, and instructions for using them are readily available elsewhere. The sixth technique, random assembly, is described in *Gaia's Garden* and several other permaculture books. Here I will focus on methods seven through ten, as they are more or less unique to permaculture. And, each is a powerful method for doing what is at the heart of permaculture design: creating connections and relationships among the parts of a design, the people who live with it, and the environment that surrounds, nourishes, and shapes it. Each of these four methods creates a specific type of relationship that is different from the others, and I will list those relationships for each method.

## PRIMARY DESIGN METHODS USED IN PERMACULTURE

Permaculturists can choose from a fully packed quiver of design methods. These include all the common techniques used by architects, engineers, and other design professionals, such as mapping overlays, bubble diagrams, scenario creation, prototyping, and mind mapping. But permaculture, not surprisingly, leans heavily on methods that focus on creating relationships among the parts of a design. The method a designer chooses will vary with the type of relationship, and also with what is being designed—you might not use the same design method to design a garden that you would use to fashion a marketing plan. The relationships themselves can be of several types, including relationships in physical space, over time, among the parts, between the parts and their users, and between the parts and the environment of the design. For creating and optimizing beneficial, functional relationships, permaculture relies on the four methods listed below. Each one helps to create a specific type of relationship in a design.

### Highest Use

When a design element has several different uses, highest use tells us how to increase yields and efficiency by arranging those uses so that after each use, as many other uses as possible remain. Here's an example of this method in action:

Every meal that is prepared generates food scraps. Often these go straight into the garbage, so there's no yield at all from them. Fortunately, many people compost, so they get one yield from their food scraps, a soil amendment. That's better than nothing, but let's look at all the possible yields we could be getting from food scraps:

- Compost
- Food for a worm bin
- Soup stock
- Seeds
- Feed for chickens, pigs, rabbits, other small livestock, or pets
- Fiber for paper or other crafts
- Dye from onion skins, mushroom stems, beet peels, and other foods
- Feedstock for biogas production
- Mulch

Now we turn this list into a plan. Of course, we're being hypothetical here. I doubt that anyone will make use of every single function of food scraps, but we're doing this as a teaching example. So what should we choose to do first in order to preserve as many subsequent uses of food scraps as possible?

The first thing to do, in fact, is to *sort* the scraps. Note once again that in whole-systems design we don't jump right into feverish activity. We first do some organizing work, such as preparing the materials for use rather than going straight to using them. In this case we sort: Pull out the seeds; set aside peels, skins, and similar materials to make paper or dye; direct items too funky to eat toward a nonfood use; and so on. With sorting done, most of the remaining items can be dropped into soup stock. Those are the first-round choices that leave many other options. Conceivably, once some of the scraps have been used for dye, they could then be added to stock (a stretch, I know, but it illustrates the theory). If not stock, then those spent dye scraps could go toward the next-lower tier of uses.

Next we could strain off the soup stock and put the solids into a biogas digester. The leftovers from the digester could feed livestock, and any inedible bits could be used for mulch. The livestock manure, then, could go into a worm bin (worms love rabbit droppings in particular), and finally the worm compost would be used as a soil amendment. The highest-use method applied to food scraps then looks like this:

Step 1. Sort food scraps into seeds, edibles, and nonedibles.
Step 2. Save seeds, and make stock and dye, paper, and other crafts.
Step 3. Put unused items into biodigester for methane or into another fermenter.
Step 4. Feed biodigester remains to animals.
Step 5. Feed suitable animal manures to worms.
Step 6. Use suitable remains as compost.
Step 7. Use compost and manures as soil amendments.

What's notable here is that the use for food scraps that most people think of first—compost—turns out, when we apply the method of highest use, to be not the first but the last of many uses.

Highest use applies to a whole palette of resources. Water, for example, is often used only once before it disappears down the sewer, but it could be used to generate energy first, which leaves it quite clean, then be used for washing and finally as graywater to irrigate. Plenty of other activities can be improved by applying highest use. In bygone, more conserving days, people wore new clothes only for formal occasions for years, and when the duds began to show wear, they became work clothes. When the clothes wore out, women sewed any intact fragments into quilts and other linens, and finally, when the quilts wore out, the pieces became rags.

Highest use also tells us how to choose the types of resources to use. Applying this method, we can divide resources into five use categories. In descending order from most to least desirable, they are resources that:

1. increase with use;
2. are lost when not used;
3. are unaffected by use;
4. are destroyed when used;
5. pollute or degrade other systems when used.

Category five resources include fossil fuels, toxic reagents, and others that cause damage when used. Industrial culture relies heavily and unwisely on this class of resources. Category four resources include most fuels, foods, and industrial feedstocks. The way we choose to use resources also affects what category they go in. Soil, for example, can be increased through wise use, but many farming methods destroy soil (category four), and industrial chemical farming turns soil into category five toxic waste, such as the sludge that is choking the Mississippi Delta dead zone. Categories four and five are ones that an ecological society would avoid, focusing instead on the first three classes.

Category three resources, those unaffected by use, can include (under some circumstances) sunlight and wind, water used for generating small-scale energy, and some kinds of information. Category two resources, those that are lost when not used, include perishable foods, manures, muscles, some skills, languages, and cultural knowledge. Finally, category one resources, those that increase when used, are often living things, such as seeds, plants that sprout when browsed or cut (maples, willows, grasses, and many more), rare breeds of plants and animals when raised wisely, and recycled biomass. This category also includes languages and other cultural knowledge, and many forms of information. As a general rule, many complex adaptive systems, living or not, increase when used. A wise culture chooses resources from categories one, two, and three, and uses them in ways that preserve or increase them.

Many people, consciously or not, apply highest use to their work habits. At the beginning of my workday, my brain is at its best, so I do the most intellectually challenging tasks, such as studying new material that I need to master. When that part of my mind feels stuffed full, I move to writing and other creative work. Once my creative juices are spent, I answer e-mails and make phone calls. It takes even less brainpower to sort, file, and organize tasks and materials, so I do that next. When my brain has deteriorated to an inert mass, I find physical work to do. Somehow that work pattern manages to preserve some social energy, which often fills my evening. We all have different work habits, so your best framework may not be the same as mine. Some people begin with the easiest tasks and save the tough ones for later. The object of highest use in this case is to organize your workday so that the tasks done first leave you with energy to do the later ones.

To sum up: Highest-use analysis builds a specific pattern of relationships among the various uses or functions of a design element, task, or resource. Highest use also creates a pattern of relationships in time; it tells us the order in which to perform a set of steps. When what's needed is a decision-making tool for choosing among multiple functions or a way to decide what tasks to do first, highest use will often give dependable guidance.

Highest use tells us how to connect design elements or activities in time by linking their functions or uses in a sequence. It tells us what to do first.

## Needs-and-Resources Analysis

Also called niche or functional analysis, this method creates relationships among the inputs and outputs of the various elements in a design. Whether we're maintaining a landscape, running a business, or supporting a family, all the parts or members have needs that must be met for them to function, and they have outputs, too, whether they are physical products, behaviors, or effects on their surroundings. In good design those needs and resources are connected to other elements; that is, the product of one activity is arranged so it provides something that another element requires. This reduces work, resource use, time, and energy. It also forges beneficial working relationships among the parts, which is the hallmark of whole-systems design.

Here's how needs and resources analysis works. Returning to our goal above, of obtaining reliable, ethical food sources, some obstacles were "a short growing season, limited land, a busy schedule, and other challenges to growing it all ourselves." One design element that could help overcome the short growing season and increase yields is a greenhouse, so let's say that we plan to build one. First, we make an inventory of the various needs and yields of a greenhouse.

The focus of this method is to look for ways to meet the needs of each design element by connecting them to the yields of other elements in the design. In the greenhouse example, some of the materials may simply need to be bought or imported, but we can be clever about it. Can we link up with someone tearing down or remodeling an old building as a source of windows or glass doors? Energy upgrades to houses often generate single-pane windows and sliding doors, and those are fine for greenhouses. Is there a salvage business in town—for example, one of the many Habitat for Humanity ReStores?

If we have an established garden, some of the plants and seeds needed for the greenhouse are already a yield in our system. Water could be pulled from a pond or a rainwater tank, which means that the water source should be connected to the greenhouse in some way—near it or uphill from it. Some of the greenhouse yields include rainwater harvesting, heat, and temperature fluctuations. This suggests that the greenhouse itself could be the water source, by harvesting rainwater. A pond could go inside the greenhouse to act as thermal mass, which could warm the greenhouse at night. A needs and resources analysis on a pond, in turn, tells us that ponds reflect light, so having the pond inside or just to the south of the greenhouse would be a useful location for bouncing light to the plants.

**TABLE 2-1.** Greenhouse Needs and Yields

| NEEDS | YIELDS |
|---|---|
| Transparent covering | Food and other plant yields |
| Frame | Organic matter |
| Door/entry | Heat |
| Ventilation | Humidity |
| Soil | Interior space |
| Fertilizer | Light |
| Plants | Wind protection |
| Water | Mushrooms |
| Pots and seed trays | Spent potting soil |
| Harvesting and transplanting | Rainwater |
| Starting seeds and cuttings | |

Noting that the greenhouse produces humidity suggests that we could grow mushrooms as well as plants inside it. We also could use the greenhouse as a windbreak to shelter tender plants, animals, or a house from chilly breezes. Since it generates organic matter, we might want to locate the greenhouse near a compost pile or poultry. Speaking of poultry, one need for the greenhouse is fertilizer, which chickens provide. Poultry also generate heat and carbon dioxide, leading some permaculturists to keep chickens inside their greenhouse, especially during the chilly season.

In this way, needs and resources analysis of the major elements in our design or activity points out useful connections to make. This method, like all those in permaculture, applies to much more than landscape design. What people should you be in better relationship to in order to meet your own needs? What "yields" of your personality can connect you to others? How can the needs and products of your business or job be taken care of better by linking them to the by-products of other commerce?

We practice needs and resources analysis, then, by listing the needs of each design element in one column and the yields of each in another column, then making connections. Sometimes those connections are best made by putting those elements near each other physically (pond *inside* greenhouse); other times the connections are nonphysical (phone your friends when you feel low). Note that yields include not just physical products but also less tangible products such as heat or shade, and behaviors such as a chicken's habits of eating insects and scratching soil.

Like each of our major design methods, this one creates a particular kind of relationship:

---

Needs and resources analysis tells us how to connect the parts of a design to one another.

---

## Zone Analysis

This method creates a third form of connection, one between the user or central activity area of a design and the parts of the design. How often have you seen a vegetable garden tucked away into the back corner of a yard—usually because one of the residents considers it ugly? So of course the zucchini swell to the size of watermelons before they are noticed, the strawberries get fuzzy with mold, and the whole affair becomes untidy and neglected, just as that person feared. That's because it's in the wrong place. It's not on the way to anywhere; thus it takes a special trip to get there, something that in our busy lives we don't do nearly enough for healthy gardening.

Permaculture's zone system is based on a very simple concept: Place the things you use the most or that need the most attention near where you are. This seems like common sense, but it's astonishing how often it's not done.

I'll briefly describe zone analysis in terms of landscape design, but I also want to show how it applies to manifold other cases where frequency of use—how often we use or need to be around something else—is an important factor.

In home landscape design, permaculture defines six zones, labeled 0 through 5. The house is zone 0. Zone 1 holds the things we use roughly every day, and it encompasses the part of the yard closest to the doors, walkways, and other easily accessed areas around the house. For most people, those everyday things include a deck or patio, herbs and salad greens, some attractive plantings to gaze at, perhaps a dwarf fruit tree, cherry tomato, or berry bushes to provide snack fruit, a bird bath or small water feature, and similar oft-used items. It may also include nursery or seedling beds, because a single hot day can fry a six-pack of plant starts, so they need close watching. Table 2-2 shows the contents and functions of landscape zones 1 through 5. Most small yards won't contain much more than zones 1 and 2, but ideally every yard will hold a patch of wild zone 5 to let nature show us what it can do.

What I love about the zone system is how much work it saves. The parts of my yard that need the most maintenance are closest to the door and path that I use the most often. When I lounge on our front deck with a cup of coffee in the morning, I will spot a weed, bend over and pluck it, and consider my weeding done for the day—because for each of the previous many days I have done the same thing. The weeds don't have a chance under that kind of easy vigilance. The same goes for watering or mulching. This putting things where they belong avoids the unpleasantness of marathon maintenance sessions by simply fitting most needed activities into the flow of my life.

When using zones, remember that they are not perfect concentric circles but are shaped by access such as paths and doorways, topography,

**TABLE 2-2.** The Zone System for Landscape: Functions and Contents

| | FUNCTIONS | STRUCTURES | CROPS | GARDEN TECHNIQUES | WATER SOURCES | ANIMALS |
|---|---|---|---|---|---|---|
| **Zone 1: Most intensive use and care. Zone of self-reliance** | Modify house microclimate, provide daily food and flowers, social space, plant propagation | Greenhouse, trellis, arbor, deck, patio, birdbath, storage, potting shed, toolshed, worm bin | Salad greens, herbs, flowers, dwarf trees, snack fruit, low shrubs, lawn, trees for microclimate | Intensive weeding and mulching, dense stacking, square-foot and biointensive beds, espalier, propagation | Rainbarrels, small ponds, graywater, household tap | Rabbits, guinea pigs, small poultry, worms, bees |
| **Zone 2: Semi-intensely cultivated. Home production zone** | Home food production, some market crops, plant propagation, bird and insect habitat | Greenhouse, barns, workshop, wood storage | Staple and canning crops, multifunctional plants, small fruits and nuts, fire-retardant plants, natives | Weekly weeding and care, spot mulch, cover crops, seasonal pruning | Well, pond, large tanks, graywater, irrigation, swales | Rabbits, fish, poultry |
| **Zone 3: Low-intensity, extensive methods. Farm zone** | Cash crops, firewood and lumber, pasture, | Feed storage, field shelters | Cash crops, large fruit and nut trees, animal forage, shelterbelts, seedlings for grafting, natives | Cover crops, coppicing, light pruning, movable fences | Large ponds, swales, storage in soil | Goats, pigs, cows, horses, sheep, other large animals, free-range poultry |
| **Zone 4: Minimal care. Forage zone** | Hunting, gathering, grazing | Animal feeders | Firewood, timber, pasture plants, native plants | Pasturing and selective forestry | Ponds, swales, creeks | Grazing animals |
| **Zone 5: Unmanaged. Wilderness zone** | Inspiration, foraging, meditation | None | Native plants, mushrooms | Unmanaged, occasional wildcrafting | Lakes, creeks | Native animals |

Increasing Wildness ↓ Decreasing Maintenance

and the habits of the user. When someone asks, "How big is zone 1?" the answer is that it depends on how far you comfortably stray from your doorstep and how easy it is to get around your yard. A fruit tree 50 feet from the house but next to the front walk might be in zone 1, while a similar tree 30 feet away but down a slope on a doorless side of the house could be in zone 3.

The zone system has countless other applications besides home landscape design. Any physical site that has a center of activity can be fruitfully organized according to zones. Think of a city, with its downtown core as zone 1, the surrounding dense commerce- and apartment-filled ring as zone 2, urban neighborhoods of single-family dwellings as zone 3, suburbs as zone 4, and rural and farm surroundings as zone 5. Each zone has a different set of functions, needs, and activities.

Closer to home, rooms in the house naturally organize into zones. In a kitchen zone 1 includes the countertops and work areas such as sink and stove, the most used spaces. Frequently used items—flatware, knives, spices—go in the top drawers and most convenient upper cabinets, where they are instantly at hand. We don't need

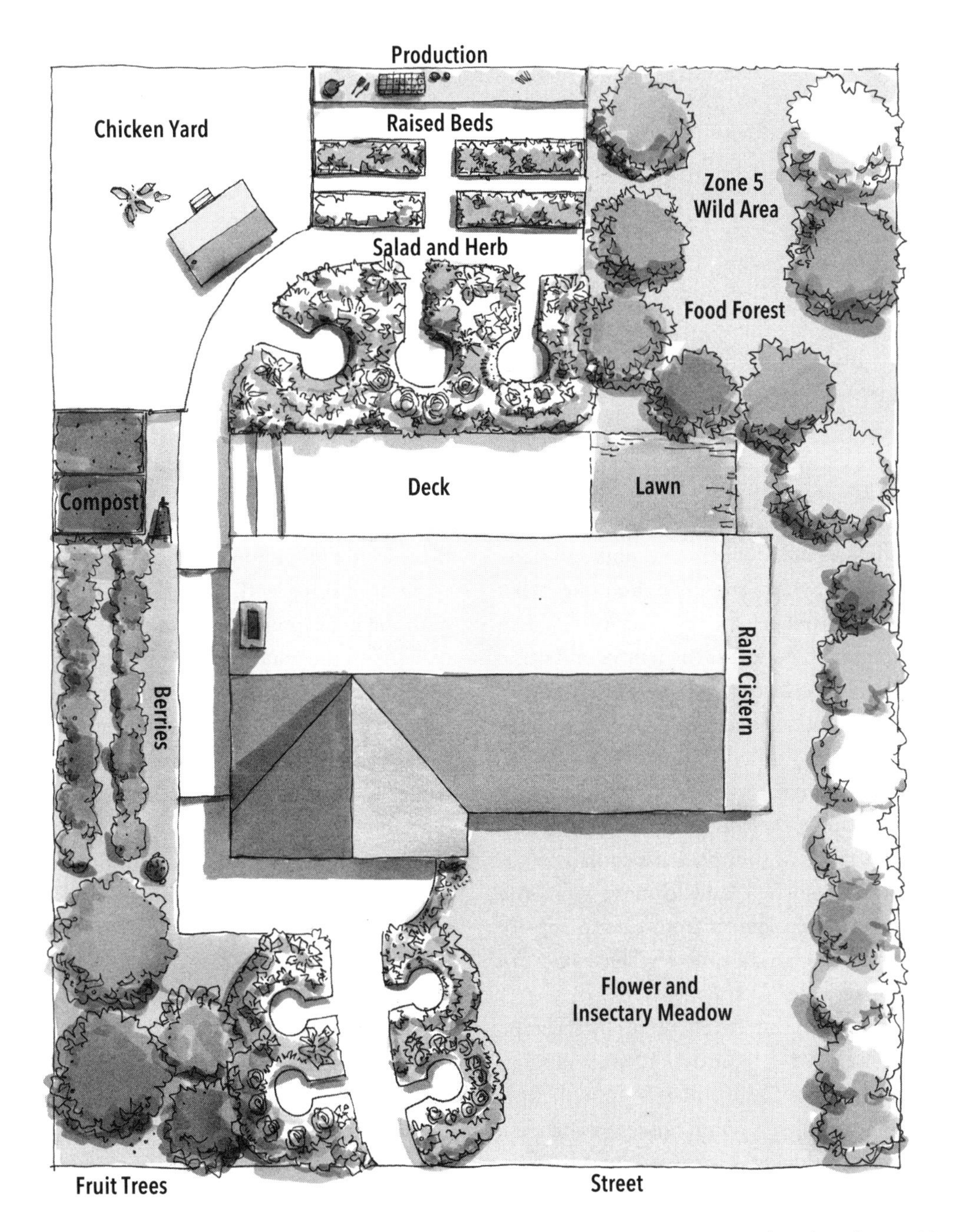


**FIGURE 2-3.** Zones in a typical small yard. Areas near paths and doors are the easiest to reach and can be considered zone 1, which is used nearly every day. Most other spaces and items in a yard this size will be in zone 2, somewhat less frequently used. The chicken coop in this case is placed near but not in zone 1 because although the birds need care every day, it may not be desirable to have poultry too close to the house. In the back corner lies a small zone 5 wild area. Illustration by Elara Tanguy

**TABLE 2-3.** Zones of Organizational Structure

| | TYPE OF ORGANIZATION | | |
|---|---|---|---|
| | **Business** | **School** | **Nonprofit** |
| **Zone 1** | Executives | Administrators, school board | Directors |
| **Zone 2** | Employees | Faculty | Board members |
| **Zone 3** | Suppliers and competitors | Students | Staff |
| **Zone 4** | Customers | Parents | Volunteers |
| **Zone 5** | Noncustomer population | District residents | Community served |

to bend or reach to use them, so I would call that zone 2. Kitchen zone 3 comprises the lower drawers and cabinets, which hold pots and pans, mixing bowls, wraps and foils, and the other supplies used frequently but less often than knives and forks. Zone 4 is the pantry and other less accessible food storage, and zone 5, the wild zone, might be that hard-to-reach cabinet over the fridge, where, buried in back, is the fondue set that Aunt Harriet sent as a wedding gift. Ordering a kitchen by zones helps suggest better placements that suit your personal cooking style. To my wife's dismay, I tend to move a favorite whisk or coffee implement from a zone 3 lower drawer to one in handy zone 2 when I find I'm using it more often.

Most offices are intuitively set up by the zone system as well. Sit in your desk chair, and what is in reach without stretching is zone 1. This usually includes a computer, note pad and pen, phone, and coffee mug. My own zone 2, because I write, holds a set of reference books—thesaurus, dictionary, usage guides—behind my laptop, which I can grab with a long arm reach. It also includes the top desk drawer. Zone 3 encompasses files in a lower drawer, and so on. Again, the beauty of the zone system is that it places the elements that are used the most often closest to where the user spends the most time.

We can also usefully apply zones to organizations of people. In a business, school, or nonprofit corporation, people could be placed into zones according to the intensity or frequency of their involvement or their relative proximity to or influence on the power or decision-making center. I've included one example of how this could be done in table 2-3.

Each of the zone communities has different needs, varying frequencies of contact and intimacy with the others, and dissimilar amounts and kinds of power and responsibility, and each is affected in different ways by the actions of those in the other zones.

Relationships with others fall into zones as well, and in chapter nine I offer some examples of how to use the zone system this way. We can use zones in several other ways. In our example on meeting food needs above, we began to touch on zones in looking at where our food comes from. If we think of food sources in terms of zones, those sources that we have the most control over and work the most intensively toward are in the inner zones. Thus our garden is food zone 1, neighbors and community gardens are in zone 2, CSAs and buying clubs in zone 3, locally owned food stores in zone 4, and chain stores in zone 5. As with all zones, we want to meet as many needs as we can in the inner zones and rarely visit the outer ones. This shrinks our food footprint, gives us influence over how our food is grown, and lets those with specific skills or better growing sites provide specialty foods for us.

In *Gaia's Garden* I described the urban zone system conceived by Resilience.org editor Bart Anderson. During a permaculture design course, Anderson was grabbed by the idea of zones, but his urban life didn't include a yard. Instead of just

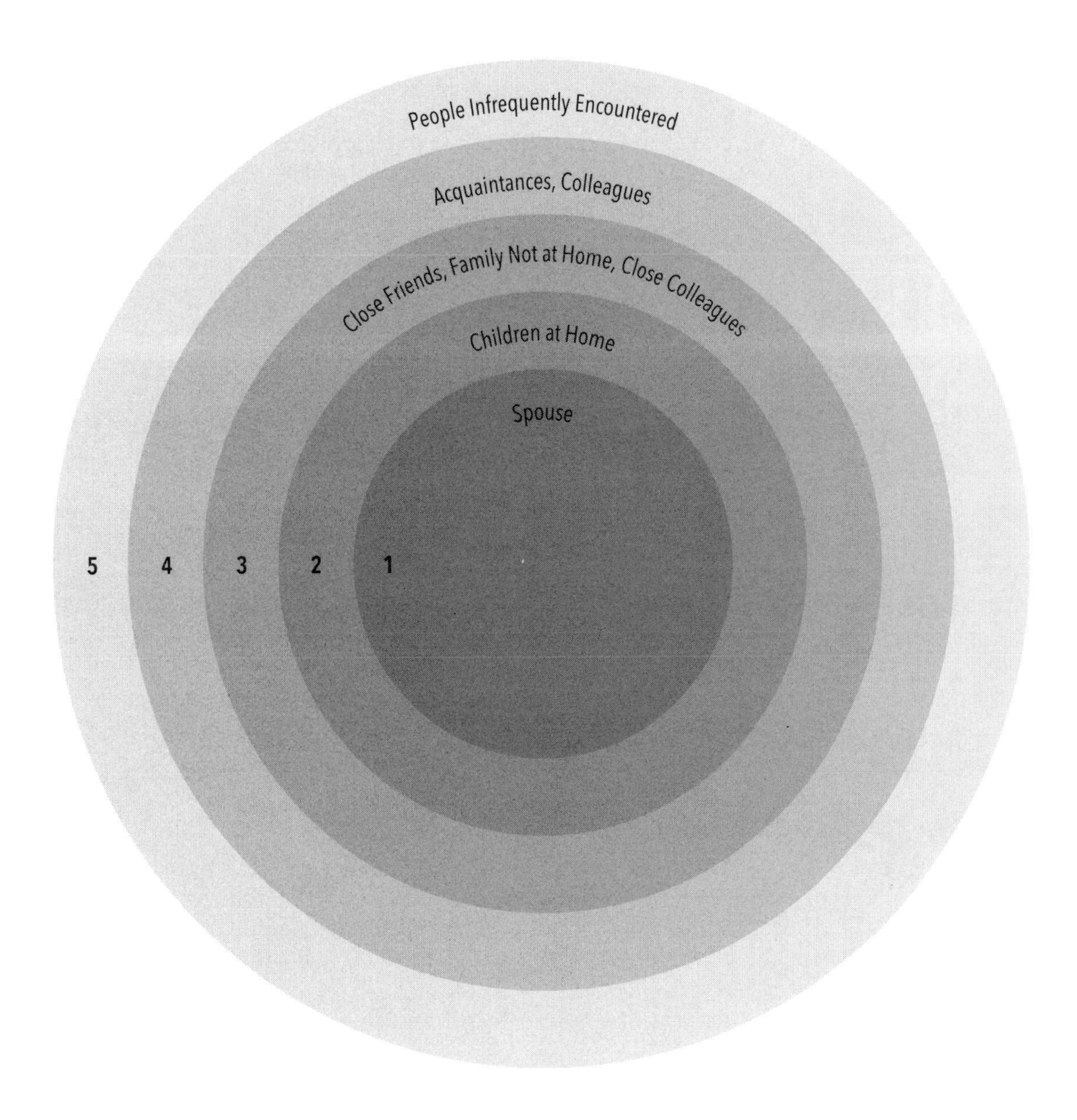


**FIGURE 2-4.** Zones of relationships. The people you see and are affected by the most are in the inner zones. Your personal zone system may differ substantially from this one. You might consider your children to be in zone 1, or perhaps your parents are living with you and are thus in your zone 1 or 2.

**FIGURE 2-5.** Urban zones by the type of transportation used to reach each. Ideally, we would arrange our lives to meet most of our needs in the inner zones, rarely needing fuel-consuming modes of transportation. Concept by Bart Anderson.

dropping the idea into the "I'll use it someday" pile, he realized that in his case the core concept of zones—frequency of use—could be linked to the type of transportation he used to get places. He sketched out a diagram in which zone 1 included all those spots that he regularly walked to. Zone 2 encompassed places he rode a bicycle to, zone 3 covered the areas he used public transit to reach, zone 4 comprised destinations he could get to only via car, and zone 5 held the far-off spots best reached by plane, train, or boat. His goal was to arrange his life so that most of his movements were in the walking and biking zones 1 and 2 because that would shrink his energy and carbon footprints. The zone system not only made Anderson's goal easier to achieve but enabled him to envision and organize it in the first place by providing a framework to map his regular destinations by highlighting the concept of frequency of use.

Remember that zones are a flexible thinking and designing tool, not a set of hard-and-fast categories filled with never-varying elements. When someone asks if the compost pile—or parking lot or vice president of sales—goes into zone 1 or 2, the answer is always, "It depends." Where is the relevant center of use, who is affected by the placement, and in which zone does it form the most useful relationships? Zones can shift, too, depending on where the users are. When I'm at my writing desk, my music system is across the room in zone 3, which is where that distracting device belongs. But when I'm in my comfy reading chair, the stereo is next to it, in the chair's zone 1.

Organizing the relationship between the user and the item used is the zone system's forte. Zones apply whenever we have a collection of things—objects, people, ideas, places—that we encounter, connect with, or use with different frequencies. The less contact needed between the user and the element, the farther from the user is the zone it's located in.

The zone system organizes the parts of a design in relation to the user or center of use.

## Sector Analysis

The final method in this list of permaculture tools is perhaps the most powerful, and it's one that should never, ever be skipped. A *sector* is a bit of Mollisonian jargon for an outside influence that can't be directly stopped or controlled but that affects the site or design area. The classic examples of sectors in landscape design are sun, wind, fire, wildlife, views, and similar forces. Sectors are more powerful than we are, so we must deal with them by creative design work. We can't turn off the sun or move it; our options are limited to choosing and placing design elements so they can use the sun (for example, a solar panel in a sunny spot), block it (a shade tree), or be unaffected by it (a shade-loving plant under a mature conifer). A detailed sector map of a portion of Naropa University in Boulder, Colorado, can be found in the color insert, pages 2 and 3.

Town dwellers must work with natural sectors but are also buffeted at least as much by not-so-natural-sector energies. These include:

- Noise
- Pollution (soil, air, water)
- Smells (restaurants, traffic, fumes, neighbors)
- Views (attractive, ugly, and for safety and observation)
- Traffic (vehicle and foot)
- Zoning and codes

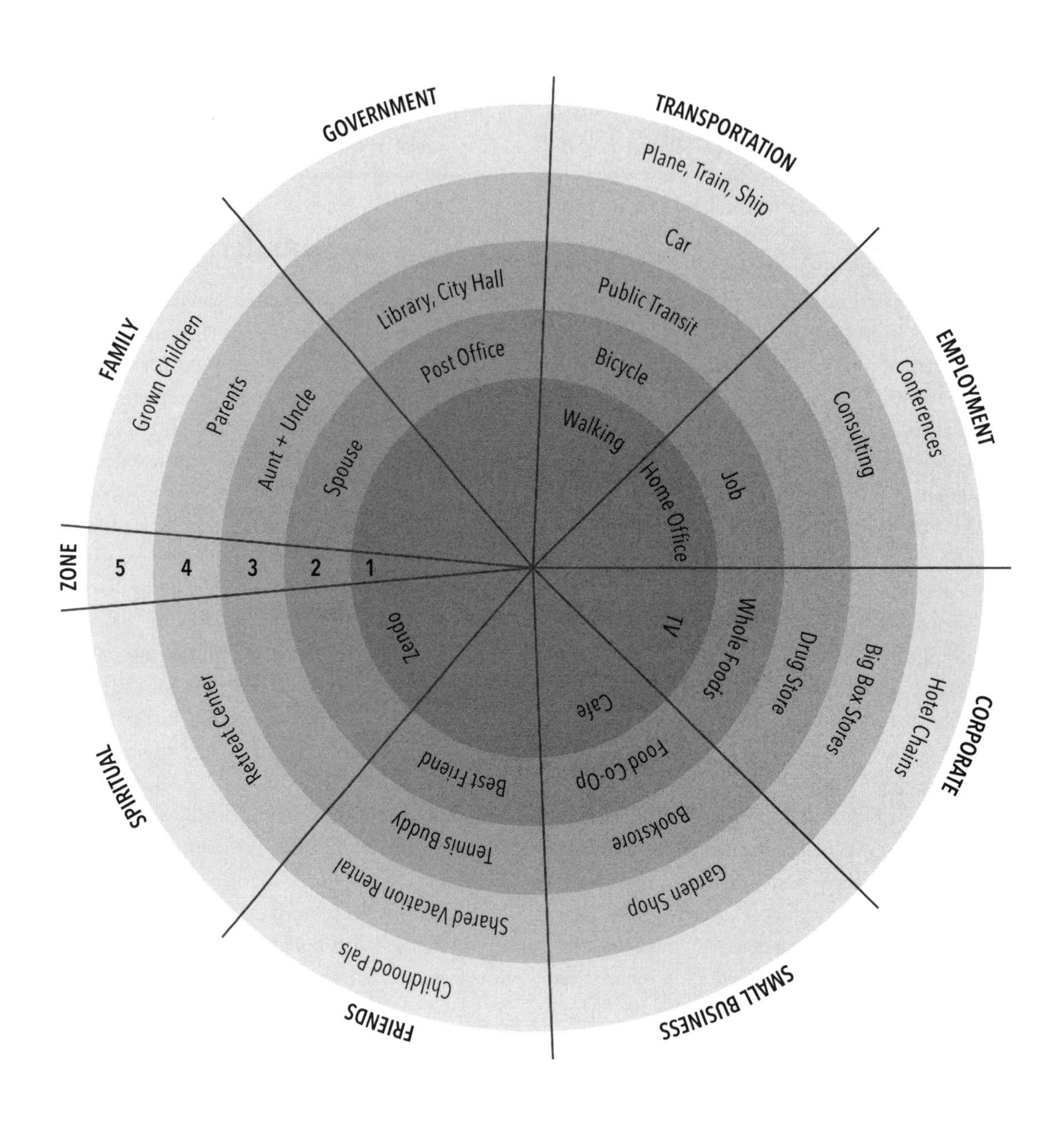


**FIGURE 2-6.** Urban sectors, zones, and some of their contents. Concept by Bart Anderson.

- Homeowners' associations
- Easements and utilities (must provide access for repairs)
- Neighbors
- Passersby (neighbors, guests, children, transients)
- Commerce (deliveries, solicitations)
- Local customs
- Pets
- Burglary and street crime
- Arson
- Law enforcement

Los Angeles permaculture designer Larry Santoyo, highlighting the critical role of good sector analysis, tells his students and clients, "Sectors trump everything." He means it. If a design gets the sectors wrong, it may fail utterly. Obvious examples include not obtaining a permit for a building and experiencing the misery of an inspector's stop-work order or installing plantings along a fence that are quickly pounded to mulch by your dog's vigorous border patrolling. Friends of mine built a sauna during the summer dry season, facing its door to the south, away from the cool northern breeze. But they missed some critical points: The winter rainstorms, which they hadn't thought of during the summer construction, blasted in from the southwest through the open door and soaked the clothes hung outside it. The sauna turned out to be in a low spot that filled with water from those same rains. As a result my friends faced tedious, expensive retrofitting: windbreaks, an enlarged roof overhang, enclosed clothing storage, and drainage trenches dug all around the site. None of that would have been necessary with thorough sector analysis. This is typical of incomplete design analysis, where one poor decision leads to a cascade of makeshift compensatory repairs—and those repairs, because of their haste, have their own repercussions.

On the other hand, good sector analysis can turn otherwise wasted or destructive forces into assets. In Tucson during the summer monsoon storm season, rainwater expert Brad Lancaster glumly watched deep rivers of rainwater surge down his street, heading for storm drains to be piped away. Irked that there, in the desert, this most precious resource was treated as a problem and unable to rebuild the city's streets or water-wasting attitudes himself, he struck on a solution. Lancaster chipped out a section of curb and diverted the street's stormwater into a set of basins on the parking strip that flowed to his trees and yard, irrigating them. This was years before this now fashionable technique, called a curb cut, became part of many a dry city's water-saving repertoire. Lancaster may have started the trend by viewing stormwater as a positive, not negative, sector.

Good sector analysis can be subtle. In *Gaia's Garden* I told the story of the ecovillage residents in Los Angeles who, because of heavy shade in their backyard, planted their tomatoes by the sidewalk. The tomatoes never had a chance because of the schoolchild sector: The kids from the elementary school across the street would nab any fruit showing the slightest blush of ripeness. The solution? The ecovillagers planted varieties of tomato that don't turn red upon ripening. The schoolchildren never really caught on.

Sectors, like zones and the other methods, don't apply only to landscape. We can also consider the influences upon our daily patterns, our jobs, and our relationships as sectors. For example, we can take our map of urban zones and overlay sectors onto it, using the influences on urban life, and populate it with some sample elements. See figure 2.6 for an example.

I regularly apply sector analysis to my own life. Doing a sector analysis on my current living situation helped my wife and me immensely. We rent the main house on a site that also holds a cottage where the owner lives, and her grown son lives in an apartment in the barn. Also on the site, in a beautiful redwood building we call Redheart, is a business run by a neighbor. The land itself, which the owner calls Amazing Grace, has its own powerful influence. Not long ago my wife and I were struggling with the increasing effect of the growing business at Redheart and of a number of other projects going on. Kiel and I had been feeling that our own interests and projects were being overtaken and undone by some of the others. Kiel, smart woman that she is, suggested I map out the situation using permaculture design. I realized that all these influences could be thought of as sectors: The Toby and Kiel sector, the owner sector, the son sector, the Redheart sector, the Amazing Grace sector. By mapping them out and seeing where the various intersections were, I quickly saw where the imbalances lay. This helped me realize that the solution was not to try to lessen the other influences—to fight the other residents and activities—but to strengthen our own influence by moving some of my classes and off-site projects to Amazing Grace. The sector analysis of our living situation also let me spot some areas of friction and potential for positive interactions that I hadn't been aware of.

Every living situation and workplace has sector influences. We can fruitfully think of relatives, friends, colleagues, staff, and supervisors as sector energies. In this book's chapters on community and livelihood, I cover several methods for applying sector analysis to human relations and other nonlandscape arenas.

Each sector has its own qualities and patterns. The sun has properties of heat and light, patterned in the sun's rising and setting and its changing elevation over the seasons. The zoning department of your town has specific qualities and patterns, too: personalities of the staff, the ordinances they must abide by, the unwritten rules that the department in fact operates by, the heavy workload, and so on. Learning the qualities and patterns of each sector influence will let us become not just better designers but more effective people, able to work smoothly and constructively with the influences in our lives. Do we want friction and conflict with the larger forces in our lives, or do we want those energies to align with our own and enhance our actions? Sector analysis is a powerful tool for organizing our landscapes, plans, workplaces, and lives into beneficial relationships with the influences we cannot directly control.

Sector analysis organizes design elements into useful relationships with outside influences that we cannot directly affect.

These four design methods offer a powerful suite of tools for making that most intimidating of all design decisions: Where does stuff go? The heart of whole-systems design is placing design elements in effective, useful relationships, and these design methods do just that, each in its own way. In combination, these methods act like a multistage filter, winnowing out the unproductive placements of the parts of a design, letting the most functional arrangements percolate through. When done well, these methods combine to create synergies on many levels among the elements. Look what we've got going: Highest use places elements and processes in an efficient sequence in time; needs and resources analysis connects the elements to each other; zones arrange elements in elegant relationships with the user or center of activity, and sectors align design elements in strategic relationships with outside influences. These four methods create four different, interacting types of useful relationships in a design, letting us be confident that we are arranging the elements in optimal places and with beneficial connections.

## THE STEPS OF THE PERMACULTURE DESIGN PROCESS

We now have a design toolkit in hand that is general enough to be applied to nearly any situation while containing enough specific directions to tell us what to do at each step. Permaculture's ethics and principles are the guides and touchstones that help our decisions to be ecologically and socially sound, while the remaining concepts and tools provide the framework for the questions that we need to answer, such as, What is our mission? What methods will work best?

To use any toolkit effectively, we need to know when to use each tool. Although the four levels of design give us a structure to fit each task into, we still need a process that guides us through design from start to finish. Permaculture designers give their design processes an assortment of names and acronyms, but they all follow the same pattern: observation, research, conceptual design, master planning, implementation, evaluation, and tweaking.

### Observation

The foundation of good design, observation is the process of collecting information about the project. That information breaks down into three primary types:

- What conditions, challenges, opportunities, and constraints exist at the site? (We're using the term "site" broadly, as a physical place, a business, a neighborhood, or any other type of project base.) These conditions include the physical layout of the site; what exists there, including vegetation, materials, and soils; the sector energies and existing zones; and the structure of its activities.
- What resources are available for the design, including money, skills, labor, and physical resources at the site?
- What are the client's needs and interests? This is where the designer and clients fine-tune the goals of the project. Observation helps us shape our mission and goals by determining the needs of the site, project, and clients and whether we have the resources to achieve the project's goals. It shows us where we are, which in turn lets us see the distance we need to cover.

The information we're gathering in this step will also be used in the diagnosis step of the strategy structure process, because it informs us about challenges and obstacles.

### Research

Here we assemble the measurements, surveys, lists, client interview results, and all the other data we collected during observation and organize it into a package of maps and other documents that forms the baseline on which the design builds. We then do further research on critical components that we've identified. Are the soils suitable for the new design's activities? What are the qualities and needs of the plants? What skill sets will the staff or clients need?

## Conceptual Design

In this step we hold our mission and goals in mind and begin developing the strategies needed to accomplish them. Here we are using the diagnosis work we did in the observation step to support the guiding policy of our strategy structure. We can ask a series of questions to work through this step:

- What systems and elements must we develop to support the goals and smooth out the challenges? (These systems can be physical, such as water harvesting, or conceptual, such as a decision-making process for a business.)
- What permaculture principles guide the choice of the systems and elements?
- How can the needed systems be made compatible with the needs and resources of the site, project, and client?
- What is the general structure of those systems? That is, are we doing graywater reuse via a laundry-to-landscape system or by gravity flow or a mix? Are we setting up a 501(c)(3) structure or a for-profit, benefit-oriented B corporation?

## Master Planning

This is the step that most people think of as design, where the locations and relationships of the systems and elements are put on paper and the organizational structure is laid out. We use the permaculture principles and design methods to guide these steps, to help us create beneficial relationships among the parts, and to give us confidence that we are making smart decisions. This is the coherent action phase of the strategy structure and the beginning of the final technique phase of our four levels of design.

## Implementation

Here we plan the sequence of tasks that will make the design real, then implement them. Each type of project has its own best sequence of implementation steps; in landscaping we do major earthmoving first, while in business design raising money is often the initial tangible step after planning. Here we are deep in technique and action, but because of the effort we've spent on observation, goals, and strategy, we can move forward with confidence.

## Evaluation

This step is missing from many traditional design processes. Often, architects and designers move onto another job by the time one project is done and don't hear whether their concept actually worked. In permaculture design it's an integral part of the design process. It creates a feedback loop, a defining hallmark of any whole system.

This part of the design process connects back to mission, goals, and the observation step. What's working? What isn't? It's also a good idea to add a brief evaluation session at each stage of the design process: Does our research raise any red flags? Is the master plan still in alignment with the mission?

## Tweaking

We're prone to thinking that a design has failed if it needs revision after it's built. But as every gardener knows, a few things always turn out not quite as planned, and they need some tweaking. Did that semidwarf pear just love your rich, loamy soil and grow twice as big as you had planned, shading the raspberries? That's life, and that's fine. Move the raspberries or prune the

## A Summary of the Design Process

Foundations of the Design Process
- Core Principles for Ecological Design
  - Observe
  - Connect
  - Catch and store energy and materials
  - Make each element perform multiple functions
  - Have each function supported by multiple methods
  - Make the least change for the greatest effect
  - Start with small-scale, intensive systems
  - Optimize edge
  - Collaborate with succession
  - Use biological resources before technological ones
- Principles Based on Attitudes
  - Turn problems into solutions
  - Get a yield
  - Recognize that lack of creativity is the greatest limit
  - Learn from mistakes
- Using the Permaculture Principles

The Four Levels of the Design Process
- Mission
- Goals
- Strategies
- Techniques

The Central Role of Strategy
- Diagnose the Challenges
- Create a Guiding Policy
- Organize Coherent Actions

Primary Design Methods Used in Permaculture
- Highest Use
- Needs-and-Resources Analysis
- Zone Analysis
- Sector Analysis

The Steps of the Permaculture Design Process
- Observation
- Research
- Conceptual Design
- Master Planning
- Implementation
- Evaluation
- Tweaking

tree hard. If you did the design process well, the changes will be modest tweaks, not wholesale revisions. Also, make a complete plan for integrating the tweaks rather than dealing with each as if they are unrelated. Gently guide the design back toward the desired trajectory, applying the permaculture principles (such as least change for greatest effect) to your tweaking strategy.

The evaluation and tweaking steps of permaculture design, as well as its reliance on ecological principles and connection-building methods, help it avoid the failures of top-down, central planning by allowing feedback and adjustment over time. The high-modernist schemes of Le Corbusier and Robert Moses mentioned in chapter one failed in large part because they treated land, people, and culture as rigid cogs in a machine and assumed that design was simply a matter of locking the gears into one another and turning a crank to churn out a happy society. Design of complex systems doesn't work that way. The parts—the people, the plants, the communities—are not gears that ratchet together but dynamic, living entities. They connect by relationships, not by metal teeth. Relationships are softer and more fluid than gears and tie-rods. Rather than bolting the elements in a design rigidly together, permaculture design connects them by placing them into flexible relationships that can adjust, change, and grow. In this way permaculture can create the conditions that let whole systems evolve, adapt, and self-organize, which is what brings them to life.

Now that we have an overview of the steps of the design process and a format for developing intelligent strategies, it's time to move on to specifics. We will explore how permaculture design is applied to life in cities, towns, and suburbs, starting with the familiar territory of food and land use as an easy beginning to developing the concepts further. But where I'm excited to go after that—and I hope you're just as eager to go there with me—is how permaculture design is serving so many other human needs as well.

# CHAPTER 3

# Designing the Urban Home Garden

The home is much more than simply where the heart is. Creating a comfortable home is what much of human endeavor strives toward. This also makes the home and its needs one of the principal drivers of the global economy. It's a critical leverage point for ecologically sound living.

To meet human needs, the worldwide economic network and its underlying national, regional, and local economies act as colossal wealth pumps. They vacuum up chunks of inanimate and living treasure from the earth's carpet of rocks and life, grind up these riches in the industrial machine, rework it all with intricate hardware and software, and spew out a cornucopia of goods. And where does the business end of this global wealth pump point? Straight at the home. If we trace the path leading from even the largest, most industry-oriented machinery—the coal mines, steel mills, tractor factories, and the rest—nearly everything they make or what their products make ends up coming into your front door, yard, or garage. Even our naval fleets and Predator drones exist largely to ensure a steady flow of oil, ore, food, and other resources toward your car, furnace, toolshed, closet, and table. The consumer is god in our economy, and the temple of consumption is the home.

The home's central place is told in the roots of the word "economy": the Greek *oikos*, or "household," and *nemein*, "to apportion or manage." Managing our households spins the wheels of nearly all economic activity. Our homes are the building blocks that construct the economy, and we've forgotten that mainly, I think, because most households have been run by women, who are unpaid and otherwise not considered part of the workforce, and this traditional women's work is nearly invisible to economists. Managing a home is an oft-ignored balancing act that demands enormous skill. Flooding through every home are innumerable streams of goods (and people) that all must be selected, cared for, kept in tune, and made the best use of. The home is where we bring the things we buy and make, and where we use what we have to support the people we're most connected to. It's the heart of our lives.

Not long ago most of what people used in the home, and often the house itself, came from the yard and from the small plots most villagers worked just outside the town walls. In many places the yard still provides for many of the home's needs. Those productive home economies are packed with lessons about designing yards that can unplug us a bit from the industrial leviathan, improve the quality of our lives, and reconnect us to land, family, and community. Some of the best examples of these are in the tropics and the Global South, where industrialization and development have not yet thoroughly transformed traditional societies and the role of the home as a place of production, not just consumption, remains. The diverse home gardens found in villages, towns, and cities in the lower latitudes not only furnish food but also fulfill many other home needs. They are true home ecosystems. We can learn plenty from them, even if we need to change the species mix to suit our climate and tweak the functions to match our culture.

## ➡ THE TROPICAL ⬅ HOME GARDEN MODEL

Modeling our yards on the tropical home garden can show us how to move some of the pieces of our globe-spanning economy out of far-off factories and offices and reinstall them to benefit us more directly, and with far fewer toxic consequences, near home. This pares down our need to earn, to commute, and to be dependent on surges of the employment index for our well-being. It keeps money and resources in the local economy and shifts us from being passive consumers to producers. And, creating this kind of diversity in our yards and homes builds resilience and abundance for us and for the rest of nature. After we take a quick look at how tropical home gardens work, we can explore how their design principles can be translated for our own yards.

The tropical home garden is densely multifunctional. A Maya farmer named Ignacio I came to know in Belize once told me, as he swept his arm across what looked—and functioned—like a wild jungle surrounding his house, "I've been to your Walmart in America. My garden is my Walmart. Everything I need is here." Ignacio showed me plants not only for food but for medicine, income, fuel, mulch, containers, rope, glue, bug repellent, and more.

Anthropologists have studied tropical home gardens intensely and find them far richer than the typical function-poor, diversity-lacking North American yard. As many as 200 species of plants might be packed into a space the size of a city or suburban lot. One study in Veracruz, Mexico, found 338 species across eight gardens that generated a constant stream not only of food but also medicine, fuel, and many other useful goods such as baskets and brooms, canoes and jewelry.[1]

Many home garden plants are multifunctional as well. The leaves of a tree used for roof beams ooze an insecticidal sap; an edible tuber can be mashed into a glue. This stacks more into a small space. Beauty, too, plays a key role in these well-used yards. In Veracruz 38 percent of all home garden plants were ornamentals, though many of those had other functions, too, such as attracting insects with nectar-rich flowers or providing fiber for rope and twine, paper and cloth.

I'm inspired by the creativity of tropical home garden designers and believe that we can do as well in our own yards. In cultures with

## Plant Products from Tropical Home Gardens

| | | |
|---|---|---|
| Fruits | Animal food | Sealers, varnishes |
| Vegetables | Cash crop | Brooms, brushes, combs |
| Tubers/root crops | Cooking fuel | Paper and cloth |
| Nuts/seeds/grains | Charcoal | Canoes |
| Seasoning | Green manure | Furniture |
| Beverages | Mulch | Stakes and posts |
| Stimulants/intoxicants | Honeybee fodder | Rope, twine |
| Medicine | Baskets | Tool handles |
| Cut flowers | Boxes | Dye |
| House construction | Smoking mix | Glue |
| Fencing | Insecticide | Jewelry and ornament |
| Firewood | Fish or rodent poison | Ceremonial goods |

these kinds of gardens, we're no longer separating consumption and production, an inefficient and often soul-killing setup where we leave home to go to work to earn the money to shop for the things that we bring home to use. Here we're putting consumption and production under one roof—or at least in one yard. And most things grown or made at home have a far smaller ecological footprint than industrially manufactured goods that are shipped long distances to your door. No, you probably won't make everything that you use, but harnessing the power of the sun via plants to let them build some of those things, beyond food, reduces the need to earn and connects you to place. And, not having to be away from home at work creates a virtuous spiral. Being home more means you spend less on daycare, fancy clothes for work, retail therapy and vacations to recover from job stress, restaurant meals because you're too tired or busy to cook, and so on. The builders of tropical home gardens worked all this out long ago.

There's some design work involved in transferring all that hot-climate experience into temperate yards, but by using the needs and resources method of good permaculture design, we can figure out which plants and other design elements offer the resources that can easily fill critical needs. We're looking for the low-hanging fruit: It's neither realistic nor desirable for an urban wage earner to expect to grow all her own food, weave the fabrics for her clothing, and reach some supreme level of self-sufficiency. (Self-reliance done as a community is another matter.) But a fertile, well-designed yard can easily produce some otherwise expensive goods or other items that, for many reasons, are wise to generate at home.

We'll start this design process by setting some goals; that is, by assessing which needs make sense to satisfy in our own yards. Then—and this part is easier, since it's just technique—we will figure out how to do that. So this section on urban yard design begins with some big-picture questions, then dives down into specifics.

## RESOURCES ABUNDANT, RESOURCES SCARCE

First, what is best for us to do ourselves, as opposed to what we'd be better off buying with our job-won cash? One way to decide is to look at which resources are most abundant in towns and which are scarce. Knowing this tells us which endeavors will be easily supported and which will take some scrambling to achieve. *Abundant* in towns are:

- **People and their output.** Human labor, an educated and skilled populace, cultural resources such as art and music, teachers and other trainers, accomplished policy implementers, and social capital are all heavily concentrated in cities and towns.
- **Built material.** There is a surfeit of manufactured goods, recycled material, Dumpster contents, castoffs left at the curb, garage sales, neighbors' giveaways, and everything purchasable in stores. Buildings themselves are abundant but often expensive unless you can use human connections, such as sharing, trading, or using a friend's space.
- **Money.** Although cities are expensive and poverty abounds, in 2011 the average income in urban areas in the United States was 32 percent higher than in rural areas.[2]
- **Commerce and jobs.** Cities are founded on the exchange of goods and labor for income. Unemployment in rural areas averages 10 percent higher, says the US Bureau of Labor Statistics.
- **Innovation.** As we learned in chapter one, innovation per capita is far higher in cities and towns. We're stimulated to be creative in our solutions and approaches.

*Scarce* in towns are:

- **Free time.** Being grossly overscheduled has become a status symbol, but in cities it's not hard to do. Time is the ultimate nonrenewable resource.
- **Land.** Population density and intense use make land expensive, and yards are small or nonexistent. Public land is heavily used.
- **Raw materials.** Timber, firewood, rocks, soil, and other natural materials are rare, though their processed cousins are available for a price or to be scavenged.
- **Organic matter.** Not long ago, chipped tree trimmings and even small-animal manure was easy to find in cities, but people are wise to these resources now and waiting lists can be long, except, perhaps, for leaves in fall. Food waste still abounds, but this resource is being discovered.

One pattern here is that in cities, not surprisingly, products from the natural environment are harder to get than human-made resources. You'll be able to find people to help you with projects but may have to hunt for inexpensive places to do them. From this we can see that two skills that make for smooth-flowing urban permaculture—and town life in general—are developing good social networks and learning how to stack many functions into a small house or piece of land. We'll save the first skill, social permaculture, for later chapters and focus for now on stacking functions on the land.

## What Could You Put in Your Yard?

Salad greens and daily vegetables
Herbs
Production and canning vegetables
Potato tower
Flowers
Plants in containers
Plants for crafts, dyes, and fiber
Medicinal plants
Nursery/seedling beds
Medicinal and mycorrhizal fungi
Plants to build soil fertility
Fruit trees
Nut trees
Berry bushes
Compost bin
Potting bench
Coppice trees for firewood
Firewood storage
Fire pit and seating ring
Patio or deck
Small lawn for kids and lounging
Dining area
Barbecue
Pizza/bread oven
Outdoor kitchen
Hot tub
Sauna
Insect and wildlife habitat
Poultry coop and yard
Birdbath
Rabbit hutch
Beehives
Bird- and bat houses
Dog run/doghouse
Root cellar
Pond
Water storage
Aquaponics system
Graywater system
Composting toilet
Outdoor shower
Workspace for projects
Toolshed
Greenhouse
Cold frames
Treehouse
Playhouse
Arbor
Pergola
Outdoor sleeping area
Small studio
Meditation/sacred space
Chairs/bench
Swimming pool
Sports area
Bicycle parking
Laundry line
Trash cans
Access paths
Fences

An important question to ask is, How much can we actually do on an urban lot? Let's expand our horizons beyond the tropical home garden's focus on useful plants and look at all that a yard can do. In his design classes, Washington State permaculturist Douglas Bullock asks his students, "What could you put in your yard?" He's looking for a list of all the activities and items that would conceivably fit into a home property. Predictably, the students' answers start with food, but they quickly broaden as the class sees the possibilities. The resulting tally swells to cover nearly all human endeavors. A summary of the answers to Doug's question appears in the "What Could You Put in Your Yard?" sidebar above.

What becomes clear during this exercise is that, given good weather, we can shift almost all indoor activities outside, including cooking, eating, socializing, working, and sleeping. Not everyone wants or has room or a suitable climate to do all that in the yard, but from this list and your own preferences and additions you can choose the design elements that fit your life.

## FOUR KEY DESIGN QUALITIES FOR SMALL GARDENS

The next step is to find comfortable, usable places for these elements. To do that well, we need to design four key qualities into each activity space:

1. Space-saving arrangement
2. Easy access
3. The right microclimate and setting
4. A sense of defined space for the activity within

These four qualities create the conditions for a well-made, multifunctional yard. We don't just plunk down an outdoor kitchen here, some raised beds there, and a fire ring out yonder. A design process that first assesses the sectors and prepares the space and its contents will ensure that we'll joyfully use what we've built instead of discovering that the deck broils in the afternoon and the barbecue lies in a mud wallow. It also raises the odds that our goal, instead of being a miserable uphill battle, will be almost inevitable. The soil will be fertile, the shade inviting, the space beautiful, the work minimal.

To do this we'll use the powerful triad of permaculture design methods covered in chapter two: needs and resources analysis, zones, and sectors. We've already begun using needs and resources by listing the possible uses of a yard. The next step in this method is to decide which of those options you need in your own life and which of those you have the resources to create in your yard.

There are few formal rules for what should go into your yard. It's an intensely personal decision, and although permaculture's design tools will help you decide, your own passions and needs will dictate the list of possibilities that you'll choose from. I know urban market gardeners whose yards hold nothing but food plants and a couple of lawn chairs crammed into a path widening. The entertainer's landscape will be dominated by the inviting patio, fire circle, and outdoor kitchen, all enfolded by fragrant flowers and colorful edibles to tickle the guests' senses.

To see how to apply the needs and resources method to selecting what you want in your yard, I'll give an example. The plants in my yard need water, and I must decide how to meet that need with resources I have or can afford to get. Of course, I'll start by reducing my water use in as many ways as possible, but right now I'm focusing on how to deliver water to my plants. So I begin with listing the possible water sources:

- Municipal water
- Direct rainfall
- Rainwater harvesting from the roof (mine and that of an adjoining neighbor who doesn't harvest his)
- Runoff from the street
- Graywater from laundry and shower

Being a permaculturist, I want to reduce my use of city water since it costs money, pulls water away from other users (human and wild), and relies on complex physical and political infrastructure. So I'd like to push municipal water off the list. Then, because my region has a long summer dry season, I know that direct rainfall won't be enough. I also know that to carry my plants through the dry season, storing sufficient collected rainfall will require immense, expensive tanks or a large pond. I want to do that eventually, but I need to save up the money. And street runoff comes only with rain; maybe

I can direct it into a large pond to store it, but I don't have one yet. That also goes on the "eventually" list. This hierarchy of resources and preferences is quickly pointing toward graywater as a top water source, with other methods coming later. And conveniently, my washer and shower generate tens of gallons of graywater each day. Also, the hardware to build a code-approved laundry-to-landscape graywater system (see chapter six) is less than $250—much less if I use my scavenging skills—and that's within my budget. So graywater is a readily available resource that meets my water need. That's where I'll start. Meanwhile, I'll plan for tanks and ponds, too, since I know I have the need but just lack the resources right now.

We'll use this method again at other scales as this process unfolds, applying needs and resources analysis to as many elements and activities we can think of to select and prioritize them. Once we've selected the elements we want via needs and resources (and probably other methods listed in chapter two as well) we can use the zone method to start arranging roughly where those elements and activities should go in relation to the house and to where we spend most of our time. The zone system also carries us toward the four key qualities, especially the first two: space saving and easy access.

## ➡ ZONES FOR ⬅ THE TOWN GARDEN

Remember that permaculture zones are shaped by how often we need to use their contents. In the previous chapter, table 2-2 lists some of the items found in each zone in a yard. Most town lots are large enough to hold zones 1 and 2 but rarely the outer zones. We don't simply do without the yields of the farther zones, but those things may come from outside our own yards. For example, an orchard is usually a zone 3 feature because fruit doesn't need harvesting more than once every week or two. But urban yards are rarely big enough for a zone 3. How are fruit lovers to fit in all their favorite varieties? Here's how I solved that, using the zone system.

In our Portland, Oregon, yard grew a huge European prune plum that each summer groaned under a load of fruit so immense that a marathon of frantic canning, drying, and belly-stretching eating fests hardly dented it. My neighbors also had fruit trees—peaches, figs, plums, apricots, apples, and pears—that bore overwhelming crops. To deal with this abundance, we all traded our excess fruit. To round out our collective harvest, I planted the fruit trees missing from our neighborhood trading circle: Asian pears, cherries, and persimmons.

I took care of another of my needs, firewood, which is usually a zone 3 or 4 item, by cutting a deal with a city park maintenance crew. They would leave their trimmings in a pile at the park, and I would haul it away, sparing them the work of loading and dumping it. Everyone won.

These examples show how one of the first steps in designing an urban yard is to decide which activities you've just got to have on your lot and which ones could occur—or better yet, already are occurring—elsewhere. You won't be able to fit every activity into your yard. Needs and resources analysis will help tell you which ones you have the assets for, in space, budget, and usefulness, and how they connect to the other elements, and the zone system will help you select from that list the things and activities that simply must go in your yard and which ones can be elsewhere. Zones tell you how an element connects to you, the user. Thus when that connection to a design element is frequent, deeply meaningful, or grossly inconvenient not to have nearby, you'll usually want to have that need met within your lot lines. Otherwise, you can probably rely on the overwhelming abundance of the city ecosystem to provide a need from outside your busy home.

Because we're using permaculture design, we know we must balance our own needs with those of the rest of nature. We'll choose materials that are responsibly produced and plants that provide habitat for more than just people, and we'll proceed according to permaculture ethics and principles. But within those constraints, there's a universe of opportunities and choices, and permaculture's methods help narrow those down.

Table 3.1 shows a sample zone analysis for the contents of a town yard. Zone 1 contains the elements that we use roughly every day, and zone 2 contains the elements we use on a weekly or seasonal basis.

Note that many elements—in this case, water storage, plant medicines, fungi, and access

**TABLE 3-1.** Contents of Urban Zones 1 and 2

| ZONE 1 | ZONE 2 |
|---|---|
| Salad greens and daily vegetables | Production and canning vegetables |
| Herbs | Fruit and nut trees |
| Flowers | Berry bushes |
| Plants in containers | Plants for crafts, dyes, and fiber |
| Snack fruits: Dwarf, multigrafted fruit tree or two; cherry tomatoes; favorite berries, etc. | Medicinal plants |
| Nursery and seedling beds | Medicinal and mycorrhizal fungi |
| Birdbath | Plants to build fertility |
| Patio | Compost bin |
| Immediate-use firewood storage | Potting bench |
| Bicycle parking | Greenhouse/cold frames |
| Outdoor kitchen | Chicken coop and yard |
| Hot tub | Beehive |
| Outdoor shower | Pond |
| Composting toilet | Water storage |
| Insect and wildlife plantings | Graywater system |
| A small lawn for kids and lounging | Barbecue |
| Medicinal and mycorrhizal fungi | Fire pit and seating ring |
| Water storage | Pizza/bread oven |
| Access paths | Dining area |
| | Workspace for projects |
| | Kids' playhouse |
| | Outdoor sleeping area |
| | Small studio or art cottage |
| | Access paths |

paths—may be in more than one zone. Also, some elements can be combined or used multifunctionally. We could meld the patio, outdoor kitchen, barbecue, pizza oven, dining area, and fire pit into one focal area that spans part of zone 1 and zone 2. The part that is used every day—the outdoor kitchen and dining area, perhaps—would lie close to the house, while the barbecue, oven, and fire pit would go toward the outer side.

Keep your zone 1 as small as possible, so that none of it will be neglected. How big is that? Again, it depends on the context. Is your yard flat? Then the area you'll shuffle across every day will be larger than in a sloped yard. Are parts of the yard far from doors or paths, behind hedges, walls, or large raised beds, or in some other way less accessible? That may push them into a farther zone. On a chilly morning when you're still in slippers and pj's, a few extra feet of soggy path can make the difference between picking some blueberries as a breakfast topping or not. The key factors dictating the size of inner zones are slope, access via doors and paths, and the presence of obstacles such as overhanging shrubs, steps, or anything between you and your destination other than a clear, level path.

As with all rules, there are exceptions. A few elements used every day may not belong in zone 1. A chicken coop is one example. Poultry need attention at least twice a day, to chaperone them in and out of the coop and to feed them. That would suggest zone 1. But a clutch of gabbling birds under the bedroom window is not everyone's idea of heaven. The coop may need to go farther away, but access should be straightforward and ideally near someplace that you're likely to go at the same time, such as a compost pile.

The small, dense yards of urban permaculturists may only contain zones 1 and 2, but decisions like those above show that these zones are intricately structured. I find it helps to organize these complex spaces by designating subzones, such as an inner and outer zone 1 or even zone 1a, zone 1b, and so on, because in a compact, busy urban yard the functions, feel, frequency of use, and built-versus-natural qualities of the subzones can be as different from each other as zone 1 is from zone 2 or 3. Here's what I mean by that.

Rare is the person who enjoys stepping straight from indoor heated and carpeted comfort into full sun, gusty winds, and crumbly earth. Most of us prefer gradual transitions from inside to outdoors. Good landscape design dictates gentle gradations from a space that is fully indoors, domesticated, and 100 percent constructed through a transition space in which the wild gently seeps in and the built slips away, until we are in full contact with nature. This is why we create porches, decks, and lawns. We are domesticated animals who need a little time to transition from homey security into the wild. So we design some transitions into our spaces. Especially in cities and towns, our yards need to be places to connect with nature, but our psyches prefer gradual shifts between their tame and wild sides.

The house, or zone 0, is our fully domesticated space. The concept of zone 0 is a recent development. In permaculture's early days, the zone system began at zone 1, which included both house and inner yard. I'm told that this was because founders Bill Mollison and David Holmgren lived in subtropical Australia, where the mild climate spurs an indoor-outdoor lifestyle like that displayed in enticing *Sunset* magazine photos of Southern California. It's a place where much of the year wide doors are thrown open onto broad patios, and upholstered furniture sits outside on shady decks and lawns. But in temperate climates, the indoors and outdoors are sharply distinct. Much of the year the environment and activities inside a temperate house are very different from what is—or isn't—happening in the frigid realm outside, and the houses are more or less impermeable. So it makes sense in temperate climates to distinguish the house as zone 0, while zone 1 begins at the doorstep. (British and some other permaculturists sometimes refer to zone 0 as the inner state of the designer or client. We won't use that meaning here.)

In town yards that hum with varied activities in the inner zones, it can help to further divide zones 1 and 2 according to their increasing movement away from the tame indoors toward wild nature. Zone 0 is an insulated, heated, fully electrified and plumbed, secure, closable, built environment. Outdoor spaces can retain many of those properties but will gradually lose them as we move farther from the house.

Architectural theorists tell us that human shelter needs to provide a balance between two qualities. One is *prospect*, the ability to see what's approaching and on the horizon. This gives us a sense of freedom but also of knowing our environment and what's coming next. The other quality is *refuge:* a sense of protection, security, and being enfolded. A good house provides both prospect and refuge, and this explains why we like looking out our picture windows as we sit by our fireplaces.

As we move outdoors, we like to retain some sense of refuge until we are fully comfortable with what's outside. All animals do this. You can see it when a cat hesitates at the door before padding outside, or when a deer eases nervously into a clearing. They pause and look around before abandoning the protected space. Subzones help us design this important feeling into our own yards. Here's how it can work.

The most houselike outdoor space is a screened porch, which is no longer insulated or fully protected but can be plumbed, electrified, and partially heated. We'll call this zone 1a. Next out is a covered deck or other floored space protected by a roof overhang. Call it zone 1b. Open decks and patios are zone 1c. These are all in inner zone 1. Outer zone 1 holds places where your feet touch earth rather than processed materials, so a lawn or fully mulched open space is zone 1d, and a planted inner garden is zone 1e. To some, this may seem like excessive dividing, so if you're a lumper rather than a splitter, just call zones 1a through 1c the inner zone 1, and label 1d and 1e as outer zone 1. But in the fine-grained intensity of an urban yard, each of the five subzones has distinct qualities, activities, and microclimates, and we design for them differently. Tables 3-2 and 3-3 shows their activities and contents, how they differ, and the gradients that distinguish their aspects.

With this information in mind, we can distribute many of the activities in table 3-2 into their subzones. Given what we now know about the space, we may think of a few more activities, too. Here's one way to apportion them, though your own choices may vary.

The town yard often must be a richly multifunctional place with many users, human and otherwise. To use a yard permaculturally, we need to think beyond vegetables, flowers, and lawns and consider all the activities that go on, both indoors and out, and add in the many needs of wild nature—habitat, soil building, water and air cleaning, and so on—as well. Zone design helps us arrange all this efficiently, and by telling us which elements might be placed in the same zone, it can also point us toward consolidating several different features

**TABLE 3-2.** Properties of Inner and Outer Zone 1

| | ZONE | | | | |
|---|---|---|---|---|---|
| | **1a** | **1b** | **1c** | **1d** | **1e** |
| **Dominant feature** | Screened porch | Covered deck or stone patio | Open deck or patio | Lawn or similar open space | Highly managed garden |
| **Activity** | Sitting, eating, social, crafts, many indoor activities | Sitting, eating, social | Sitting, eating, social, sunning | Play, eating, social, sunning | Plant tending |
| **Privacy** | Most private → | | | | Least private |
| **Overhead** | Roof that matches house | Light roof of wood, textile, or metal | Arbor or none | None | None |
| **Underfoot** | Finished or smooth wood | Wood or stone | Wood or stone | Grass, mulch, walkable plants | Path mulch or grass |
| **Furniture** | Interior or exterior | Exterior | Exterior | Light exterior | None or portable |
| **Exposure to wind** | Little wind → | | | Most windy | |
| **Temperature** | More controlled → | | | | Least controlled |
| **Lighting** | Artificial and natural | | | Natural with spotlighting | |
| **Plantings** | Houseplants | Containers for house- and outdoor plants | Containers | Containers and in ground | In ground |

**TABLE 3-3.** Zone 1 Subzones and Their Contents

| | ZONE | | | | |
|---|---|---|---|---|---|
| | **1a** | **1b** | **1c** | **1d** | **1e** |
| **Dominant feature** | Screened porch | Covered deck or patio | Open deck, patio | Lawn or similar open space | Highly managed garden |
| **Furnishings and small structures** | Seating, tables, lamps | Outdoor furniture | Outdoor furniture, fire pit, birdbath | Lawn furniture, fire pit, birdbath | Lawn furniture, raised beds |
| **Water systems** | Household | Spigot | Spigot, outdoor shower | Spigot, small pond, irrigation, graywater | Spigot, pond, irrigation, graywater |
| **Plant types and functions** | Houseplants, container ornamentals and herbs | Container herbs, greens, ornamentals | Container herbs, fruit, greens, ornamentals | Ground covers; containers and in-ground plants for beauty, fragrance, microclimate, snacks | In-ground plants for food, fiber, medicine, microclimate, beauty |

into one, such as a combined outdoor shower/graywater system or an integrated kitchen/dining/barbecue/pizza oven/fire pit space.

Zones are a key tool for creating beneficial connections and multifunctional spaces by placing design elements in relation to the user or the center of activity. Now that we've examined how to use this method in the small yard, let's look at how sector analysis helps us decide what goes in the yard and where.

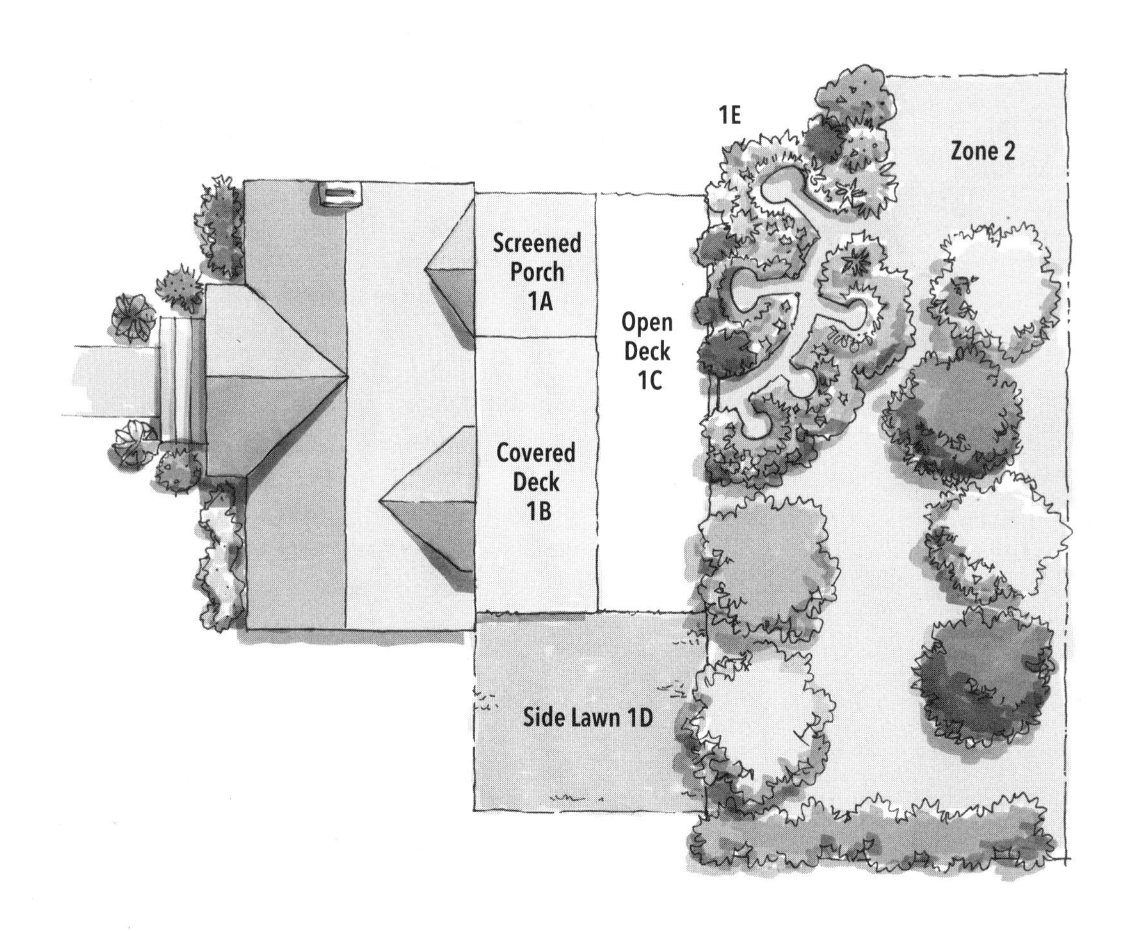


**FIGURE 3-1.** Schematic of zone 1 subzones for an urban backyard. The transition from fully indoors to fully outdoors is gentle. Illustration by Elara Tanguy

## SECTORS INFLUENCING THE TOWN GARDEN

We saw in chapter two that sector analysis is permaculture-speak for arranging design elements to be in useful relationships with influences or forces coming from off the site, usually ones that we can't directly control. The classic sectors are forces of nature: sun, wind, wildfire, flood, and so on. However, in towns and cities, not only are these forces often altered by human patterns—fire is more likely to come from a neighbor's faulty space heater than from a forest blaze—but many urban influences are not found in the wild at all. In towns these purely human sectors include traffic, codes and covenants, neighbors' sensibilities and customs, crime, utility easements, spouses and children, artificial light, shade or reflections from buildings, and a host of other built-environment influences. All of these are crucial to take into account when designing the town yard. Remember designer Larry Santoyo's warning: sectors trump everything. A disgruntled neighbor wielding an antiquated ordinance forbidding front-yard vegetables, a feral cat using your seedbed as a litter box, day-length sensitive plants confused by a neighbor's security light: All these can stop a project in its tracks, and each is preventable by good sector analysis.

Every site will have a unique combination of sectors. For example, at the Los Angeles Eco-Village a nearby billboard throws deep shade across their backyard, an unusual but powerful sector influence. Thus it's important to observe carefully and think hard about what real and potential influences sweep onto and across your site. Remember the three possible ways to work with a sector energy. The first is to harvest or collect it, as with a solar collector or a curb cut to sink rainwater from the street. The second is to block it via, say, a windbreak or privacy fence. And last, allow it to flow by untouched, as with a wildlife corridor.

As the list in chapter two (pages 41 and 42) attests, many urban sectors are unpleasant, human-spawned ones such as crime, pollution, or a stop-work order tacked to the door. This means that a variant on the third strategy—avoiding a sector altogether—is often the best to use. For many unpleasant sectors, we're better off if we design to never have to deal with them. It's an aikido approach to sectors. You don't want to block a mugger or unhappy code enforcer. You don't want to see one at all.

How do we become skilled in eliminating negative sectors before they reach us? Earlier I gave the example of how the Los Angeles Eco-Village evaded the schoolchild sector by planting tomatoes that didn't turn red. To forcibly block that sector would have meant building a fence or running a constant tomato patrol, which is extra work for little return. The ecovillagers' strategy vaporized the schoolchild sector with no additional work while preserving the yield of sweet, ripe tomatoes. The key to good strategy here is to change the rules of the game and use subtlety rather than brute force.

Another example: Portland architect Mark Lakeman designs innovative structures, and it can be expensive or even impossible to obtain building permits for his outside-the-box designs. One code that he constantly bumps into demands that roofs more than a few square feet in size must be permitted, and his outdoor seating spaces in rainy Portland need a roof. Instead of flouting the code or otherwise fighting city hall, Mark builds several tiny, overlapping roofs instead of a single large one. This way, he avoids needing a permit yet remains legal. (See photograph in the color insert, page 4.) As with so much of permaculture, when we learn to think on the high level of strategy, instead of the lower level of techniques and objects, our designs flow more smoothly and efficiently. Don't pack a gun in a high-crime area; stay out of it, travel through it with a posse of friends, or find some other way simply not to be a target.

This shows that after listing the sector influences, a good next step, as a high-level intervention point, is to find strategies like those above to "disappear" the negative ones. By using strategies that eliminate the sector altogether, we don't have to figure out where to place design elements in the best relationship to it. That's one less thing to think about. After the "disappearing" step is done, the usual practice can begin of arranging elements to harvest a sector, block it, or let it pass by unmolested.

As I've said, applying multiple methods to place elements lets us be confident that we have found beneficial, functional places for them to go. It also helps create the conditions for a yard not only to save labor, money, and resources but also to be a dynamic, richly interconnected place that acts like an ecosystem.

## DESIGN AT THE EDGES OF TOWN

Permaculturists are fascinated by edges. Edges—the transition from one set of conditions to another—are often full of life, busy, and where innovation and novelty blossom. The classic definition of edge is the place where two ecosystems meet. The species from each of the two ecologies gather there, as well as a unique set that thrive in the special conditions of that edge. Thus edges can be nexuses of diversity. They can also be places where resources collect—think of a fence that stops blowing debris, which then rots into topsoil, traps seeds that sprout, and catches droppings from perching birds. They can also be places of tension (say, cultural boundaries between ethnic groups) and of original ideas (when musicians from two cultures meet, jam, and invent a new musical form).

In design we manipulate edge to save energy and work and to enhance desired qualities. The decision to increase or decrease edge depends on what lies on either side of the edge and what our goals are. We're looking for the optimum amount of edge. Edges allow us to define spaces, to see their boundaries, what flows across them, what accumulates at them. We work with these flows. Edges are places where matter and energy change direction and speed or stop altogether and often change into something else.

Out in the countryside, edges are often soft, gradual, and fairly permeable. Think of walking from full sun toward the trunk of a tree. You traverse from full-strength daylight to speckled light, then deep dappling, and finally full shade. It's still bright enough under the tree for some plants to grow, and the gentle gradient from sun to total shade spans many feet.

But in a city, in one inch you can step from blazing sunlight, with the high-albedo glare of a light-colored sidewalk and dazzling sunbounces off glass towers, into the deep, chilly gloom of a building's north side where the sun doesn't shine, literally, ever. Edges in the built environment are hard, abrupt, frequent, and often impenetrable as well as unpredictable. Also, edges in cities are not just in space but also in time, such as the edge between my busy calendar and the schedules of the people I want to see. Towns are full of invisible edges: cultural, legal, linguistic, and economic. But edges are often places of opportunity, because the resources of two neighboring systems are available. At edges, materials and energy leave one place and flow into another, where they have new potential in a new environment. To help us be aware of their value and effects, some of the edges in a typical town yard are featured in the "Edges in the Home Landscape" sidebar.

Notice in the sidebar the frequent overlap between edges and sectors: Sun, wind, privacy, view, pollution, and many others are both. This is because edges are where something from outside comes inside (and vice versa), and that something is often an influence and thus a sector. Some edges can mark the limit of a sector's influence, such as a fence that screens the view into a yard.

Spotting and controlling edges help us gather resources we want, block potential troubles, and preserve what we have. Deepening our understanding of edges lets us find benign microclimates, safe or quiet spots, places need-

## Edges in the Home Landscape

- Sun/shade
- Wet/dry—e.g., under the eaves of a house where rain can't reach versus wet soil
- Wind/calm
- Hard/soft—e.g., from sidewalk to lawn or house foundation to soil
- Warm/cold—often at sun/shade and wind/calm edges but also near heating vents, doorways, fire pits
- Level/sloped
- Private/public
- Above/below ground
- Seasonal edges
- Built/natural
- Open view/blocked view
- Accessible/inaccessible
- Fenced/unfenced
- Less regulated/more regulated (in backyards we can build or plant most anything, in front we're more scrutinized, and parking strips may be highly regulated)
- Polluted/unpolluted
- Native soil/imported or reworked soil
- Child-impacted/Adults only
- Accessed by animals/protected from them
- Planted/unplanted
- With utilities/without (such as plumbing, electric, heat)
- Covered/open
- Resource producing/ornamental or nonfunctional
- Tended/untended
- Irrigated/nonirrigated
- In your yard/outside it

ing improvement, and opportunities to harvest resources or enhance habitat. We'll look at a couple of edges in detail as a template for understanding how our yard's edges work in general.

### The Unregulated/Regulated Edge

We live in a regulated culture. The old (albeit patriarchal) adage "A man's home is his castle" has been perforated by accumulating laws and customs, and our designs must acknowledge these intrusions upon our autonomy. The backyard is still a private space where inner zone 1 creativity can reign, but the front yard is visible to neighbors and authorities and subject to their biases. News stories tell of wildflower gardens mowed down by fire regulations, curbside planter boxes cited as impediments, and other testimonies to the supremacy of the lawn as the accepted front-yard icon. A few decades of edible-landscape education have nibbled at these prejudices, but front-yard food and habitat may still require finesse to achieve. Thus it's important to be pattern literate here and create a front yard that is pleasing to passersby, educational to the curious, yet still following permaculture principles. An unprovocative front yard also raises the odds that your backyard experiments will thrive unchallenged.

If your neighborhood already appreciates front-yard food and wildlife gardens, you're lucky. For the pioneers on their blocks, here are some hard-won tips for breaking the Kentucky bluegrass barrier.

- Learn the local ordinances and homeowners' association rules. How are parking strips regulated? Must lawns (which may stretch to include wildflower meadows) be kept below a certain height? Is it legal to keep bees, chickens, or even pygmy goats?
- Get to know your neighbors. It's easy to report a stranger's transgressions. It's less likely that a neighbor will bust "that nice Alice, who gives us her extra eggs."
- Make it pretty. Keep mulch piles, salvage collections, and other hallmarks of the reuse/recycle ethic out of sight. Prune near sidewalks; remove overtly visible dead growth.
- Place innocuous or beautiful species on the margins. Your outer zones are the neighbors' inner ones, so those edges should be acceptable to them. A screen of flowers or common shrubs ringing a less conventional center will help. What's normal for the hood? If boxwood or rhododendrons are de rigueur, a buffer of these will deflect criticism.

Obviously, if your intent is to challenge the outdated laws and habits that force our yards to be ecological deserts, then some of these rules don't apply. But please don't tick off people unnecessarily—it gives permaculture a bad name—and pick the battles you can likely win, one at a time.

## The Wind/Calm Edge

Even light breezes can alter the climate of an outdoor space for people and plants alike. Air movement sucks away both heat and moisture, making it a key element in managing microclimates, irrigation, and plant placement. In cities buildings form a jagged windbreak, so average wind speeds are lower than in rural areas, but winds in town are more turbulent and varied because the urban windbreak is randomly sized and spaced. The best windbreaks are slightly permeable, but the solid walls of our cities deflect wind around them and squeeze it into the narrow channels between buildings. This causes what is called a Venturi effect that can turn light breezes into blustery gales that dry plants, increase frost damage, and create a harsh checkerboard of windy spots interspersed with still ones.

We tend to avoid producing Venturi winds and drafts in our yards, but soft air movement is a plus on hot days. Luckily, plants are the perfect solution. Good windbreak design lets 40 to 60 percent of the wind through, which won't create a turbulent low-pressure zone behind it or Venturi winds around it but breaks the gusty force of the wind. Plants happen to create that level of permeability easily.

Like almost everything else in cities, urban windbreaks must be compact. On farms windbreaks stretch for hundreds of yards and may be single purpose: shrubs and trees chosen primarily for having the right size and permeability for slowing wind. They are often tall, too, since the area to be protected behind them may be hundreds of feet. (A windbreak stills a downwind distance equal to five to ten times its height.) In town windbreaks ought to be multifunctional, providing shade, beauty, products, and habitat. And, they will be small, often simply plugged into a gust-producing gap between nearby structures. Since yards are petite, the plants don't need to be more than 6 to 10 feet tall to shelter the whole place.

For example, here's how we use the wind/calm edge in our yard. The sitting spots are a front deck, facing south and sheltered east and west by structures, and a backyard that

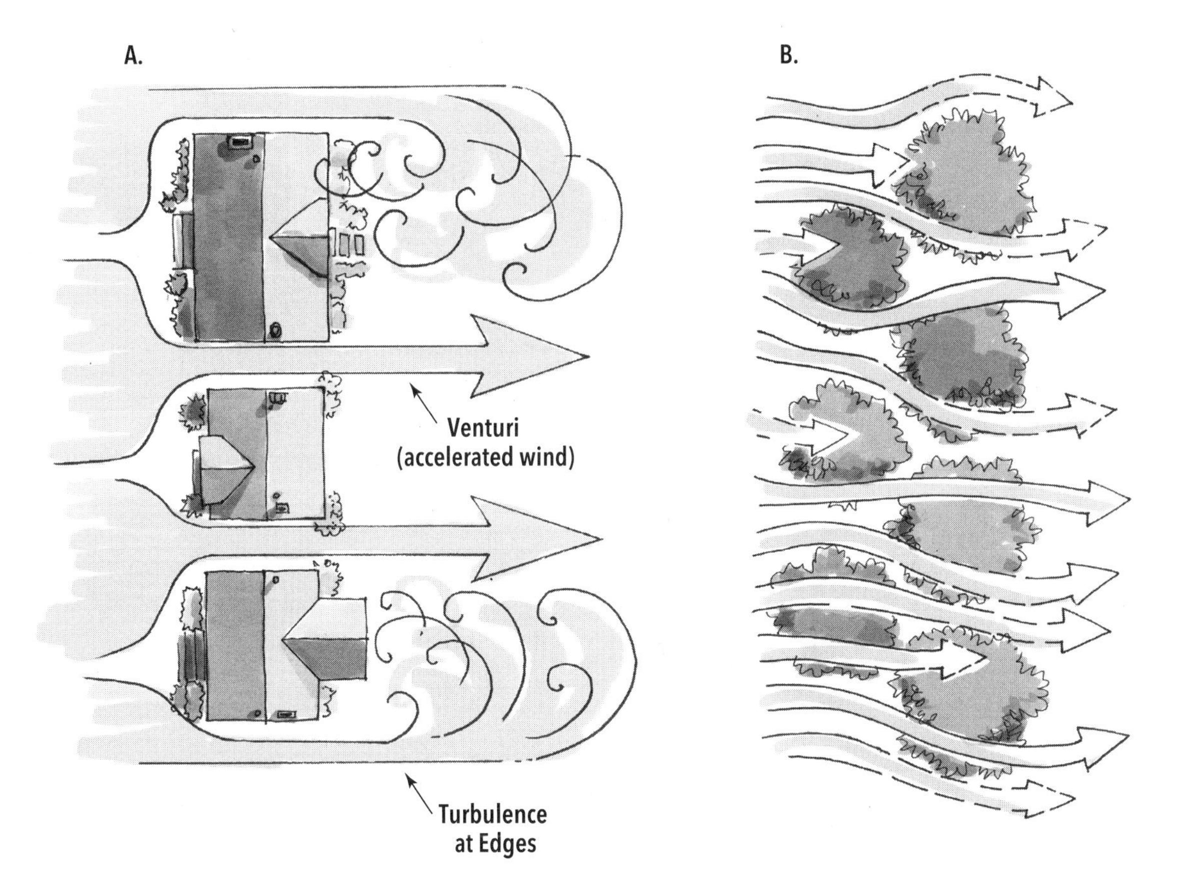


**FIGURE 3-2.** A well-designed windbreak allows 40 to 60 percent of the wind to pass through it, avoiding turbulence and low-pressure areas behind it. *A*, Buildings make a poor windbreak because they channel wind around them. This concentrates and accelerates wind, creating strong gusts and Venturi effects between and behind buildings and turbulent swirls at their edges. *B*, Trees and hedges on the windward side of buildings dissipate and slow the wind, eliminating Venturi winds and creating comfortable places behind them. Illustration by Elara Tanguy

is open to both sides. Thus the front deck is protected from wind, and the backyard is not. On any cool day—and there are plenty in coastal California—any breeze pulls the heat out of us. Because it is sheltered, the front deck is perfect for lounging and becomes uncomfortable only on the hottest days, and that's when—you guessed it—we retreat to the backyard and its cooling zephyrs. So having a wind/calm edge in our yard increases its usefulness.

## MANAGING YOUR MICROCLIMATES

Working with several of the edges I've listed—wind/calm, level/sloped, wet/dry, sun/shade, and others—helps to create and maintain benign microclimates in the yard. Microclimates are smallish, local nooks that are warmer, cooler, wetter, drier, or in some other way different from climate conditions in the larger surroundings. They form where influences such as sun, breezes, and moisture meet the features on the ground: buildings, slopes and hills, plantings, roads, and so forth. The more features, the more varied are the microclimates. That's why towns are festooned with microclimates that range in scale from the entire metropolitan area down to a sheltered corner of your yard. Designers work with the whole array of them to extend the growing season, stretch the plant palette into less hardy species, ripen tricky fruits, reduce heating and cooling costs, and make more of the yard pleasant in unfriendly weather. Microclimate management is a powerful leverage point in the town yard because it can pump up yields as well as biodiversity by creating more variability in a small space that otherwise might be boringly uniform. It lets us fit more into a small space. Two of the main elements for microclimate design, structures and plantings, are already abundant in urban yards. The secret to creating blissful microclimates is knowing how to arrange these elements to work harmoniously with the physical forces that shape microclimates.

Most of those forces involve heat transfer, the movement of heat energy from one place to another. In a landscape this happens in several ways.

## Radiation

Radiation is a powerful force in shaping microclimates. Although the word may conjure up thoughts of atomic bombs, that's a special case; here we mean the general radiating of electromagnetic energy: particles or energetic waves emitted, usually, from the sun or from something warmed by solar rays. That's what you feel in full sunshine, even on a chilly day, as radiant solar energy strikes your body and heats it. Solar radiation is the primary driver of climate, both global and micro. Capture sunlight to warm up; block it to cool off.

## Air Mixing

This is the second big factor. This occurs in two ways: convection, where heated air rises and is displaced by settling cooler air, and what is called fluid flow streaming, which in this specific case becomes just a fancy term for wind. Moving air carries heat from one place to another, which cools warm things and heats up cooler ones. Managing air movement allows us to hang onto the benefits we've accumulated when we capture or block the radiance of sunlight.

## Evaporation

This factor is important in cooling off. Water evaporating from an object carries heat away from it. That's why mist sprayers and swamp coolers abound in hot, dry climates.

## Conduction

Conduction, heat transfer by direct contact, is a lesser factor for us. It shows up mainly in warming soil, where solar radiation heats the soil surface, and this warmth is conducted deeper into the soil and to roots. That's one reason some gardeners strip mulch from soil in spring. Mulch insulates soil from the sun's rays, and pulling it away lets the sun warm the upper layer of earth via radiation. That heat gets conducted to the frigid dirt below.

Besides heat transfer, there are physical factors that influence microclimate, outlined in the following sections.

### Thermal Mass

Thermal mass refers to a material that stores heat. A massive, dense substance, such as rock or water, can hold more heat than a lightweight one such as wood. When a dense mass warms up, it reradiates that heat back out over time. A rock wall that heats in the hot sun will keep anything near it warmer at night. Water also has serious thermal mass and will influence the temperature of anything close to it. In hot climates we can shade thermal masses to keep them from heating, or we can minimize thermal mass and use; say, metal, fabric, or light wood for fences and walls. These can't store much heat and will cool off fast once the sun is off them.

### Slope

The slope of land, especially with respect to the sun, influences its temperature, moisture, and many other important qualities. That's why slopes facing away from the sun are the classic gardener's challenge: They are colder, darker, and often damp. Also, cool air can pool on flat, open hilltops and in basins or where it is blocked from draining. So where we are on a slope matters.

### Precipitation

Rain, fog, and snow all affect microclimate. Keeping dry, blocking snowdrifts, and protecting plants from hail are a few of the ways that we can work with this major force. Specific methods are below.

### Albedo

Albedo is otherwise known as reflectivity. Light-colored surfaces reflect light and thus stay cool (but can warm adjoining darker areas by that reflection), while dark surfaces absorb the energy from light and convert it to heat, which then can radiate to warm people and other objects nearby.

Manipulating these factors gives us huge control over microclimate. Whatever the general climate, we have strategies for arranging the objects in the yard so that those microclimate forces combine to produce the conditions we want. See the sidebar for some examples for working with microclimates in different conditions.

## SETTING THE SCENE FOR SUCCESS

By this point in a chapter on gardening you may be asking, "Yes, but where are the vegetables?" All this talk of edges, zones, and microclimates is not just throat clearing; it's necessary guidance for creating the conditions for healthy plant growth, nutritious food, comfort, low maintenance, and thriving habitat. Just as the key to a great garden is not just to toss down some seeds on any old dirt but to do comprehensive soil preparation, at the heart of a successful whole-systems design is learning to arrange all the large systems—soil, plants, water, energy, compost, graywater, animals wild and domestic, human spaces, habitat, and so on—in the best possible relationships. Without that, we'll be constantly intervening, laboring, buying inputs, carting off unused outputs, and spending much more time strong-arming the yard into being the place we want rather than watching it naturally move in that direction while we enjoy it.

By understanding and managing sectors and edges, we control the flow of resources and influences into and across our design, put them to best use, and avoid the disaster of having some powerful force—such as feral animals or contaminated soil—lay waste to our plans. Gaining microclimate savvy speeds plant growth, reduces heating and cooling costs, and ensures that we will spend more happy time in those outdoor spaces we have worked so hard to create. All these strategies are critical parts of good design, and good design lets us make our mistakes on easily correctable paper instead of less malleable reality. It paves the road to success. The rest—implementing the design—is just technique and sweat. Compared to the intense, focused brainwork that goes into the planning, building out the design is straightforward. Careful design gives us the confidence to build our plan on the ground quickly, without hesitation or second-guessing. Next, we'll look at techniques that are part of doing just that.

## Creating and Enhancing Your Yard's Microclimates

### Warming a Cool Yard

- Identify and preserve sunny areas, and use them well. Place sun-loving plants and warmth-requiring activities in them.
- Use pavement, masonry, earth, water tanks, and other thermal masses in sunny places to store heat during the day. Afternoon sun will warm decks and patios on the south and southwest sides of the house, which will then radiate their heat in the evening to warm the air and the house. Deciduous trees and vines provide shade in summer for potentially hot places, if needed, but will let in light during the winter to warm them. Patios used in the mornings should be on the east or southeast of the house. But remember that massive objects (patios, stone walls, water tanks) in shady places will stay cold until the sun can warm them up. This means that morning sitting spots should be built of light materials that will warm fast, while afternoon spots can be of more massive materials so they won't get too hot too fast and will radiate heat in the cool evening.
- Create sun bowls—U-shaped areas open to the sun and protected on the sides by plants or walls.
- Use dark colors for surfaces and structures. Dark stones, bricks, and rock mulches can store heat and warm soil.
- Use windbreaks (plants, fences, or walls) to block and divert chilly winds. Watch where the wind is being channeled to avoid placing activity areas in Venturi zones.
- Don't put evening activity areas at the bottom of slopes or where air drainage is blocked; those will be cold pockets. They will also be chilly in the morning.
- Dark vertical walls facing the sun collect heat and are warmer places for seating and gardening.

- Sun-facing niches and protected pockets in walls will be several degrees warmer on sunny days than the ambient air temperature and often warmer on cloudy, breezy days.
- Small ponds in sunny areas will store heat and warm the surroundings, but in cold, cloudy weather these will be cool spots. It's generally cooler near large ponds, so place these away from cold-weather activities.
- Prune trees to allow sun to reach thermal masses and activity areas.
- Tree canopies, roofs, and fabric covers will reflect heat from the ground and will be warmer than open areas. They can be shaped so that low winter sun will reach under them. Tucking tender plants close to this cover will protect them.
- A light dusting of charcoal (not additive-laden briquettes but pure charcoal or biochar) on snow will speed melting and can then—stacking functions!—build soil carbon after the thaw.

### Cooling a Hot Yard

- Create lots of shade. In particular, shade the south and west sides of buildings and activity areas, especially stone and concrete walls, patios, and other thermal masses. Start with temporary shade structures such as shade cloth or lattice, meanwhile planting trees around the house and activity areas. Build trellises and arbors and run pole beans, vining squashes, and other fast-growing vines up them while waiting for larger, perennial vines to mature. Grow vines on sun-facing walls to reduce thermal gain. Remember that in northern climates, hot, late afternoon sun in summer comes not from due west but from the northwest.
- Cover nonplanted activity areas, but leave the sides open for air movement.
- Place activity areas in shade. In the Northern Hemisphere, use the north and northeast side of structures.
- Shade ponds near activity areas and the house to keep the water—and thus the surrounding air—cool.
- Avoid large paved areas. If you must pave, use paving blocks or strips of pavement and grow plants between them to reduce their exposed, heat-collecting surface.
- Take advantage of roof overhangs, arbors, trellises, gazebos, and other covered areas. These are cooler in the day but at night will be somewhat warmer than their surroundings because heat from the ground reflects back from them and can't escape easily.
- Use light colors for surfaces, structures, and even plants (dark foliage absorbs more heat than light-colored leaves).
- Plant in layers so that little sun reaches the soil. Minimize large areas of short cover. Even small shrubs and large herbaceous plants will keep areas cooler than a lawn.
- Mulch soil with light-colored, loose mulch such as straw. Almost any kind of mulch will keep soil cooler than if it were bare.
- Where water is ample, use mists and sprays for cooling.
- Prune low branches to increase air movement.
- Identify cooling breezes, and remove obstacles that block them.

- Identify sources of warm winds, such as breezes heated by passing over pavement, and block them.
- To open a view, rather than cutting whole trees remove lower limbs to preserve the shade of the canopy.

### Increasing Humidity and Moisture

- Preserve tree canopies and cover. This holds moisture in the air below the cover and slows evaporation from soil.
- Plant thickly and in layers; transpiration increases humidity, and layers hold that humidity close by.
- Use plants instead of pavement as ground cover, and let them grow tall unless it will create a fire hazard.
- Contour the soil with swales to prevent runoff, and create basins to collect and hold rain.
- Where sensible, use mists and sprinklers as well as ponds and fountains. Just the sound of water helps us feel cooler and wetter.
- Use living roofs.
- Block hot, dry winds.
- In areas of fog and dew, create fog brooms: tall trees (especially conifers) and vertically mounted screens or shade cloth that harvest significant amounts of water vapor.
- Catch and store the rain. Build sunken, not raised, garden beds, and raise paths to shed rain into planted beds.

### Decreasing Humidity and Moisture

- Let the sun in. Prune trees to let light penetrate, and minimize covered areas.
- Keep grass short, use ground cover plants instead of taller ones in some areas, or replace grass with gravel. Since this may increase runoff, be responsible about where the runoff goes.
- Move water off the site via drains and ditches, also prevent external water flows from entering the site.
- Allow airflow, especially warm winds, by removing obstacles.
- Design any required ponds to have small surface areas (depth rather than width) to reduce evaporation.
- Plant on raised beds and berms.

### Decreasing Wind and Its Effects

- Learn your seasonal wind patterns: Where do unpleasant winds come from and when?
- Create windbreaks. Agricultural Extension service websites have many excellent resources on windbreak design.
- Build temporary windscreens of wood or fabric until vegetation grows in.
- Locate activities downwind of buildings, fences, and other protection.
- Excavate sunken areas for activities, or build earth berms around them.
- Encourage low branch development on trees and shrubs.

### Increasing Breezes and Benign Effects of Wind

- Learn local wind patterns, especially those of cooling and warming winds.
- Remove plants, structures, and other obstacles between the ground and 10 feet high in the path of winds.

- Prune low branches.
- Locate activity areas on high ground or on mounds; build raised decks, raised beds, and elevated sitting areas.
- Create Venturi effects: Channel wind via solid structures from where it is not needed to where you want it. When a wide volume of moving air is constricted through a narrow channel, its speed picks up, creating a stronger breeze.

**A Few General Tips on Dealing with Wind**

Observing wind at different times of day will help with windbreak planning. From midday to early evening, winds usually move upslope,; they often flow downslope at night and in early morning. Most winds blow from lakes and oceans toward land in the day and from land to water at night.

# CHAPTER 4

# Techniques for the Urban Home Garden

Sector, zone, and needs-and-resources methods, as we saw in the examples in the last chapter, give us a toolkit for choosing from a near infinite palette of possible activities and functions that we could have in our yard and for narrowing this almost bewildering array down to the selections that work for us and for our conditions. These methods also help create beneficial connections among our choices. Then, getting a handle on our yard's edges and microclimates gives us the power to shape the many flows of resources entering, leaving, and already present there.

Now it's time to drill down to specific techniques and elements to answer two more big design questions: What should we do to meet the goals of our design, and what living and nonliving elements do we need? This chapter looks at developing the key elements needed for a town yard to function like an ecosystem—the soil, plants, and animals—and shows how to integrate those elements into a whole-systems design from the perspective of the land and its roles. Later we'll see how water, energy, structures, waste, and even livelihood and community can be brought into this picture, with the home ecosystem as the focal point. But for now, we'll stick to the land itself and how it can provide for us and for the rest of nature. That brings us to the interconnected cycles of soil, plants, and animals both wild and domestic, which all need to be elaborated thoughtfully and fully in themselves, as well as integrated in a complete design, so that each supports the other elements and the highest functions of a yard.

## THE SPECIAL CASE OF URBAN SOIL

An old gardening adage counsels us to wait one year when gardening in a new place, meaning it's best to observe your yard and climate through a full round of seasons before making major changes. That's a wise piece of advice but one that's rarely followed, because most of us simply must get out there and *do something* well before a year has dragged by and usually before we've finished the full garden design. A good first step that will exploit some of that gardening ardor without risking disastrously wrong moves is to get our hands dirty, literally. No matter what else we do, we'll need healthy, nutrient-rich soil as the garden foundation, so we can safely expend bottled-up energy on soil building while we polish the design and learn our local conditions. Besides, soil enhancement is the leverage point where a little up-front work causes countless benefits to unfold spontaneously.

Sometimes city dwellers are blessed with fertile soil, since in a natural if unfortunate evolution, many towns agglomerated and fissioned on top of prosperous farmland. In spite of this, we can't count on having fertile earth. Soil in town yards is notoriously patchy. Housing developers often scalp off the original rich loam, sell it, and sprinkle down an inch of industrial topsoil before rolling out a carpet of sod. And in many cases history has left its confused mark on the soil. My former yard in Portland featured a central core of delicious Willamette Valley loam, one of the finest agricultural soils on the planet, but that had been surrounded by a hodgepodge of wildly varying soils. The basement of the 1885-vintage house had been enlarged in the 1920s, and that project's clay subsoil filled the north border of the yard. A different clay fill, trucked in to flatten a slope, capped most of the front yard. Decades of trash burning had mutated a corner of the back into a caustically alkaline ash pit. An abandoned gravel driveway, now buried, cut across the back edge, and at some point an ancient cesspool had been filled with yet another soil type.

That's typical of any yard having a past, and the only way to understand those complex patterns is to dig test pits around the lot and see what's there. A couple of hours of backyard archaeology will spare you unpleasant surprises such as stricken seedbeds and fruit trees dying in an ash pit.

Two factors predominate in urban soils. One is the need for intense use and production. We don't have a lot of space, so we're often coaxing as much yield—food, habitat, flowers, biomass—from the soil as is feasible. To do this, we need to generate high fertility. The second big factor is contamination. Old yards can have a nightmare history as the locus of ecologically horrific activities—home-mechanic oil changes, furniture stripping, pesticide disposal, trash dumping—while newer yards are often topped with soil from chemical-laced farms and municipal sludge.

## Toxic-Soil Remediation

Let's look at toxic-soil remediation first, since it pays to eliminate the negatives before (or while) building up the positives. The most common soil contaminant in any city yard is lead, from two sources. Houses built before 1978 were often painted with lead-based pigments, which can leach from the walls into the soil near the house. Lead doesn't move easily in soil, so it's usually restricted to close to the house, but it may have been spread by tilling and digging. The other source of lead is the fine dusting from leaded gasoline upon any land within reach of the internal combustion engine; the more engines, the more lead, so cities have high lead levels, and parking strips are often lead hot spots. Many cities offer free lead tests, and they're well worth doing, as lead is conclusively linked to birth defects, lower IQ, and a broad spectrum of health problems.[1]

## Good Sources of Organic Matter

- Arborists, electric utilities, departments of transportation and parks. Their tree trimmings and chips may be had with just a phone call, though in progressive areas where organic matter is rightly valued, there's a waiting list.
- Restaurants and coffee shops. Their food waste is nitrogen rich (which means you'll need a source of carbon to match it). These businesses need reliable people to do pickups, as a pileup of rotting food is disastrous for them.
- Groceries, breweries, wineries, food processors and distributors. Health laws may prohibit pickups, but the flow of materials from these businesses can be immense.
- Urban poultry, rabbits, pygmy goats. Usually their owners are gardeners, too, but if not, the manure needs disposal, and that's where you come in.
- Pet stores and rabbit rescue centers. Some pet stores separate rabbit, guinea pig, and other vegetarian manure from undesirable carnivore manure. I've gotten great used bedding from rabbit rescue operations, and I've heard of poultry rescues, which would be another good source.
- Leaves and grass clippings. Many cities require leaves to be bagged for pickups, making them easy to collect and haul home. Neighbors who don't compost may give you their lawn clippings and other organic matter for the asking. Just be sure they don't use pesticides.
- Municipal compost programs. The quality of this compost is highly variable, so it's worth doing some research.

The most effective remedy for lead-contaminated soil is removal, which usually means hiring a specialist to strip off the top foot or so of fouled soil and haul it to a toxic waste site. For low, widespread levels of lead, several tricks will reduce its mobility and availability to plants:[2]

- Most fruiting crops—tomatoes, peppers, squashes, and so forth—and those that produce seeds and seed heads such as beans, peas, and sunflowers don't readily accumulate lead. These are safer to plant in lead-tainted soil than those with edible leaves and roots, which can sequester lead from soil.
- Keeping soil pH between 6.5 and 7.5 reduces lead's mobility.
- Organic matter such as compost, decomposing mulch, and manure binds lead, keeping it out of plant roots. Leaves from street trees may be high in lead, so those are not ideal for mulching urban food crops.
- Supplementing soils with iron or manganese may reduce lead uptake and flush it from soil.[3]

Various other metals show up in urban soils, and city agencies often record which ones and where they occur. Specific metals have their own remediation techniques, but in general adding organic matter along with healthy minerals such as calcium, potassium, phosphorus, and essential trace elements can help bind or flush toxic metals. If you know your soil has a particular metal contaminant, research specific methods for getting rid of it.

Petrochemicals and other toxic hydrocarbons, including fuels and pesticides, are the other major soil contaminants. Backyards were once the favorite place for changing engine oil, dumping oil-based paints and thinners, and burning all manner of trash, all of which left a shadowy legacy of hydrocarbon-filled soil. Researchers have cataloged many species of bacteria and fungi adept at degrading petrochemicals, which is another argument for keeping soils rich in organic matter and minerals: They bolster microbe populations. Mushroom guru Paul Stamets, in his eye-opening book *Mycelium Running*, shows how oyster mushrooms can convert fossil-fuel waste into harmless compounds. Other fungi, such as king stropharia and various mycorrhizal blends, can also break down petrochemicals.

If your soil is seriously contaminated with hydrocarbons, have a specialist look at it. For typical soils near roads or those only lightly contaminated, my personal treatment is to amend it with high-quality compost, add minerals, mulch deeply, keep soil and mulch moist, and let the microbes flourish in this now-conducive clime. This can break down many carbon-based toxins, but it will not reduce the amount of lead or other heavy metals in the soil. A test after treatment is always wise.

While we're shrinking the toxic burden of our yards, we can increase soil fertility. We're not only building a verdant paradise that will sooth our frazzled urban nerves but also pulling immense amounts of food and other products from our yard. So we'll need a lot of inputs, at least at first, to boost productivity and to replace what our heavy harvests are taking from the soil. Our goal, eventually, is to close the nutrient loops with self-renewing fertility plants and by returning wastes to the soil, but we're rarely lucky enough to have all that in place by the time our first harvests start making hefty withdrawals from the soil bank. So to start, we need to make some large deposits into that bank and create enticing working conditions for the bank's staff—the soil life.

I'll stretch that bank analogy a little further. Although having a large balance sitting in an account makes us feel secure, what really engenders a feeling of wealth is the flow of money: having plenty pouring in, and spending it with lavish abandon. The same goes for soil. Almost any soil holds more nutrients than plants can possibly use, but most of it is tied up in compounds that the plants can't access, leaving the plants in the same bind as trust-funders restricted to a miserly allowance trickling out of a fat but locked-up inheritance. The life in the soil is like a team of shrewd lawyers able to break the trust and dig into the capital (okay, I'll stop beating the analogy now—but it works). Soil life converts stored, bound-up nutrients into available ones, and the more soil life there is, the faster the nutrients flow rootward.

Many key plant nutrients—phosphorus, potassium, calcium, magnesium, and a dozen-odd more—are held in rocks. Ground-up rock in the form of sand, silt, and clay makes up most of the solid content of soil, but plants can't digest those rock-bound nutrients easily. To become plant food, these nutritious minerals must weather out of the stone. Once freed from their rocky prison, most of the liberated minerals stick to the surface of clays or are captured by humus and other dead organic matter. These bound nutrients are somewhat available to plants, but only with effort and must be etched loose by root secretions of mild acids or enzymes. It's hard work for the plants. Microbes, armed with sophisticated chemistry sets, are more adept than plants at teasing minerals from their clay and humic confines and usually snag them first. Once the microbes have had a go at the mineral morsels, they pass them on to plants through excretions or death or by being eaten, dissolved, and excreted.

Thus most available minerals are either cycling in living things, bound lightly to clays and organic matter or dissolved in soil water. It's that last source that plants largely rely on for food: the soil nutrient solution. But hungry plants can very quickly use up the nutrients in the soil water, and those nutrients must diffuse back in from soil storages or be pumped in by living things, most freely in the form of manure. By "manure" I don't mean only cow pies and other macropoops. Two major sources of nutrients in soil solution are insect droppings, or frass, and excretions from microbes and larger soil critters such as worms. And there's a lot of that excreting going on. Below the soil surface, microbes reach densities of several billion per teaspoon, and animal life there can reach 2 to 4 tons per acre, which may be double or triple that above the soil. These immense populations of microherds produce vast amounts of microturds packed with water-soluble nutrients for plants. If the soil life is abundant, those droppings will constantly renew the fertility of the soil moisture.

That means that feeding the soil life is a key to healthy plants. If nutrients are scarce, hungry microbes will greedily grab them and hang on until they die. If microbes are scanty, they won't have the numbers to pull more than a few nutrients from the soil bank and cycle them. In either case, there won't be much for plants to eat. So boosting soil life and getting those swarms to feast at overflowing tables like Edwardian royalty is the strategy we need to follow. The key is ample organic matter (the energy source and nutrient storage bank) and abundant minerals (the nutrients for life).

Fortunately, we live in a world—and in towns—awash in both. I have been surprised to find, on my various migrations between city and country life, that organic matter is often easier to get in cities than in rural places. One reason is that farmers need organic matter in huge volumes and make arrangements with local stables and animal operations that often squeeze out little guys like me. Rural tree-trimmers often just blow their wood chips back into the forest, which is a good place for them but means I can't share in the bounty. But in town, organic matter is a problem to be gotten rid of. Restaurants can't dump their food scraps just anywhere, the fall leaf crop must be disposed of, arborists have to haul away their chips—and don't get me started on the mountains of coffee grounds. You can intercept these flows usually for the asking.

The quantities of soil minerals needed in a town yard are small but critical. I'm a believer in the value of soil tests to see what's needed. There's almost no way to know what's been done to the soil in a town yard, so simply dumping a universal fertilizer mix into soil that has been worked, amended, and altered in countless ways by previous residents can bring on toxicities or deficiencies. Home soil-test kits give varied results, so have it done at a professional lab, which local garden stores, agricultural Extension offices, and the Internet can recommend for you.

In garden soils, nitrogen, calcium, and phosphorus are the most common nutrients in short supply. Manure will supply nitrogen; lime (for acid soil) and gypsum (for alkaline soil) add calcium. Soft rock phosphate is my favorite for phosphorus, as it also adds calcium, but if I need phosphorus in a hurry, guano and fine-ground bonemeal deliver it fast. I also value rock dusts such as azomite and basalt dust for their trace elements that release over a long period. Seaweed is another source of trace minerals.

To dive deeper into soil building, a topic that rightfully fills many volumes, you can read chapter four in *Gaia's Garden* or peruse some of the excellent books listed in the resources section.

## Closing the Human Nutrient Loop

Many mineral fertilizers are mined via extractive industries, so any good permaculturist will phase out their use quickly. Ideally, we would use these products only to jump-start the soil into high fertility while we're developing perennial-based, fertility-renewal landscaping and tight composting cycles. That still leaves one important source of mineral loss: the food you harvest and eat. Those nutrients literally go down the toilet and out of your home ecosystem unless that broken loop is closed via composting toilets, urine collection, and similar methods. Not all towns permit composting toilets, especially homebuilt versions, although many towns allow commercial models. Good resources for building composting toilets abound, from complex, virtually maintenance-free designs to the simple 5-gallon-bucket method in Joe Jenkins's classic *Humanure Handbook*. Fully composted human waste can be used like any compost, although some feel it is safer to use it on trees (including fruit and nut trees) or nonedibles rather than on vegetables.

In this fecal-phobic culture, many people are not comfortable dealing directly with their own feces—and using the bucket method, they'd be doing some of that. Until that cultural squeamishness abates, some may find it easier to use urine for fertilizer. Fresh urine from a healthy person is nearly sterile and, if religiously collected for the garden, can supply about 80 percent of the nutrients needed to grow that person's food, much more than feces, so it's an effective leverage point. Urine holds enough nitrogen, phosphorus, and potassium to be equivalent to 11-1-2 fertilizer.[4]

I've used urine via two techniques. One is to collect it in a bucket filled with sawdust or wood shavings. Being male, I find it simple to use the bucket; women might want to set a toilet seat on the bucket or build a frame to hold a seat. The wood shavings prevent odors. When the shavings are saturated, empty the bucket into a compost pile and mix it in. A full 5-gallon bucket takes a few days to fill, so you'll notice some fragrance upon emptying. If that bothers you, use fewer shavings and empty more often. There is enough nitrogen in urine to balance all the carbon in the shavings and give your compost pile a boost as well. The minor drawback to this system is that you aren't applying the fertilizer directly to the soil but indirectly via compost, where there will be some losses. The advantage is that you aren't moving, pouring, and mixing volumes of urine.

The direct method is to collect the urine in a suitable container (without sawdust), dilute it 1:5 to 1:10 with water, and apply it to the soil, avoiding splashing on edible plant parts. You can collect it indoors or create a private outdoor urination station. The width of the mouth of the collection container depends on your sex, agility, and comfort with transferring urine. You probably aren't going to want to take it to the garden after every pee, so you'll need a way to store the liquid in a closed container. Some folks design their urination station to hold a container prefilled 80 percent with water to dilute the urine immediately, then apply this when the diluting container is full. This also minimizes smell. Fresh urine starts out clean and sanitary, but after a day or two the rich nutrient solution will grow various bacteria, and that can get odorous. Sealing it or diluting are solutions to this. As with any fertilizer, even diluted, don't overdo its use. A couple of applications per growing season is plenty.

I realize that many of us would prefer not to deal with our bodily wastes. But ecological responsibility demands that we do, and these methods offer a clever stacking of functions in which the problem becomes the solution. Not only do we avoid polluting clean water, but we also send nutrients to the garden without importing them specially for that use. It's a double closing of cycles. I've also learned that most of us are more interested than we let on about human waste and its consequences. In my permaculture courses I used to allot fifteen minutes to discuss composting toilets and urine collection, but I found that the ensuing lively conversation always expanded to an hour or more. As our ecological consciousness evolves, we see that this is one place we can all make a direct difference. That and the freedom to talk about a taboo topic pulls everyone into the discussion.

## HIGH YIELDS IN THE MICROYARD

The small size of most town lots means we need to use intensive, small-scale growing methods rather than extensive, large-scale ones. I've covered many of these strategies and techniques in detail in *Gaia's Garden*, so I'll stick mostly to summaries of them here, though I've added some new ones. Details on how to implement and use them can be found in my previous book and in many other sources in print or on the web.

### Packing Plants Horizontally

Getting big yields from a small space isn't just a matter of squeezing plants tightly together. In fact, that may reduce yields because of nutrient and light competition. We need, as always, to use diverse and complementary strategies. Here are a few.

#### Grow Perennials When Possible

Though everyone loves annuals such as tomatoes and zucchini, perennial vegetables often yield over a longer season and lessen the time the plant is not productive as a seedling or bolted, weakened plant. Perennials reduce the work of annual seeding and cultivating, and their richly developed root systems make them drought tolerant. Perennial greens such as French sorrel (*Rumex scutatus*), Good King Henry (*Chenopodium bonus-henricus*), and sea kale (*Crambe maritima*) will poke up shoots as soon as the weather permits, eliminating possible delays from too-late seed starting, and yield until winter descends in earnest. In mild climates they grow year-round. For these reasons I recommend finding perennial substitutes as well as broadening your taste to new varieties that have no annual counterpart, such as Jerusalem artichokes and scorzonera. Eric Toensmeier's book *Perennial Vegetables* covers over 100 species of these useful plants.

#### Use Space-Saving Bed Patterns

Gardeners know that raised beds of solid plantings use much less space than single-row

**FIGURE 4-1.** Improved keyhole bed design for small yards. The circular area, 7 to 9 feet in diameter, is the most easily reached part of the bed. But squaring the circle lets the beds pack together more efficiently, especially in rectilinear yards. The back corners can be planted with insect-attracting flowers, cover crops, or vegetables that are harvested only once. The front corners that are along paths can be planted with more vegetables. Illustration by Elara Tanguy

**FIGURE 4-2.** Hugelkultur. Mounds can be built on top of the soil or laid into trenches. They can be anywhere from 2 to 6 feet tall, depending on the space and materials available. The woody debris holds moisture between rains, releases modest amounts of mineral nutrients, provides organic matter, and generates heat as it decomposes. The raised surface increases the planting area, useful for small yards. Illustration by Elara Tanguy

plantings separated by paths. Another pattern, the keyhole bed, takes raised beds one step further by bending a rectangular raised bed into a horseshoe shape with a short path leading to the center of this now-circular bed. This reduces the area lost to paths to 60 percent of that in conventional raised beds. Herb spirals, which coil a linear path of herbs or any small plants around a 2- or 3-foot-tall mound, are another space saver. Hugelkulturs, raised piles of soil and yard trimmings mounded on top of brush, also increase the surface area available for planting.

### Plant in Compact Patterns

The common square-grid pattern of laying out seeds and plants wastes space. Planting in hexagons, triangles, and sine waves fits more veggies in a given area. See figure 4-3 for examples. Mel Bartholomew's classic *Square Foot Gardening* and related books detail other close-packing methods.

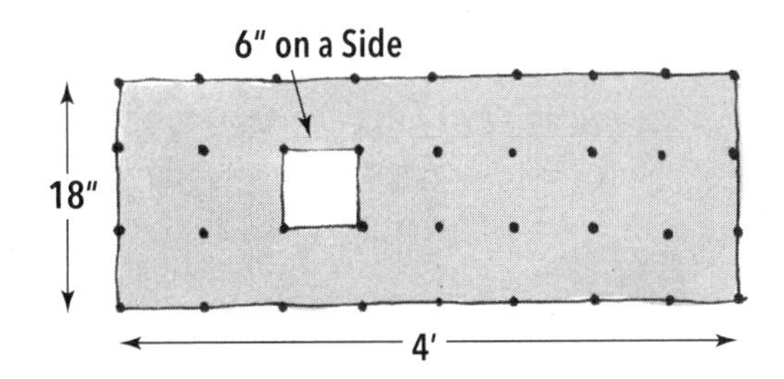


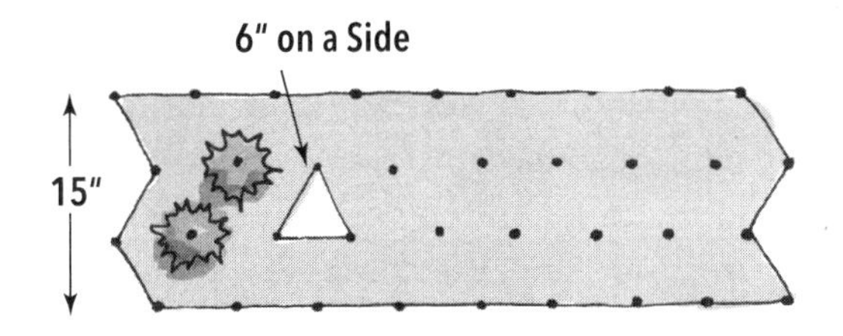


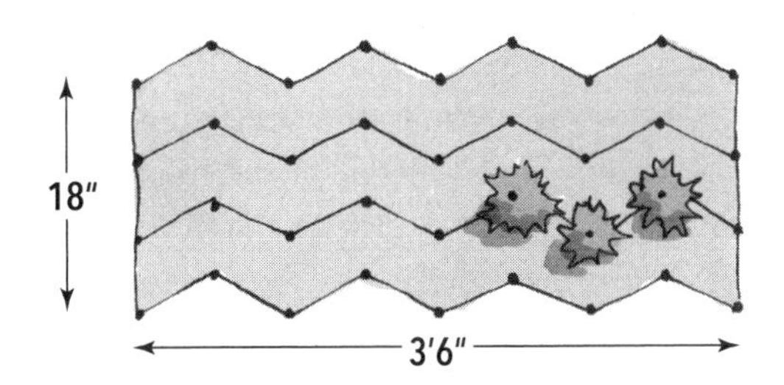


**FIGURE 4-3.** Planting in compact patterns for small yards. To save space, instead of using rectangular seed or plant spacing, plant in triangular or zigzag patterns. Illustration by Elara Tanguy

### Maximize Fertility

Soils that are light and fluffy, packed with organic matter and minerals, optimized for pH, and properly watered will usually yield heavily. A small area of intensively managed, extremely fertile beds will provide more food than a larger area of less fertile ground. We're taking a lot from the soil, so we need to give a lot back.

### Use Small Plants with Big Yields

Think small and harvest big with dwarf varieties, multiply-grafted fruit-salad trees, and microgreens. Fruit trees can be kept under 8 feet tall and wide by selecting dwarf rootstock or heavily pruning larger cultivars, yielding more fruit from a given space. One tree often bears more than a household eats, so grafting several varieties onto one tree is a good strategy. A small apple tree can hold four or more varieties: early, midseason, and late ripeners for the table and other cultivars for pies, long storage, or cider. Plum rootstock can be grafted with multiple plum varieties as well as almonds, peaches, nectarines, apricots, plumcots, pluots, apriums, and, with more difficulty, cherries. Greens such as chard, kale, lettuce, and many others can be sown densely, about one seed per square inch, and the harvest, via cut-and-come-again methods (snipping only larger leaves), can begin when leaves are 2 or 3 inches long. As the greens mature, the plants can be thinned to eventually reach the usual full-sized spacing.

### Grow in Polycultures

Seeding an array of complementary vegetables and greens into the same space boosts yields. One polyculture mixes cabbages, radish, dill, parsnip, calendula, lettuce, and beans together, producing an astonishing amount of food from a tiny space and extending the production over a long season. Polycultures are detailed in *Gaia's Garden.*

## Stacking Plants Vertically

It's easy to forget about the third dimension when growing food. Outside of the tropics, the sun is never straight overhead, so we can use methods that take advantage of that by training or stacking plants vertically.

### Support Vines and Trailing Plants

Use trellises, arbors, walls, poles, teepees, netting, stakes, and other vertical supports. Many sprawling vegetables such as squash, melons, and cucumbers can run vertically, saving valuable soil space. Grapes, hardy and fuzzy kiwifruit, runner and pole beans, peas, and passionflower can fill other vertical space.

### Use Intensive Fruit Methods

Fruit trees can be grown via espalier, cordon, Belgian fence, fan-training, column, and other heavy-pruning methods that shrink the space used by one tree. This allows more varieties and species to fit in a small space.

### Garden in Guilds

A central tree can be ringed and underplanted with other food plants, as well as insectaries and fertility plants that may also produce food. Guild design and use are detailed in *Gaia's Garden* and some recent permaculture books.

### Grow Vegetables in Stackable Containers

Many tubers, such as potatoes, yams, and Incan root crops, do well in barrels or wire baskets that can be filled with soil as the tubers grow, stacking the harvest upward. Strawberries, greens, and other small plants can be planted one above the other in holes cut into soil-filled towers and tubes.

### Stack the Food Forest

A forest garden blends and stacks relatively large and small tree varieties, tall and short shrubs, large and small herbaceous plants, root crops, vines, and even fungi into the same space. Excellent books such as David Jacke and Eric Toensmeier's *Edible Forest Gardens* and Martin Crawford's *Creating a Forest Garden* give a wealth of detail on food forests, and *Gaia's Garden* covers this subject as well.

## Stretching the Harvest in Time

Now that we've packed our space with food, we can extend the harvest by moving into the fourth dimension, time. Here are some ways to lengthen the growing season and use clever timing strategies to boost yields.

### Use Relay and Succession Plantings

Follow early spring crops such as peas and hardy greens with beans, tomatoes, peppers, and other warm-season varieties. In mid- or late summer, start fall cool-season broccoli, cabbage, cauliflower, and greens in the same bed as the hot crops. The next relay can be started among the maturing earlier crops, which will be removed as the following crop grows. The optimal combinations will depend on your climate. In regions with hot summer nights, successions will move quickly, and three or four series of crops may be possible per year, while cooler coastal and mountain climates may allow only two relays.

### Find and Create Warm and Cool Microclimates

The surest way to make a warm microclimate is with a greenhouse, cold frame, or cloche. These will give your plants a head start on the season. Use your microclimate savvy to locate and enhance natural microclimates, too, and in summer plant heat lovers in the warm spots and cool-season crops in the chilly ones. In fall and spring use the warmer sites for hardy veggies to keep them growing in cold weather. You'll be extending the harvest on both ends of the season and increasing yields.

### Use Perennial Plants When Possible

They will know better than you when to emerge in spring—often earlier than you would think because of their ample root development—and many, especially perennial greens, can last well into the cold months if given some protection under plastic, glass, or in a warm microclimate.

### Use Permaculture Design

Permaculture focuses on more efficient use of resources, and those include your efforts and time. The zone system places elements where they are easiest to reach and maintain. Considering sectors protects plants from destructive forces and puts them in the path of helpful influences. Permaculture's principles and design methods will help you pattern your garden to get more from a small space with less work and less waste.

## CALORIE COUNTING, SURVIVAL, AND WHAT TO GROW

Choosing what to grow is one of the food gardener's biggest decisions, one made more exacting by the limits of a small yard. We can apply a few permacultural methods to help us make those choices, among them the zone concept of frequency of use and needs-and-resources analysis. The first guideline is purely practical: Grow what you like to eat. It's easy to be enticed by seed and nursery catalogs into planting exotic, unfamiliar species—that latest Russian berry full of antioxidants sounds tempting!—but it's wise to follow a variant of the 80/20 rule: 80 percent of what you plant should be easy-to-grow, reliable producers that you know you like to eat. Then you can indulge your speculative tendencies on a safe 20 percent. When that sexy new perennial vine turns out after a five-year wait to yield bitter, seedy, wormy fruit, you won't have to tear out vast, expensive acreages of it, and your trusted 80 percent will carry you through. Grow a lot of what you like to eat that thrives in your conditions.

Next, consider growing what is hard to find or expensive to buy. Kiel and I love pesto, and fresh basil costs a fortune in stores, so we start a sizable basil plantation each year. Some plants, such as tomatoes, never seem to taste as good when store bought as garden grown, and that's another pointer for crop selection. Flavorful, nutrition-packed heirloom varieties of favorite crops aren't often carried in stores, and they are pricey in farmers' markets, so consider growing those.

Then let's consider a recent trend in the context of the town yard: calorie crops. Many gardeners have begun calorie gardening, growing the energy-dense foods that really feed us. The familiar garden crops of greens, fruits, and vegetables are high in flavor, vitamins, and minerals but very low in calories—a pound of kale holds roughly 150 calories—and calories give us the energy we need to live. You can get away with skimping on fresh greens for short periods (though I don't recommend it), but if you fall very short of the 2,000- to 2,500-calorie daily average for more than a few days, you'll get hungry and weak fast. To truly feed yourself from a garden, you must grow calorie-rich crops such as grains, potatoes, squashes, and beans. The bible of calorie gardening is Carol Deppe's excellent book *The Resilient Gardener*.[5]

However, calorie crops need a lot of space to yield enough to truly feed a person, and space is what town gardens usually lack. Commercial yields of grain range around 5 million calories per acre, which equals about 11,000 calories from a 100-square-foot bed.[6] That's a five-day food supply. Winter squash and potatoes yield more than that in a home garden: 100 square feet can easily provide 50 pounds of potatoes, but at 415 calories per pound, that's 20,750 calories, just ten days of calories tying up critical real estate for several months. Although you'll stretch out those potato calories by adding them to other dishes, I'd rather buy my potatoes from a farmer who deserves my support, and plant my 100 square feet to a polyculture of greens that I can harvest for much of the year.

Rather than trying to fit space-consuming calorie crops in a small yard, my strategy for potential food shortages is to have several months of dry staples plus seeds on hand, a plot of land earmarked in the yard or elsewhere for conversion to calorie crops if needed, and ties deep enough to local farmers to last through hard times. I can live off our stored food for the few months it will take to grow a calorie crop. Converting a garden from veggies to energy-dense crops is straightforward enough for an experienced gardener to do easily, and we can do it when the need arises.

Calorie crops are also almost all single-function annuals, not the permaculturist's first choice. Those crops produce food for people and little else (beans produce nitrogen, but most of that goes into the bean and not the soil, since it's harvested after setting seed). Thus most calorie crops don't make it through the filter of many functions, and as I've exhorted repeatedly, most everything in a small yard should play multiple roles if we are following permaculture design.

## GUILDING THE SMALL YARD

One of the most entrancing, productive, and sometimes baffling practices of permaculturists is growing plants in guilds, or communities of species chosen to support each other, increase yields, and decrease work and resource imports. By now the guild concept has been well tested and is bolstered, too, by several observations.[7] One is that nature rarely creates monocultures. Another is that wild species often grow in guild-like ensembles of species called communities or associations. A third is that many indigenous cultures plant their crops in multispecied mixes (polycultures) that prevent many of the negative effects of row-crop farming, such as erosion and loss of biodiversity.[8] And last, polycultures often yield more than any single species planted over the same area.[9] (See photograph in the color insert, page 4.)

I've described guild theory and design in detail in *Gaia's Garden*, so I'll briefly introduce

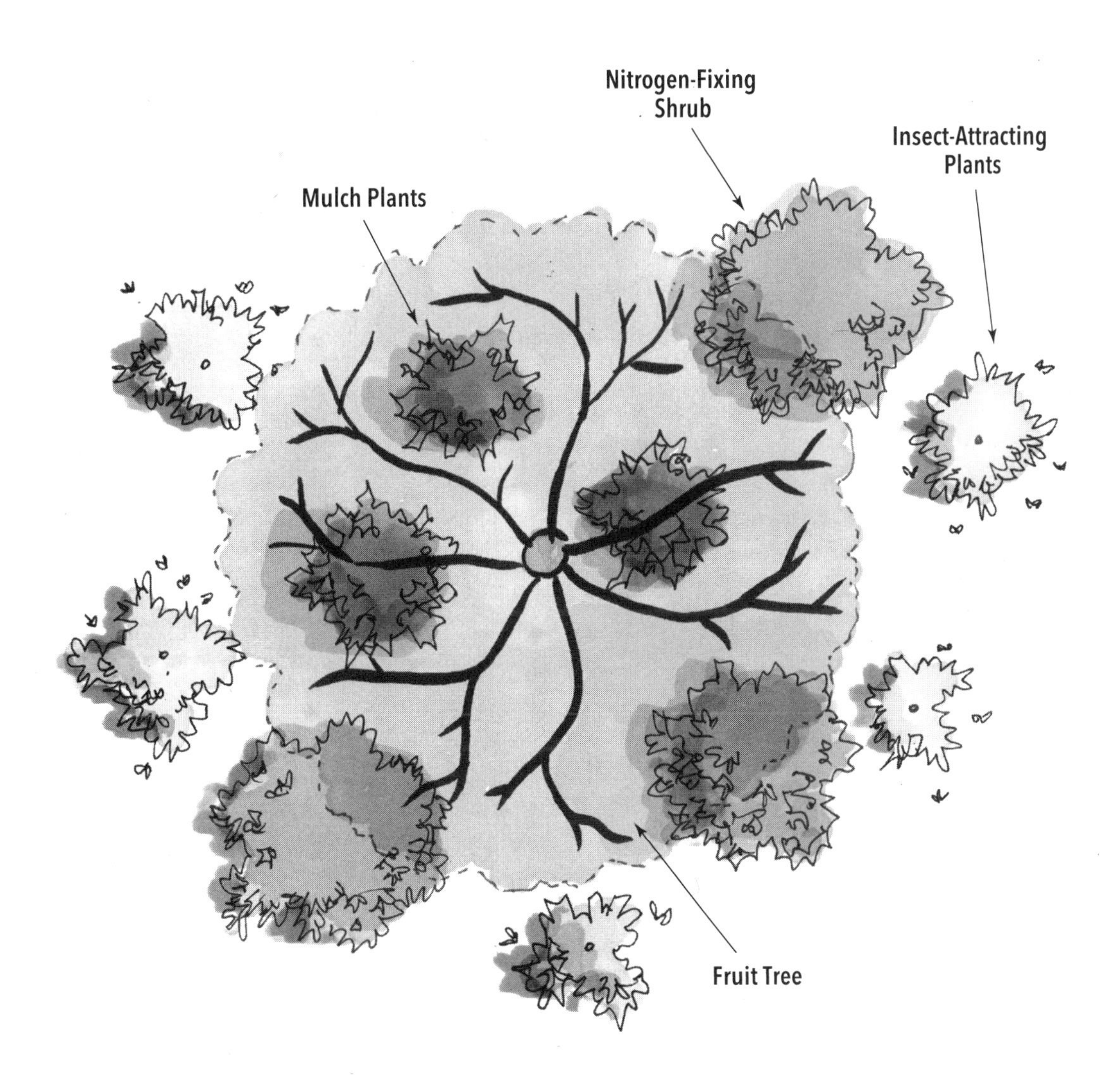


**FIGURE 4-4.** The basic elements of a fruit tree guild. Mulch plants such as comfrey should be placed inside the drip line of the tree for chop-and-drop mulching. Nitrogen-fixing shrubs such as Siberian pea shrub or goumi can be located around the mature-tree drip line and pruned heavily at least once each summer to stimulate nitrogen release. Insect-attracting flowering plants can be placed in profusion anywhere near the tree, as pollinators and predator insects can easily find the tree from some distance away. Additional guild members could include berry bushes; plants with thick root systems, such as mustard and rapeseed, for breaking up heavy soil; and fungi such as garden giant (*Stropharia rugosoannulata*) to aid soil development and add yet another edible crop. Illustration by Elara Tanguy

guilds here, then quickly move on to look at guild specifics for small yards and city settings. A guild was originally a group of craftspeople sworn to support each other (each member contributing an amount of gold, or *guild;* hence the name). Using principles learned from designing and observing plant guilds, permaculturists are again creating human guilds, but in ways that are more multifunctional and inclusive than the old, protectionist craft guilds.

Any organization has roles that, when filled by the right people, help it function in a dynamic, resilient manner reminiscent of a healthy ecosystem. Human guilds can be scaled up, too. Businesses and nonprofits can link up in guilds composed of mutually supportive organizations that reduce the need for outside capital, guarantee markets to one another, and help weather the rockier segments of economic cycles. Thinking of the household as a guild as well helps identify the important functions that need to be filled for it to run smoothly and as an integrated whole. These are more examples of how permaculture's study of natural systems has broad use far outside the garden.

We'll explore human guilds and other applications of the guild concept in later chapters. The reason I mention this now is to point out that you and your housemates or family are also part of all the plant guilds you create. Your habits will shape your guilds, and the guilds in turn will shape you. Although one point of guild creation is to reduce the work of obtaining their yields, the guilds in our yards are dependent on people to maintain them and perform the other tasks that we're not clever enough to make innate to the guild.

A successful guild is tied together through multiple partnerships among its member species. The gold-standard guild, the one that we all wish we could have invented, is the Native American three sisters, the ancient blend of corn, beans, and squash. It is a richly interconnected, calorie-packed guild. The corn links to the beans by donating a trellis for the bean stalks, the beans aid the other members via the nitrogen-fixing bacteria snuggled into its roots, and the sprawling squash leaves cover and shade the soil, keeping roots cool, smothering weeds, and slowing evaporation. This exemplar of vegetable cooperation gets to the heart of good guild design: selecting species that satisfy the essential needs of the other plants. The three sisters method is unusual because all three members are food producing and none of them dominates the guild; it's a more or less equal partnership. That kind of balance is rare, for it's much easier to design guilds that are centered around a dominant, larger plant such as a fruit or nut tree. This central element defines most guilds and guides our choices of other species to support it.

That tells us that a good place to begin in guild design is to list the basic plant needs, then pick species that can supply them for the central element and, if possible, for each other, too. Some major plant needs are:

- Water
- Nutrients
- Soil rich in organic matter
- Pollination
- Protection from predators (certain insects, birds, mammals)
- Appropriate sun or shade

Table 4-1 offers a sampling of functions needed in guilds and plants that supply them to illustrate some members of these categories, but there are many, many more.

**TABLE 4-1.** Guild Plant Functions

| ESSENTIAL ROLES IN ALL GUILDS | DESCRIPTION AND SAMPLES |
|---|---|
| **Fortress/barrier** | Help keep out unwanted plant and animal species: daffodils and euphorbia spp. (repel gophers), mustard (good smother crop for weeds), buckwheat, gooseberry |
| **Insectary** | Support beneficial insects: sweet alyssum (attracts aphid predators), golden marguerite, lavender, dill, angelica |
| **Mulch maker** | Decompose quickly, providing mulching in place: comfrey, cattail, cardoon |
| **Nitrogen fixer** | Host nitrogen-fixing bacteria: clovers, sweet pea, bladder senna |
| **Nutrient accumulator** | Draw nutrients from soil, concentrating them in tissue: eastern bracken, yarrow, sunflower |
| **Soil builder** | Produce organic matter and improve soil structure: mustard, Sudan grass, crotalaria |
| **Soil cultivator/ (Spike -root)** | Penetrate, loosen, and aerate the soil: little bluestem, daikon, fava bean |
| **AUXILIARY ECOLOGICAL FUNCTIONS** | |
| **Acting as nurse/ scaffold/chaperone** | Hardy pioneer plants support establishment of other plants: alder, acacia species, Spanish broom |
| **Air purifying** | Clean pollutants from the air: English ivy, common milkweed, chrysanthemum (known to remove benzene) |
| **Animal forage** | Provide food for domestic animals: Siberian pea shrub, buffalo grass, buckwheat |
| **Erosion control** | Hold soils in place with fibrous root systems: sea buckthorn, bamboo, sumac, many salvias |
| **Fire retardant** | Sagebrush (*Atriplex* spp.), white rockrose, jade plant, aloe vera, lavender, salvia species |
| **Flood management** | Can stand submersion in water, and promote percolation to water table: many native aquatic plants, annual ryegrass, feather reed grass, fountain grass |
| **Pest repellent** | Repel pests, eliminating pesticides: peppermint (repels insects and mice), lemon balm (repels flies and ants), garlic (repels aphids, deer, and rabbits) |
| **Scavenging nitrogen** | Remove excess nitrogen from soil: berseem clover, barley, oats |
| **Toxin absorption** | Black nightshade (removes PCBs from soil), curly-top gumweed (absorbs selenium) |
| **Water purification** | Cattail, common rush, canna lily, ostrich fern |
| **Wildlife food** | Blackberry (small birds), witch hazel (ruffed grouse, pheasant), elderberry (many bird species) |
| **Wildlife habitat** | Barberry, highbush blueberry, hazelnut, cornelian cherry |
| **Windbreak** | From small to large: Jerusalem artichoke, shrub willows, hybrid poplar |
| **HUMAN USES** | |
| **Aromatic/Fragrance** | For their fragrance alone or to mask other smells: lavender, rosemary, jasmine |
| **Basketry and weaving** | Shoots are straight and flexible: common reed, prairie willow, fragrant sumac |
| **Cleanser/Scourer** | Pennyroyal, cranberry, horsetail rush, potato |
| **Compost** | Comfrey, chicory, yarrow, stinging nettle |
| **Cut flower** | Black-eyed Susan, cosmos, yarrow, daffodil |

**TABLE 4-1.** *continued*

| ESSENTIAL ROLES IN ALL GUILDS | DESCRIPTION AND SAMPLES |
|---|---|
| **Dried flower** | Lavender, milkweed, statice |
| **Dye** | Goldenrod (mustard, brown, and yellow), carrot (orange), red raspberry (purple) |
| **Essential oil** | Sage (for perfumes and shampoos), dill (for soaps and medicines), creeping thyme (for perfumes and mouthwashes) |
| **Fiber** | New Zealand flax (for rope and simple ties), swamp milkweed (bark is used for twine or cloth), oats (hulls used in construction board) |
| **Food** | Fruits, nuts, greens, vegetables, edible flowers, herbs |
| **Medicine** | Comfrey (for wound healing), purple milkweed (to treat warts), burdock (as a general tonic) |
| **Oil, wax, resin, or polish** | Bitternut hickory (oil from seeds used in oil lamps), Japanese stone pine (pitch obtained from rosin is used in waterproofing), hazelnut (seed used for furniture polishing) |
| **Pest repellent** | Rosemary (for sachets), parsley (juice used to repel mosquitoes), lavender (repels mice) |
| **Soap** | Snowberry, New Jersey tea, wild lupine |
| **Wood** | Sugar maple (for furniture and musical instruments), Amur corktree (for cork substitute), walnut |

## Water

Sadly, no plant species will actually irrigate plants, but we can use living and dead mulches to hold water in the ground and plant in multiple layers to create shaded soil during the hot part of the season. So we will choose species and arrangements that create mulch (see below) and have varying heights and leaf density. In foggy climates, tall trees, especially conifers, act as fog brooms. Paul Stamets has told me that many fungi synthesize their own water from oxygen and organic matter, which is a great reason to make sure guilds (and the rest of the soil) are home to healthy mycelium.

## Nutrients

We can use nitrogen-fixing varieties, usually from the legume family, as well as accumulator plants, which are species that pull certain nutrients, such as calcium, phosphorus, and iron, from deep in the mineral soil and ferry them to leaves, where they land in the topsoil at leaf fall.

## Soil Rich in Organic Matter

Organic matter—the decomposing or composted remains of plants and other creatures, as well as manure—is crucial for soil and plant health. Organic matter nourishes the life in the soil, stores water, keeps soil fluffy and light, and acts as the savings bank for nutrients. Unless you've been conscientiously building organic matter over the years, the odds are that your soil could use more. Once again, nature can step in and do this for us. Here are three categories of plants that build organic matter, each in its own way.

### Mulch Makers

These sport large, soft leaves and other non-woody biomass (examples include comfrey and cardoon). To turn these into mulch, I use the chop-and-drop method of cutting them down or pruning them heavily and letting the trimmings shield the soil.

### Spike Roots

These are plants with large, deep root systems, such as daikon. These punch holes in compacted or heavy soil and, if left to rot, pump organic matter into the ground.

### Cover Crops

Cover crops are legumes, grasses, and other species that build fertility and biomass and can be tilled into the soil, chopped-and-dropped, or removed to make compost, then returned to the soil.

## Pollination

Pollination is a by-product of insects foraging for nectar, pollen, and other blossom-borne exudations to eat. So grow flowers, lots of them! The Asteraceae (composite family), Apiaceae (umbel or carrot family), and Lamiaceae (mints) are especially attractive to pollinators, but so are many other flowers. It also helps to choose varieties that are just finishing their bloom as the guild's fruit tree member is beginning to blossom. The host of pollinators lured by the flowers will then turn to larder in the blooms of the central fruit tree.

## Protection from Predators

Fruit and nut trees are prone to insect problems. The best cure for a bug problem is another bug or perhaps a bird. So we want to attract these pest predators. Flowers and other host plants will attract specific predators, such as tachinid flies, ladybug larvae, parasitic wasps, and several others. The strategy here is to learn what pests are drawn to your crops, identify the best predators, and plant the predator's host species. *Gaia's Garden*, many other books, and the Internet offer extensive lists of specific beneficial-insect host plants. For thwarting larger critters, thorny species can be woven into tight hedges.

In general, the permaculturists' approach to pest problems is to learn the habits and life cycle of the creature. Then they can use that knowledge to find ways to avoid creating the conditions that attract the pest, intervene in its life cycle when it is most vulnerable, attract organisms that prey on it, and develop effective ways to exclude and block it. You can thus fashion a complete set of strategies—not just one but several that yield all the benefits of redundancy—to reduce the pest's depredations.

## Sun and Shade

To give sun-loving species ample light, space plants widely to leave light gaps. Beginners often pack guilds too tightly, which not only creates shade but blocks access for pruning and harvest and traps moist air that causes fungal rampancy. We can plant shade-needing species in layers, which has the bonus benefit, beyond shade, of packing more into a small space.

I realize that allowing room between plants and stacking them in dense layers is contradictory, and that tells us that designing proper sun, shade, and access patterns is one of the most challenging aspects of guild design. One great aid here is to know the mature size of the plants and space them accordingly. To avoid a too-sparse look in the early years, we can plant extras that will be moved later, or intersperse annual vegetables, herbs, or flowers in the temporary gaps. Also, plants can be clustered in layered groups, sorted by light, water, and fertility needs, with spaces between.

**TABLE 4-2.** The Thirty Most Valuable Plants for Urban and Suburban Yards

| COMMON NAME | BOTANICAL NAME | HARDY TO USDA ZONE[A] | GROWTH HABIT | LIGHT[B] | EDIBLE PART OR USE[C] |
|---|---|---|---|---|---|
| **Chestnut, American, sweet, or Chinese** | *Castanea* spp. | 5 | Large decid. tree | ○ | Sd, Med |
| **Italian or other stone pine** | *Pinus pinea* | 4 | Large evergreen tree | ○ | Sd |
| **Black locust** | *Robinia pseudoacacia* | 3 | Large decid. tree | ○ | Fl, Sd |
| **Honey locust** | *Gleditsia triacanthos* | 3 | Large decid. tree | ○ | Seedpod |
| **Apple, multigrafted** | *Malus sylvestris* | 3 | Small decid. tree | ○ | Fr |
| **Stone fruit tree, multigrafted** | *Prunus* spp. | 3–6 | Small decid. tree | ○ | Fr |
| **Medlar** | *Mespilus germanica* | 6 | Small decid. tree | ○ | Fr |
| **Bamboo** | *Bambusa, Fargesia, Phyllostachys, Sasa* | 5–9 | Evergreen shrub | ○ | Sht |
| **Buffaloberry** | *Shepherdia argentea* | 2 | Decid. shrub | ○ | Fr |
| **Currant, red** | *Ribes rubrum* | 5 | Decid. shrub | ○ ● | Fr, Fl |
| **Gooseberry** | *Ribes uva-crispa* | 5 | Decid. shrub | ○ ◐ | Fr |
| **Goumi** | *Elaeagnus multiflora* | 6 | Decid. shrub | ○ | Fr |
| **Hazelnut** | *Corylus* spp. | 4 | Decid. shrub | ○ ◐ | Sd, Oil |
| **Purple osier** | *Salix purpurea* | 5 | Decid. shrub | ○ | Med |
| **Raspberry, red** | *Rubus idaeus* | 3 | Decid. shrub | ○ ◐ | Fr |
| **Siberian pea shrub** | *Caragana arborescens* | 3 | Evergreen shrub | ○ | Sd |
| **American licorice** | *Glycyrrhiza lepidota* | 3 | Peren. | ○ ◐ | Root, Med |
| **Cardoon** | *Cynara cardunculus* | 5 | Bienn. | ○ | Fr |
| **Comfrey** | *Symphytum officinale* | 5 | Peren. | ○ ◐ | Lf, Med |
| **Daylily** | *Hemerocallis fulva* | 4 | Peren. | ○ ◐ | Fl, Lf, Root |
| **Egyptian or walking onion** | *Allium cepa proliferum* | 5 | Peren. | ○ | Fl, Lf, Root, Top Bulbs |
| **Groundnut** | *Apios americana* | 3 | Peren. | ○ ◐ | Sd, Root |
| **Kale, perennial** | *Brassica oleracea ramosa* | 7 | Short-lived Peren. | ○ ◐ | Lf |
| **Maximilian sunflower** | *Helianthus maximilianii* | 4 | Peren. | ○ | Root, Sht |
| **Oca** | *Oxalis tuberosa* | 7 | Peren. | ○ ◐ | Fl, Lf, Root |
| **Russian sage** | *Perovskia atriplicifolia* | 6 | Peren. | ○ | Lf |
| **Sweet cicely** | *Myrrhis odorata* | 5 | Peren. | ◐ ● | Lf, Sd, Root |
| **Yarrow** | *Achillea millefolium* | 2 | Peren. | ○ | Lf, Tea, Med |
| **Grape** | *Vitis vinifera* | 6 | Decid. vine | ○ | Fr, Lf |
| **Kiwifruit** | *Actinidia arguta, A. deliciosa, A. kolomikta* | 4–7 | Decid. vine | ○ | Fr |

| ANIMAL USE[D] | OTHER USES[E] | NOTES |
|---|---|---|
| Ins, Hab, Chk, For | Wbr, Hr, Wd | In blight areas, use Chinese chestnut. |
| Hab, For | Wbr, Hr, Wd | Many other species have edible seeds. |
| Ins, Hab, Chk, For | Wbr, Hr, Wd | Nitrogen-fixer, may sucker aggressively |
| Ins, Hab, Chk, For | Erosion control | |
| Ins, Hab, For | Hr | Late and early varieties for eating, storage, and cider can be grafted on one tree. |
| Ins, Hab, For | Hr | Hardiness depends on what fruit is grafted. |
| Hab | | Fruits in late fall |
| Hab, For | Wbr, Hr, Poles, Fbr | Hardiness, size, and spreading depends on genus. |
| Ins, Hab, Chk, For | Wbr, Hr, Dye, N-fixer | Drought resistant |
| Ins, Hab, Chk, For | Hr | Easy to propagate |
| Ins, Hab, Chk, For | Hr | |
| Ins, Hab, Chk, For | Wbr, Hr, N-fixer | Tolerates air pollution |
| Hab, For | Wbr, Hr, Bskt | |
| Hab | Wbr, Hr, Bskt | |
| Ins, Hab, Chk, For | Hr | |
| Ins, Chk, For | Wbr, Hr, Dye, Soil stab, N-fixer | |
| | N-fixer | Can be opportunistic |
| Ins | Mulch | |
| Ins, Chk | Nutr, Biom | Spreads easily by root division |
| Hum | Fbr | |
| Ins | Nutr, Dye, Rpllnt | |
| | N-fixer | |
| | | Strong flavor that becomes mild when cooked |
| Ins | | Deer resistant |
| | | Can be grown as annual in colder zones; tubers store well |
| Ins, Hum | Wbr, Hr | |
| Ins | Plsh | |
| Ins | Nutr, Dye | |
| Hab, Food | Dye | |
| Hab, Food | | *A. arguta* is spread aggressively by birds in areas with summer rain. |

**[A] The USDA hardiness zone system represents the lowest temperature a plant will normally survive:**

| Zone | Minimum Temp (°F) |
|---|---|
| 2 | −50 to −40° |
| 3 | −40 to −30° |
| 4 | −30 to −20° |
| 5 | −20 to −10° |
| 6 | −10 to 0° |
| 7 | 0 to 10° |
| 8 | 10 to 20° |
| 9 | 20 to 30° |

**[B] Light:**

○ - Prefers full sun
● - Prefers shade
◐ - Tolerates partial shade

**[C] Edible Part or Use:**

Fr - Fruit
Fl - Flower
Lf - Leaf
Med - Medicinal
Sd - Seed
Sht - Shoot

**[D] Animal Use:**

Chk - Poultry forage
For - Forage, browse, or other animal feed
Hab - Provides habitat
Hum - Attracts hummingbirds
Ins - Attracts beneficial insects

**[E] Other Uses**

Biom - Plant produces large quantities of biomass
Bskt - Stem, branches, or root used for basketry
Dye - Some or all of plant used to prepare dye
Fbr - Leaf, stem, flower parts, or root used in paper, cordage, or other fiber product
Hr - Hedgerow species
Wd - Woody parts used for lumber, firewood, or craft wood
N-fixer - Nitrogen-fixing species
Nutr - Nutrient accumulator species
Poles - Stem or branches used for poles and support stakes.
Plsh - Used as furniture polish
Rpllnt - Used as insect repellent
Soap - Leaves, sap, fruit, or other part used as soap
Soil stab - Used for soil stabilization
Wbr - Windbreak species

## THE WHOLE YARD AS GUILD

When selecting guild plants, remember to stack functions by choosing species with multiple roles and yields. Table 4-2 will get you started. For example, goumi (*Elaeagnus multiflora*) is a nitrogen-fixer and also bears edible berries, attracts birds and pollinators, and tolerates air pollution. Buffaloberry (*Shepherdia argentea*) is another nitrogen-fixer, and it offers nectar and pollen to insects, berries for us and wildlife, leafy forage for chickens, and a red dye from the fruit, and it can be pruned heavily to create a dense hedgerow while producing biomass for compost, hugelkultur, or kindling. Using multifunctional species not only packs more yields into a yard but also shrinks the number of plants needed to fill all the niches and roles. That leaves more room for other activities and design elements.

In classic guild design, support plants (insectaries, nitrogen-fixers, mulch-makers, and so on) are arranged inside or around the drip line of the central tree, and each guild is often a circular grouping independent of any others. But because town yards are often small, there's no need to isolate guilds from each other. Guilds can blend, and pieces can be scattered around the yard. Insects attracted to a pollinator plant on one edge of the yard will most likely visit all the yard's trees, and probably the neighbors', too. This means that once you've chosen the plants that satisfy the essential functions, they can be arranged in many ways rather than be placed only with regard to their central trees. The exception to this is the fertility plants (nitrogen-fixers, accumulators, and mulch species). Some of these, though not necessarily all, should go near or within their central tree's drip line to deliver their goodness right to the tree. But insectaries, spike roots, and plants with other functions can be clustered, scattered, placed in drifts, and gathered in arrangements dictated by the patterns, flows, access, aesthetics, water and light availability, and soil types of the entire yard and design.

Arranging guild members this way, not just crammed under a tree but in other functional patterns, gives the designer enormous freedom and ensures that the plants get proper light, water, and care more reliably than if they were tethered to a central tree. And, guild plants that offer food can go close to the house or in food beds instead of out with their fruit trees, and habitat plants can tuck into quieter spots where birds and other wildlife will feel safer. This strategy also allows more variety in the way the central trees are patterned with respect to sun, access, spacing, and other factors. All the yard's fruit trees could be packed together or arranged in a U-shaped sun trap; strung in a line as part of a windbreak or for easy irrigation; placed to enclose a lawn, deck, or play area; or arranged in some other practical pattern, as their placement is not dictated by the need to fit all their guild companions close to them. And this makes for a beautiful yard. I've arranged guild elements in some yards so that flower colors combine elegantly, foliage patterns complement each other, and shapes and arrangements lead the eye to intriguing features and restful views.

## ANIMAL PARTNERS IN GUILDS

With perhaps no exceptions, healthy ecosystems include animals. Plants magically sculpt sunlight, carbon, minerals, and water into the green treasure that feeds all the rest of life, kicking off the cycle of producer-consumer-decomposer and back to producer that weaves this planet into a blue-green tapestry. Animals are the plants' big beneficiaries, harvesting the edible starch and protein made by plants via the sun's energy. Animals do far more than reconvert plant tissue back into plant fertilizer, though. They direct huge flows of energy and matter around the landscape. Spawning salmon return critical phosphorus and other nutrients from the ocean to the uplands, courtesy of the predators that haul their carcasses onto land. Birds renew entire forests with their fertile droppings. Burrowers such as worms, prairie dogs, and gophers aerate and turn countless acres of soil. Large mammals, from cattle to caribou to elephants, graze, browse, and trample whole prairies, savannas, and forests, altering the species mix, changing soil composition, redirecting water flows, and changing micro- and perhaps macroclimates. And in a dynamic coevolution, plants and animals have become intimately entwined through pollination, seed dispersal, aiding germination by gut passage of seeds, and pest control. Observing this, we can find or design roles for animals in the yard that increase yields, fill critical functions, and keep the dynamics humming. In a properly designed garden system, animals are not added on. They plug into important niches that would otherwise remain empty or be filled, tediously, by our own labor.

Guilding, then, applies to more than just plants; many guilds contain animal partners. In the American Southwest, the piglike javelina plays an important role in wild piñon pine/juniper communities. While they crunch pine cones in search of their tasty nuts, the javelina scatter the seeds, which they conveniently trample into the torn-up, freshly manured soil left in the wake of their impressive rooting. Pigs themselves are handy in rural guilds for snuffling up dropped, rotting fruit. Renowned Virginia farmer Joel Salatin has created an all-animal guild in which cattle deposit manure that is picked over by chickens looking for fly larvae. Then in spring, when his cattle leave their winter barn quarters, pigs root through the thick layer of bedding that has accumulated in the barn over the cold season, hunting for manure and for nuggets of barley that Salatin has cunningly broadcast into the layers of straw. The pigs replace machines in the work of turning the cattle bedding to speed its composting, and the bedding in turn helps grow pigs. With these farm-scale strategies as inspiration, we can find ways in urban plant guilds to deploy small livestock such as rabbits and chickens to weed, trim, eat fallen fruit, and control pest bugs. Along with people, guilds and the rest of the yard rely on nonhuman animals to support the home ecosystem. Let's look at some domesticated varieties of those.

## SMALL YARD, SMALL LIVESTOCK

Town life puts constraints on livestock. Many municipalities don't allow much beyond cats and dogs. Although I'm a dog lover, I see some irony in forbidding hens and pygmy goats, which are fairly quiet, provide food and fertile manure, weed and mow for us, and have many other benefits, while allowing an animal that injures and even kills people, can wreck relations between neighbors, produces a resounding and often incessant alarm call, offers no edible products, and deposits copious amounts of noxious, shoe-fouling feces. But our relationship with dogs predates all other domestication, and they have a powerful lobby.

Dogs and cats came with us into the first cities and stayed, while most other livestock have been relegated to the farm. For cities to behave more like ecosystems and produce food, we need a broader palette of animals. Planting for birds, insects, and other wildlife helps redress this imbalance. But livestock can also fill important niches. We'll start with a look at some easily cared for and permitted animals and then progress to those that take a bit more work.

## Worms

Charles Darwin devoted his final book, *The Formation of Vegetable Mould through the Action of Worms*, to these keystone soil creators. In it he wrote, "It may be doubted whether there are many other animals which have played so important a part in the history of the world."[10] For gardeners' purposes, worms fall into two categories: wild and semidomesticated, and both kinds have important roles in generating soilborne flows of fertility. Wild worms include the familiar earthworm, *Lumbricus terrestris*. Surprisingly, the earthworm is not native to North America but migrated from Europe in tandem with the plow agriculture that churned millions of acres of prairie and forest into its favorite habitat of tilled soil and open fields. Earthworm castings can contain five times the nitrogen, seven times the phosphates, and eleven times the potassium of the surrounding topsoil, making worms ideal hardworking partners for gardeners.[11] We can pay them back by supplying nonwoody mulches and other soft, raw organic matter. These ingredients create the conditions that worms love, so covering soil with leafy mulch boosts earthworm numbers and gets nutrients cycling.

I think of manure worms, or red wigglers (*Eisenia fetida*), as semidomesticated because they thrive in livestock manure and compost piles, which didn't appear in abundance until the dawn of agriculture. In contrast to earthworms, manure worms don't do well in straight garden soil; they need a richer meal. Red wigglers are yet another species favored by our habit of creating garbage dumps and manure piles wherever we go but are more helpful to us than other similarly inclined camp followers, such as rats, pigeons, raccoons, and cockroaches. Red wigglers are also one form of livestock that apartment dwellers and other yardless folks can raise. A worm bin, properly managed, is an odorless, bugless, low-maintenance way to convert kitchen scraps into nutrient-rich castings perfect for sprinkling on potted plant soil and window boxes or amending garden beds.

The essence of red wiggler wrangling, also called vermiculture or vermicomposting, is to provide a dark environment such as a covered bin or wooden box, a damp but not soaking wet carbon source such as shredded newspapers, food scraps, and a starter batch of a few ounces of worms, available commercially or from a friend with a worm bin. Ingenious systems exist for raising worms and easily harvesting their castings. As usual, YouTube and the rest of the Internet abound with vermiculture methods and advice, as do many books and journals. It can be done in a drawer in a tiny apartment, on a farm in Dumpster-sized bins, and at any scale in between.

Worm-friendly gardening is a potent leverage point for tipping so-so soils into ripe fertility, and it intervenes at an important, early point in the path to abundance, good health, and life-friendly towns: the soil. It's yet another way to extend a helping hand to nature, which will then pay you back by recruiting an army of millions of happy subterranean workers to aid you in these tasks. (See photograph in the color insert, page 6.)

## Rabbits

Next up on the scale of easy-to-raise animals for the town yard, bunnies can grace us with a near perfect manure for the soil; naturally shed fur for spinning; and provide weed and grass control, meat and wearable pelts if we wish, and a gentle presence that enchants children and adults. Their most important needs, beyond the basics of food and water, are a sturdy, predator-proof hutch and protection from below-freezing temperatures. My expert on rabbits is Portland, Oregon, gardener Connie Van Dyke, who has raised them in her urban yard for nearly two decades. Connie's ingenious cage system is two tiered, which saves space, and has a slanted manure collector beneath each tier that channels the droppings to a trench below. Although rabbit droppings can be added directly to soil, in some cases seeds can pass through rabbits and germinate, so Connie usually hot-composts the manure.

Her rabbits rarely free-range because of predators, but they often vacation in rabbit tractors, movable rectangles of light fencing 3 or 4 feet on a side that hold a rabbit or three for grass grazing. Shifting a tractor around the yard trims and fertilizes the lawn while growing rabbits. Connie's bunnies also dine on her yard's offerings of bamboo trimmings, garden greens, root crops, oat straw, dandelion, blackberry leaves, clover, fruits, berries, plus some commercial alfalfa feed. Rabbits have finicky metabolisms, so the commercial feed helps provide a balanced diet, but the yard's produce shrinks costs and provides fresh fodder.

In addition to manure and mowing, the rabbits produce other yields: fur for yarn and weaving, which Connie harvests at natural molting times from her Angoras; young rabbits that she trades for other goods; pelts for clothing and blankets; and, from her New Zealand red and silverfox breeds, meat. This last is a challenging edge for some, as bunnies are the very definition of cuteness, but Connie believes in taking responsibility for her carnivory by butchering her own meat. She also is able to compartmentalize her attitudes toward her rabbits: Although she regularly slaughters some of her rabbits, she considers her breeding pair to be pets. She says, a little shockingly, "I don't eat my pets. But I do eat their babies." This is one way that raising rabbits poses more challenges than raising chickens: If they breed, the result is live animals rather than eggs. Dropping an egg into a pan, for many, causes much less anguish than butchering a rabbit and is simpler than finding new homes for a string of fast-arriving litters. If you don't want to deal with all that, don't get a breeding pair.

As highly multifunctional animals, rabbits plug nicely into the permaculture yard, as the rabbit tractor shows. Another clever integrated system uses a rabbit hutch with a collection chute like Connie Van Dyke's that is set over a worm bin. Here, the droppings instantly change their role from waste to food for red wigglers. Still another useful combination is to place an elevated rabbit hutch in a chicken yard, where the birds can range underneath, pluck seeds from the droppings, and mix rabbit manure in with their own droppings plus bedding to create a potent mulch/fertilizer blend. In Jacksonville, Florida, Kevin Songer has taken these ideas one step further. Songer runs MetroVerde Florida Green Roofs, and as a living roof advocate, he's built an elevated rabbit hutch with a roof planted in garlic, a cover crop mix, and greens for the bunnies to graze on. The hutch is inside Kevin's chicken yard, offering a safe harbor for both species. The elevated hutch restores production from the lost ground beneath it while feeding the rabbits and the occasional leaping chicken, providing more bonuses for the small yard. (See photographs in the color insert, pages 6 and 7.)

## Bees

European honeybees (*Apis mellifera*) require little care and would precede rabbits in the easy-to-raise hierarchy of this section but for one thing: They can sting, and rabbits can't, which means that the neighbor and child sectors enter into our beekeeping plans, while rabbits (if they are not for meat) do not. But with proper placement and attention, these valuable microherds can easily fit into town yards. The big factors in keeping others happy with bees are letting the neighbors know (you can invite them and their kids to see the hive and learn about it), locating the hive where the main bee flyway won't be toward a concerned neighbor's yard, having plenty of flowers in your yard and those of nearby bee lovers, and sharing the honey to encourage neighbors to look kindly on your hive. It's best to have these and other relevant strategies worked out before getting a hive.

Bees are the quintessential permaculture bug, embodying innumerable design lessons. They are edge workers that may range over a dozen miles to forage, gathering dispersed resources and focusing them into the hive, where they are distilled into concentrated sweetness, which in turn is centered on the invaluable queen, who transforms the harvest into larval bees for the cycle to begin anew. Bees are multifunctional. Their honey is sweet, antiallergenic, antimicrobial, and rich with minerals, enzymes, and complex sugars and can be the base for many lotions and salves, including burn ointment. Propolis, a hive sealant that bees make from plant resins rather than from nectar, quells infections and sore throats and kills viruses. The hives' beeswax can be used in candles, salves, waxes, wood finishes, and lip balm. Even their venom has benefits: It can reduce arthritic inflammation. Without bees' pollination of wild species as well as human crops, plants would need other strategies for reproducing, and we animals would lack most fruits and vegetables.

Keeping a hive attunes us to both time and space. From the patterns of the bees' excursions and the flavors of their honey, we become aware of what is blooming where and when. Their expansive range ties to an enlargement of our own attention, from only our yard to many square miles around us. Their sensitivity to the rhythms of daily and seasonal sun and temperature, winds, and even barometric pressure helps pull us out of electronic-induced trances and financial fretting, back to something like our once-wild selves.

Like the earthworm, our familiar honeybee is a European that has thoroughly naturalized in the North American landscape, economy, and culture. The omnivorous appetite and fluid adaptability of this generalist species has, in some places, put pressure on more specialized native pollinators.[12] To reverse this, we can grow the native plants that support specialist native bugs. But in developed areas, where native plants have been replaced wholesale by our introductions, generalist insects such as the European honeybee fill important, now-empty niches. Raising bees in town thus becomes a service to all wild, domesticated, and human nature.

An aspiring beekeeper can choose among three major hive styles: the familiar stacked-box, or Langstroth, hive and its variants; the Kenyan top-bar hive; and the Warré hive, which melds features of both other types. How to decide? Here are some pros and cons of each.

### Langstroth Hives

Langstroth hives are the most common, although they have lost favor among the homesteading and low-tech crowds because they require specialized tools such as a hot knife or capping fork, capping tank, extractor, and extra boxes and frames. Bees naturally make cells of varied sizes to suit the circumstances, while Langstroth cells are preformed and uniformly large to push up yields. This exertion may stress the bees, and the jumbo cells allow room for deadly varroa mites to lay their eggs alongside the bee larvae. The honeycombs, after being centrifuged to extract the honey, are carried over from year to year, so this design won't supply much beeswax. On the plus side, not having to build new combs reduces the energy expenditure of making wax and increases honey production. Langstroth hives give roughly twice as much honey as top-bar and Warré hives but with more human work and investment. High yields and mechanization make the Langstroth design the favorite of commercial beekeepers, but I think other designs are simpler.

### Top-Bar Hives

These hives, from Africa, are an ancient, traditional tool. The most common design is the Kenyan top-bar hive, a trough-shaped box with sloping sides. Strips of wood hang across the top, and from these the bees suspend their combs. Honey harvest from top-bars is easy: You just pull out a honey-filled comb by lifting the wooden bar, then crush the comb and strain out the honey. Honey squeezed from the comb picks up some pollen and propolis, which gives it a richer flavor than Langstroth honey and, some say, stronger antibacterial effects. Top-bars require some attention while the hive is growing because new empty bars need to be slipped in periodically to give the growing brood new homesites.

People who want a close relationship with their bees prefer top-bar hives because the bars can be arranged to suit the way the hive is growing, allowing, for example, more protection of the queen or segregation of workers. Newer designs have a viewing window that lets the beekeeper watch the hive's progress without disturbing the bees.

### Warré Hives

Warré hives use boxes that stack vertically like a Langstroth hive, but each box holds eight or so bars across the top as in a top-bar hive. The stack is capped by a wide roof atop a quilt that insulates the bees, and new boxes can be added from the bottom as the hive grows. The inventor, Abbé Émile Warré, was an early twentieth-century French priest who trialed over 350 hive designs and settled on a simple but versatile style that he called "the people's hive," tailored especially for impoverished villages. The straight-sided boxes are easy to build, but the comb is more likely to stick to these than to the angled sides of the Kenyan top-bar hive, making it harder to harvest. The abbé was trying to produce a hive that would stay warmer than other designs, would reduce condensation on the comb walls, and could fill completely with the bee cluster, thus producing more bee scent. This last lets the bees find the queen more easily, keeping them calmer. As with the top-bar hives, honey can be harvested with minimal tools. (See photographs in the color insert, pages 10 and 11.)

Bee enthusiasts can debate the merits of hive styles endlessly. In my own experience, the top-bar hive is simplest to make and use, the Warré seems to have the fewest disease problems, and the more expensive Langstroth yields the most honey and, being the most familiar to other beekeepers, gives you access to their extensive knowledge pool. One of these styles may suit your own circumstances more than the others.

### The Mason Bee

If dealing with hives, honey harvest, and potential stings is not your cup of tea, you can invite another common bee into your yard that requires nothing more than habitat: the mason bee (*Osmia lignaria* and other species), a solitary, that is, non-hive-building bee native to North America that is nearly impossible to anger into stinging. Mason bees nest in small holes and hollow reeds or other tubes—I've found their dried-mud nests in keyholes, bamboo sticks, and even the hole in the end of a bicycle handlebar grip. They are outstanding pollinators: In a mason bee's six-week lifespan, she will tickle the anthers of about 60,000 flowers. Nests for them are easily built by drilling ⅜-inch diameter holes spaced ¾ inch apart, 3 to 5 inches deep in an 8- to 12-inch-long block of untreated 4" × 4" or 4" × 6" lumber. Don't drill all the way through; leave ½ inch or so of wood at the back of each hole. The block should be mounted on a wall, fence, or post firmly so that it won't swing, in a protected spot such as under a house eave, and facing southeast to catch morning sun.

An even simpler design, shown to me by permaculturist Rick Valley, is to find pieces of bamboo with ⅜- to ⅝-inch inside-diameter openings, and cut them into 8- to 12-inch lengths so that the septa (the dividers between hollow sections) are in the center of each length. This leaves a two-ended divided tube perfect for mason bees to build nests in each end. Then tie a bundle of these together and hang them on a southeast corner of a building. Or you can use plastic straws or paper tubes of the right diameter.

Some mason bee professionals, such as Dave Hunter of Crown Bees, say that permanent holes and tubes are prone to disease and parasite buildup. The best bee homes contain slightly larger holes lined with a paper tube. This tube can be changed every year, giving the bee a clean new home each season.

Fancier designs, such as a tray-style mason-bee house that allows viewing, cleaning for removal of mites, and storage through bad weather, can be found on the web. Mason-bee tending is fun for kids, fascinating and educational, and benefits plants, too, in an easy stacking of functions. One of my favorite books, *The Mason-Bees*, a freely downloadable, enchantingly written masterpiece from 1914 by the consummate French insect observer Jean-Henri Fabre, offers a wealth of personal observations about these important pollinators. (See photograph in the color insert, page 10.)

## Ducks

Ducks are considered slightly easier to raise than chickens, partly because they don't scratch. (A few minutes of untended chicken scratching in tender seedlings will trash a season of vegetables.) Yes, they need water, but a simple kiddie pool or tub can be enough, and these have been used by countless urban duck-raisers. Ducks are also more docile and quieter than chickens and, if I may make a judgment call here, better than chickens at staying out of trouble. Farm-unfriendly neighbors are also less likely to be irritated by ducks than by an obvious livestock animal such as a chicken. Many breeds, such as Campbells and Indian Runners, are prolific egg layers. The latter is a superb forager and can be free-ranged on pasture. Heavier breeds such as the Pekin are meat birds. The eggs and meat are stronger flavored than those from chickens, however, and may take getting used to. Ducks became a permaculture icon in the wake of Bill Mollison's famous remark that you don't have too many slugs but a deficiency of ducks, since the birds will feed on that slimy nemesis of gardeners. Although Muscovies and Indian Runners are good slug-eating breeds, they may not take naturally to gulping down these naked mollusks. Most ducks must be trained by hand-feeding them slugs when young. They can be pastured free-range or in duck tractors the same way that chickens can.

Like all poultry, ducks need a secure yard and house, protected from predators. They don't necessarily need a large pond, but they do require enough water to rinse their eyes and bills and to splash in, which stimulates their preen glands to release the oil that helps keep their feathers clean, intact, and waterproof. A small wading pond with sloping sides or a stepped ramp for easy duck entry may be enough, but they do appreciate enough water to swim and dive in.

A clever idea for an integrated duck pond is to channel rainwater from the roof of the duck house to their pond, then run the overflow from the pond to a swale or drainage ditch that runs between fruit trees or garden beds. This carries duck-fertilized water to the garden. Some people use an old bathtub as the pond, since it is already plumbed with an overflow drain, making it cleaner and easier to direct the outflow where you want it. Others set a small kiddie pool, light stock tank, or other easy-to-move basin

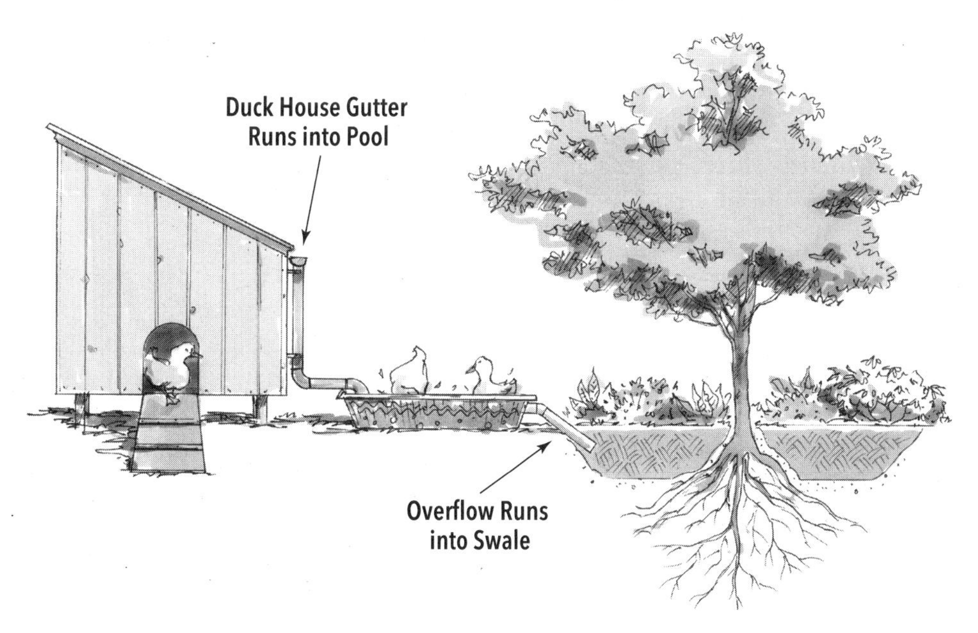


**FIGURE 4-5.** A swale-fertilizing duck pond. Rainwater from the duck house roof drains into a duck pond, in this case a kiddie pool. The manure-rich overflow from the duck pond feeds a planted swale. Neighbors of the author in Portland had such a setup, and the vegetation along the pond provided copious vegetables and biomass. Illustration by Elara Tanguy

near a tree or bed they want "fertigated" with duck water, let the ducks splash (and poop) in the movable pond for a day or two, drain the enriched water to the plants, and move the port-a-pond to a new location. The ducks may need a ramp to get into the basin if it's more than 6 inches high. Again, emptying a container via a built-in drain is less messy than dumping over a tub full of duck-pooped water.

Duck house designs are easy to come by, but they can always be augmented with permaculture variants. British garden writer Wendy Ogden houses her two ducks in a small shed topped by a living roof, on which she overwinters potted tender perennials. She says that the heat from the ducks spurs the plants' growth, giving them a running start in spring, and the plants, in turn, insulate the birds from the cold. A few people have experimented with duck houses that float in ponds or are on islands, with the aim of better predator protection, but they are inconvenient for egg harvest and maintenance, although a floating house can always be hauled to land with a mooring rope.

For those wishing more information, a permacultural approach to raising ducks suffuses Takao Furuno's book *The Power of Duck*, distilled from his fifteen years of observing how ducks boosted his rice and vegetable yields.

## Chickens

Chickens are so multifunctional and plug into so many niches in the urban yard that they are often the first animal that a gardener considers, in spite of their needing more care than bees or rabbits. Their material yields of eggs, meat, feathers, and manure are all useful, but it's their activities that fit them into the garden so seamlessly. They eat bugs, weeds, and seeds; scratch the soil to remove weeds and till in their manure; forage for their own food; sound the alarm at any unusual event; and provide endless hours of entertainment and interest. The now-classic tool for stacking many chicken functions into one place is the chicken tractor, a small, bottomless movable pen in which the birds can pick a garden bed free of weeds and seeds, loosen soil and mix in manure, and in a day or two move to a new bed to prep that one for planting as well.

And, chickens can be raised in so many different ways that one of the methods is likely to suit your lifestyle. The least permacultural approach is the conventional fenced chicken yard plus coop, feeding them with a commercial mix and some kitchen scraps. This can be improved by allowing avian escapes into a chicken tractor, where the birds can forage on more nutritious wild food and encounter greenery instead of the compacted bare-dirt desert that concentrated chicken yards can rapidly become. Another improvement is the deep-litter chicken yard, where layers of straw or similar mulch are constantly added to the bird pen along with kitchen scraps, shredded newspaper, weeds, and leafy trimmings. These regular mulchings, combined with the birds' scratchings and manure, will turn the yard's soil into fluffy, fertilizer-rich compost that can be harvested for the garden.

Next up the scale of good poultry habitat is the pastured poultry pen. These are open-bottomed, covered pens 10 to 15 feet on a side that are semipermanent homes for the birds, moved every day or so. These are more suited to the farm scale because of the size of meadowlike pasture they require, but in special circumstances they could be used in a quarter-acre yard where most of the land is dedicated to the birds. Well-respected farmer Joel Salatin's book *Pastured Poultry Profits* describes this method in detail.

Free-ranging your birds approaches the natural habitat, behavior, and diet of chickens, which are descended from Asian tropical-forest birds. Free-ranging lets the fowl perch in trees and shrubs, a natural need and comfort that pens don't permit. Many permaculture plants can feed your poultry, which broadens the birds' diet but has a downside: Free-range birds may eat or uproot your plants. Thus the method requires fencing off or covering newly seeded or transplanted crops. But once that's done, the roaming birds will range in search of larvae and veggie-eating bugs. The most pest-insect-free yards I've seen are those patrolled by free-range chickens.

A method that lets you raise a large number of chickens while giving them a natural diet and habitat but also spares plenty of the yard for other activities is rotational, or paddock, grazing. A stellar example of this is at the home of Corvallis, Oregon, permaculture designer Andrew Millison, who has divided the backyard of his ⅕-acre urban lot into four fenced quadrants. In the center is a chicken coop with four doors, each opening onto a different paddock. Andrew runs his birds in one paddock at a time, then follows them with a crop, rotating through the four paddocks over a year or so. Here's a typical planting and rotation pattern for his paddocks:

September: The chickens are in paddock 2. Andrew plants garlic in paddock 1, fall broccoli in paddock 3, and a cover crop of fava beans in paddock 4.

February: Chickens move to paddock 3 after the broccoli harvest. The birdless paddock 2 is planted to onions, peas, and potatoes.

Mid-April: Chickens leave paddock 3 and move to 4, where they forage in the fava canopy. Summer squash goes in paddock 3 when it's warm enough.

July: Chickens switch to paddock 2 after the harvest there of onions, peas, and potatoes. Favas in paddock 4 are tilled in, and the paddock is seeded with fall broccoli, carrots, beets, and onions. Garlic is harvested in paddock 1 and replaced with fall crops. (When I interviewed Andrew, he had just moved his beehives into paddock 1 at this point in the rotation and would keep chickens out of it for several rotations.)

September: Chickens move to paddock 3 after the summer squash harvest. Paddock 2 is planted with favas, winter onions, and broccoli. The harvest of fall veggies begins in paddock 4. (See photographs in the color insert, page 12.)

Millison grows fruit trees and berry bushes in the paddocks as well as many other useful plants, giving additional yields. The paddock method lets the birds forage and hop into the trees as they would in the wild, while in the birdless paddocks seedlings grow robust before the chickens wander through, stalking bugs. If you don't want permanent fences dividing up your yard, a variant of this system is to use linked panels of 4-foot-high portable dog fencing to build a single movable paddock that can be rotated around the yard.

Folks who have used paddocks tell me that their crop yields increase 20 or 30 percent and all plants grow faster and are more lush than before they used the method. It reduces reliance on purchased feed, too.

## Goats

Goats are the latest livestock to break the urban barrier, and they have vigorous advocates. In 2007 the Seattle City Council voted to allow small goats in town after the motto of the local Goat

Justice League, "I'm pro-goat and I vote," garnered national attention. Numerous other cities allow pygmy and dwarf goats, often with requirements for numbers, lot size, and distance from homes and neighbors. Goats are usually gentle and always entertaining, and they provide milk that is less allergenic than cow's milk and can easily be turned into soft cheese. Their manure arrives in tidy pellets and is a favored worm food. A loose goat is a gardener's nightmare, debarking trees and chomping shrubs and branches down to nubs, so they must be tethered or penned. Many urban goat owners take their animals for leashed walks in the neighborhood or use them to pull carts for children, and these are great attention-getters that will introduce you to your neighbors instantly.

Most people raise only does (females), since unneutered bucks are famous for their strong, unpleasant odor and during rut become aggressive and randy toward people as well as does. Milk goats must be milked once or twice a day, and they require a high-protein, mineral-rich diet. To keep the does giving milk, they need to be periodically "freshened," or impregnated. Goats can give milk for up to roughly ten months. Then they need a break for two months or so at minimum. They can be milked for the first three months of their five-month-long pregnancy. This once-yearly (at most) freshening means you'll be dealing with birthing goats and finding homes for the goatlings. Roughly half the births will be males, which almost no one wants. Obviously, you won't be abandoning bucks at the shelter just to preserve your chèvre's mild flavor, so you'll be faced with butchering or keeping (and castrating) them. Goats are herd animals, so keeping only one may leave her lonely. All this makes goats a serious commitment, so although some say that goats are the new chickens, I don't think so: Chickens are orders of magnitude simpler to raise. For a busy urbanite with conventional responsibilities, such as a day job, goats would be a stretch. But for a more independent urban homesteader willing to deal with the work, goats can fill an important niche. I've lived with goats, and they are wonderful, gentle companions.

Somewhere in this livestock list is at least one creature that can link your yard to the animal kingdom in ways that lighten the weight of your own tasks and fit into your routine. The right animal will tuck into the voids in your yard rather than just be added on as more work and money. And perhaps a combination of animals will work best. As I've related, livestock combinations such as rabbits and worm bins dovetail into each other's needs in functional relationships. A good grasp of permaculture principles will suggest other networks of species, such as bees in the same yard as worms and chickens, that can be patterned together to recycle nutrients, transform waste into food, reduce pests and predation, increase yields, provide constant fascination, and even build interspecies friendships. I have a friend whose pit bull sleeps curled up around her chickens; no raccoon will harm those birds! Animals are a lively, engaging, nurturing way to turn the parts of your yard that you can't use yourself—grass clippings, weed seeds, dead bugs—into things that you, and other parts of the web of life, can.

## Providing Homegrown Food for Livestock

Given both the space and the time, we could grow crops that feed ourselves and all of our livestock, untethering us from commercial sources and ensuring good nutrition. But growing staples for animals, just like growing all your own food, is just about a full-time job and requires ample land. Still, there are a few ways to provide some homegrown food for your animals to supplement purchased feeds:

- Kitchen compost is one of the best sources of animal food. Worms, rabbits, poultry, and goats will happily pick through food scraps.
- Livestock enjoy many of the same annual and perennial fruits, vegetables, nuts, and greens that you do. You can skim the best pickings for yourself and feed the less perfect ones to Squirmy the worm, Fluffy the bunny, Quackers, Henny Penny, and Bucky, and they'll automatically convert your scraps into fertilizer.
- Letting a few veggie plants set flowers, especially lettuces and the brassicas, will benefit bees.
- Converting part of the lawn, or the garden paths, to red or white clover offers a forage that poultry, goats, and bees will all appreciate, and it pumps nitrogen into the soil.

Because animals, wild and tame, play such a critical role in ecosystems and have so many useful products and behaviors, keep an eye out for ways that they can plug into your circumstances. Many common food plants can feed two-, four-, and six-legged beings, so choose vegetation that supports many species. Animal housing can also play multiple roles, as some of the clever ideas above indicate.

Everything in the dense environment of the town yard—the land, plants, and structures—will have demands loaded on it from multiple sources. Strive to make each piece and the whole design as multifunctional as possible. Walls of sheds can serve as tool storage, privacy screens, trellises, and windbreaks; roofs triple up as water-collection sites, growing space, and forage zones; and various parts of outbuildings can house different animals under one roof. Rather than seeing the need for intensity of use of urban spaces as a constraint or deficit, we can view it as encouragement to stretch our design muscles and be innovative.

## CHAPTER 5

# Strategies for Gardening in Community

How much land do I need to grow all my own food? It's one of the most commonly heard questions among permaculturists, urban and rural homesteaders, and other sustainably-minded people. I've asked it myself, though not for many years now, not since I tried it and learned what it truly entailed. I've come around to thinking that trying to grow all your own is not a very permacultural approach to obtaining food. It's a noble project, but the less-often-talked-about potential for isolation, monotony, and even disaster looms large. Meeting a need as critical as food in only one way—all on your own—violates the permaculture principle of supporting important functions in multiple ways. Diverse strategies will build much more resilient personal and regional food systems. If you love to garden, then, sure, grow lots of food. But a strong food system, whether personal or regional, is a network, designed in depth, built in community with others.

This chapter applies permacultural strategies and methods to the challenge of food security in ways that go beyond the home garden. Permacultural thinking can help us identify and reforge the weak links in how we obtain our food. Throughout human history, we've grown, harvested, and eaten our food in community far more often than in solitude—the image of someone always dining alone is not a happy one—and here we will look at a few ways to build a richly interwoven community through food and how to strengthen our food system through our community.

## ➡ WHAT ABOUT GROWING ALL YOUR OWN FOOD? ⬅

Let's return to that initial question, How much land do I need to grow all my own food? because it helps us unpack many preconceptions around sustainable ways to get our food. The problem isn't that growing all your food takes an unreasonable amount of land. In theory, one person should be able to grow all her food on ½ to 2 acres of fertile land, though that's more property than most of us have, and that doesn't include the land to grow the needed fertility crops and manures rather than importing them.[1] And the problem also is not that it is a lot of work, although it certainly is.

Over the couple of years I spent growing a lot of our vegetables and fruits when Kiel and I lived in rural Oregon, I found it a lonely way to spend much of each day. Many other enjoyable and important parts of my life had to be put on hold or dropped altogether to make room for the work of growing my food. Those accumulated solitary hours gave me plenty of time to think about why I was doing it and what it was doing to me. I began to ask myself, Why does the idea of growing all our own food have such allure?

For several hundred million people or more in the world, growing their own food—subsistence farming—is the principal way to feed themselves. But that's rarely true in the developed world. For the last half century, food in the developed world has cost less, in terms of how much time is spent working to pay for it, than at any other time in history. Yet the desire to grow all their own is often the default response of North Americans who want to get back to the land or to unplug from consumer culture. One reason is that once we realize that we're dependent on systems that we no longer believe in and that do immense harm, pulling entirely out of the commodity-food culture seems the quickest, easiest solution. When we first consider living more sustainably, self-sufficiency is an attractor that sucks us powerfully toward it. But another reason that Americans in particular default to going it alone is that the United States was settled in a unique way that idealizes that path.

Most of this planet's land was settled either by an expanding population pushed to adjacent unpeopled places in their search for new resources or by conquest and subjugation of the people living in a desirable spot. In either case the new settlers were usually groups such as tightknit tribes or armies and their camp followers. But once the United States was founded, much of the migration into new land was done by government edict, and the land was deeded to individuals and single families in a singular, atypical process. To encourage settlement as well as to raise money, in 1785 the US Continental Congress passed the National Land Ordinance, which plastered a grid over what was eventually three-quarters of the lower forty-eight states, turning millions of square miles of open land into a checkerboard of rectilinear townships 6 miles on a side, each divided into square-mile "sections" regardless of geography, natural boundaries, or indigenous inhabitants. Settlers bought or were given sections that were often far away and deeded unseen.

This process of individual rather than collective settlement and ownership was bolstered by the Homestead Act of 1862, which gave heads of households or individual adults title to parcels of federal land if they met certain requirements. To win the title to land, settlers had to "prove" the parcel: live on it, build a home, add improvements such as wells and barns, and farm there for five years. Homesteads were almost always settled by single families, each driven by the mandate to make a go of it themselves. This virtually unique aspect of US history, of individual rather than collective land development, driven by the demand to prove the land (and themselves), is embedded deep into our psyches. It has helped make self-sufficiency a national fetish.

Of course, there are real and valid reasons to want to grow your own food. It's immensely satisfying to see seeds sprout and become delicious nourishment. Growing food connects us to the cycles of life. Homegrown food tastes better than the stuff from the store, and it's almost always fresher. And whatever your politics, supplying your own resources offers emancipation from a system that seems, for so many of us, no longer to reflect our values or support the things we want to see in the world. The skills for self-reliance are gratifying to develop and give us a sense of security and strength in a world that frequently does not feel reliable, safe, healthy, or sensible.

But does that mean that it makes sense to grow *all* your own food? I can attest that growing just a small portion of it will earn you the same skill set that you'd gain from growing all of it. And if your hard work is the sole source of your food, how safe or reliable will it be if you get hurt or sick? You could be watching from your sickbed as your crops wither and die. As mentioned, having only one source of food violates the essential permaculture principle of meeting essential needs in multiple ways, and that's a serious error. And yet the desire to grow much of our own food won't go away so easily. So how do we work with that?

## ➡ REACHING FOR THE ⬅ HIGHEST GENERALIZATION

Let's parse this subject by using another permaculture thinking tool, *moving to a higher level of questioning*. When I find myself racing to a foregone conclusion such as, "I must do it all myself," or "I've got to buy a bigger machine," I try to catch myself, to back up and restate the question in a broader, more inclusive way. In this case, what need are we really trying to satisfy—what question are we actually asking—when the answer we slide toward is "grow all my own food"? Permaculture strategy advises us to move to a higher generalization of our question, to ask it in a larger and deeper fashion that opens us to possibilities missed by an immediate, narrow question and reveals the underlying need and goal we're trying to achieve. This also helps meet design needs in ways that preserve the most options to work with. That last phrase—preserve the most options—tells us that we're working with a larger application of the technique of highest use.

If we know that our desires are to develop skills, to unplug from consumer culture, to eat healthy food, to see that crops are grown in ways that help rather than harm the earth, then what question can we ask about getting food that keeps all those needs met and our many options open? When we automatically default to, "I must grow all my own food," we've channeled our thinking to a single option. In truth, our possibilities are far richer than that.

When we move up a few levels to look at needs, one larger way to ask that question, one that preserves all those best practices and goals, is, "How do I meet my food needs sustainably?" Now, suddenly, we have many choices. And since permaculture is at heart a decision-making system, we can apply its methods to threshing out the best options for us. So which design method will help craft a way to meet these needs? The different design methods work with different types of relationships, so what relationship are we working with here? We are trying to arrange our food supply in a beneficial relationship with ourselves, thus the permaculture zone system, which organizes design elements in relation to the user, should be a fruitful tool to apply.

## ZONING THE FOODSHED

The concept of the foodshed can help us apply zones to meeting our food needs. A foodshed is analogous to a watershed, a term made familiar over the last few decades by activists and policymakers who found it a potent tool for raising people's awareness of the way that water quality and ecosystem health depend on how we treat a region's water at every step from rainfall through soil, creeks, and reservoirs to the faucet. Foodshed is an equally powerful concept. Just as your watershed is the area of land that supplies your water, your foodshed is the area within which your food is produced. Our foodshed today is often something we know little about. For those of us in the developed world, our foodshed is pretty much the whole planet. Winter fruit is grown in the opposite hemisphere, beef operations sprawl across former rain forests, grain is imported from whichever country is selling it the cheapest, and the globe is crisscrossed with delivery trails that end in our kitchens.

Making the whole world our foodshed triggers a cascade of downsides. The fuel and carbon footprint of shipping food is one, but it's not the worst feature of the global foodprint. In fact, transportation makes up only 11 percent of the energy used in industrial food's journey from farm to table; most of the energy in food is used on the farm and in processing.[2] Reducing that transportation trail by growing and buying locally is still helpful, but there are other powerful leverage points to work with. Shipped-in food is grown, processed, packaged, and transported in ways that we have no control over or even knowledge about. A substantial percentage of our food rots or is wrecked during shipment. The money spent on nonlocal food and processing bleeds out of our community and often supports labor and environmental practices we don't believe in. Shrinking the scale of the foodshed increases our power over how our food is grown because we can directly reach the producers, processors, and retailers. The farms that produce local food are often smaller than the industrial operations that are scaled for global shipment, and small farms tend to use much less energy per calorie of food than huge agribusinesses and are far kinder to the environment.

Growing all our own food is an intuitively appealing alternative to buying commodity food because it gives us virtually perfect control over it, but it has limited reach: It won't foster an alternative to the current global system. Going it alone doesn't build on the power of connectivity and functional relationships that are at the heart of good design. Thinking of foodsheds in terms of zones can correct that. And for those who, after reading this section, don't plan to implement the foodshed zone system in its entirety, the pieces of it still offer many possibilities for obtaining healthy food, growing community, and reducing resource use. Given that we must begin from where we are, if we earn money in a way that feels ethically and ecologically sound, then using that income to support food growers who are in alignment with our values is an excellent way to build a more just food system.

I'll illustrate one way to use zones to develop a personal, small-footprint foodshed. Remember that zones are based on frequency of use. The things you use the most often belong in the inner zones, and arranging the elements in the design to minimize trips to the outer zones reduces work as well as diminishes our impact on those outer, wilder zones. Here's the list of foodshed zones in brief form. The pages

**FIGURE 5-1.** Foodshed zones. By meeting as many of our food needs from the innermost zones as is reasonable, we shrink our food footprint and gain control over and information about the sources of our food.

after that provide detailed how-to for designing your own foodshed system.

- **Foodshed zone 1** is your own garden. If you're not a gardener, don't worry—this plan still lets you meet your food needs from the other zones. As I've mentioned, you can use permaculture's tools to help decide what to grow yourself in order to create a garden that is tailored to your own needs, food preferences, conditions, and workload. The idea here is to use your own yard to take care of as many of your food needs as makes sense for you and let other places—the other foodshed zones—take care of the rest.
- **Foodshed zone 2** contains other nearby gardens that you can use: those of neighbors that you share with, community gardens, edible schoolyards if you're lucky enough to be eligible for their harvests, and wild foraging from the many edible plants that horticulturists have sprinkled, knowingly or not, around our cities. Ideally, these would be gardens that are within walking or short biking distance. In zones 1 and 2 you would have many of your fruit and vegetable needs met, at least seasonally, though perhaps you would still need to buy staples such as grains, dairy, and meat from an outer zone.
- **Foodshed zone 3** contains farmers' markets and CSAs. If you can't get all your food from zones 1 and 2, then zone 3 can fill most of those remaining needs. Policies at farmers' markets vary widely. Some allow only what the sellers have grown themselves, while others permit produce from far-off sources, so if you want to stay within the minimal footprint, you'll need to inquire. Once you've expanded to the farm scale of zone 3, the selections enlarge to include most of the foods you use, as farmers' markets and CSAs carry far more than just produce, often supplying meat, dairy, eggs, staples such as bread and grains, and value-added foods such as preserves and honey. This means that much of your food can be supplied within the first three zones. That keeps your foodshed very local.
- **Foodshed zone 4** holds locally owned supermarkets that favor nearby growers over far-off sources. If you're adhering to the foodshed zone system, even trips to these worthy sources won't be very frequent.
- **Foodshed zone 5** comes last and definitely least. It includes chain and big-box stores. In this system those furtive trips to Costco, where you pray your green friends won't see you, will be rare or nonexistent.

The foodshed zone system is packed with benefits. It encourages you to grow as much or as little of your own food as you want. It connects you to your neighbors, other gardeners, and your community via frequent visits to zones 2 and 3. It reduces travel for both you and your food. It supports local business and keeps many of your food dollars in the community. It makes it easy to know how your food is grown and encourages you to make suggestions to farmers about their practices and varieties (try doing that at a big-box store!), giving you some control over this important aspect of your life. It builds local food sources and regional self-reliance, which increases food security. It helps develop locally appropriate practices and diet. And most important, it points your resources, money, and energy toward building, supporting, and enhancing exactly the food system that you want to see in the world. That's something that trying to grow it all yourself won't do.

## ➡ STRATEGY BUILDING ⬅ USING FOODSHED ZONES

In keeping with the focus of this book on strategies as ways to help us assemble effective patterns of techniques, I want to use this issue, meeting our food needs, to deepen our exploration of designing strategies. We started off this chapter with the question, "How much land do I need to grow all my own food?" In a way, we're still answering it, but we've reframed it now, twice. First, we enlarged the question, finding that a more useful, broader generalization was, "How do I meet my food needs sustainably?" This let us consider all the possible ways of getting food that are in alignment with our values and the practices we want to support. The benefit of this enlarged reframing was that it encouraged us to brainstorm a set of options to choose from. This helped us see that we can meet our food needs in multiple ways and from many places. And that in turn showed us that a logical follow-up question was, "How can I keep my foodshed at an earth-friendly size?"

Notice how we're developing a chain of inquiry here, which is a smart approach to solving any complex problem. Each new way of reframing the question brings with it a few built-in suggestions and slants, which means that each time we do it, we get some fresh perspectives. Also, every question carries its own set of biases and assumptions (something that polltakers take advantage of). It's important to spot those biases before going too far down the road that the framing points us toward, to be sure that we've chosen a suitable way to narrow down our options rather than one that leads us toward poor choices. For example, the bias created by thinking in terms of the foodshed is that when we notice that our foodshed is currently the whole planet, this innately, quietly suggests that our foodshed is better designed smaller than larger. That seems like a reasonable bias. It means we use less energy, know more about our food, and support a local economy. Asking the question that way also suggests some methods for answering it, another useful thing a carefully framed question will do. In the case of the foodshed, the method that stood out was the zone system, because we were focusing on the relationship between the user (us) and a resource (our food).

The thought process above is an example of a useful strategy for finding and selecting options. We went through this process to explore the specific question of food needs. The steps for answering strategy questions in general are these:

1. First, move to the highest generalization of the design problem. Look at it in a way that encourages unrestrained, big-picture thinking—a free-ranging brainstorm that corrals a diverse herd of ideas and possibilities. We did this when we abandoned the narrow question, "How much land do I need to grow all my food?" for the broader "How do I meet my food needs sustainably?"
2. Next, work through this untamed menagerie of possibilities, and cull it by applying a set of appropriate filters or other criteria, which we do here by limiting ourselves to the real possibilities for getting food: which neighbors are growing produce, where are the local CSAs, and so forth.
3. Then use a design method to rank priorities and build approaches for organizing and implementing the chosen options. Here the zone system does this by telling us which options for food getting to use first and

most often. In other cases we may want to use other methods, such as sectors, needs-and-resources analysis, or conventional tools such as bubble diagrams or flowcharts.

4. The final step is to go through the solution options and determine how to implement each one. Each option may require some design work of its own. So this process is iterative (we repeat it) and fractal (we repeat it at different levels). In this case we would look at the food sources (neighbors' yards, farmers' markets, and so forth) that we've placed into the various zones and figure out how to draw on each one wisely. In the section above, "Zoning the Foodshed," we've listed the major food sources and put them into zones, so now let's examine each zone and see how we can assemble their contents into a whole system that provides both food security and a stronger, more resilient community.

## A DETAILED LOOK AT DESIGNING OUR FOODSHED ZONES

I'm going to stack functions here. While the following may appear to be simply a list of the contents and functions of the various foodshed zones, I'm also using it as a way to describe the many opportunities for creating community around food and to show how we can enhance food security in our towns and cities.

### Foodshed Zone 1

We've spent the previous two chapters exploring how to design foodshed zone 1—the home garden—and we've seen that there are plenty of resources of many kinds for doing that, so we can move on. The focus of this section is on the remaining choices, which all involve other people and land we don't own, from neighbors and community gardens to the farms and workers that supply local food stores. All of those mean working in community with others, and that's a big theme of this book. Cities and towns are places where we are surrounded by people, and if we're going to make a go of it on this planet, we need to develop the skills to collaborate with them in ways that improve the lives of everyone and everything we come into contact with.

### Foodshed Zone 2

In foodshed zone 2 we want to use nearby land for growing, harvesting, trading for, or buying food. In cities and towns, land access is a big issue, so this subject is worth spending some time on.

Urbanites often have no land of their own. Condominium and apartment dwellers may not have a private yard at all, and renters who have a yard often aren't allowed to garden there. Thus, for many city dwellers, gardening in foodshed zone 1 isn't even an option; they must start in zone 2, on someone else's land. Even those with their own land are growing only as much food as is sensible given their conditions, and zone 2 will supplement those yields. Zone 2 strategies for meeting food needs include:

#### Neighbors

Options here split into two main categories: land for growing food that belongs to neighbors you know and to neighbors you don't know. If you know some of your neighbors, I'm hoping you have the chutzpah to ask them to collaborate in growing food in each other's yards. The typical trade here, and in any of the neighbor options, is to share the harvest with them, as well as obvious arrangements such as setting up optimal times to be in their yard, offering to help with the water bill, and in general showing appreciation for their kindness in sharing their expensive real estate and private space with you. Neighbors with large fruit trees will often be overjoyed to trade their surplus with you; few people can eat several bushels of pears in a couple of weeks. Elderly neighbors in particular may be open to a yard-share, since so many people love to garden but may be unable to do it comfortably because of age or injury. Those folks will be delighted to have their gardens rejuvenated and homegrown produce pouring out of them.

To get neighbors you don't know to open their gardens to you, you have several options:

- **Putting up flyers** around the neighborhood and in nearby stores or leaving notes on front doors. Mention not only your request and contact information but also what you can offer in return.
- **Attending a neighborhood association meeting** and pitching your idea. Other neighbors may be interested, and more than one yard-share may result.
- **Signing up at yard-share websites,** which connect would-be gardeners with landowners. National sites that cover multiple US cities include hyperlocavore.com, sharingbackyards.com, and y2g.org. Many cities and towns have their own yard-share websites as well. Surprisingly, landowners looking for gardeners often outnumber garden seekers in these programs.
- **Common community-building methods** can open up yard-sharing possibilities, such as organizing a neighborhood potluck meal or block party and mentioning your idea to the attendees.
- **Gardening collectives and micro-CSAs** are local food-sharing networks that link gardeners, urban land, and the tendency of vegetables and fruit trees to pump out bumper crops that risk going to waste. In Portland, Oregon, the Urban Farm Collective (www.urbanfarmcollective.com) has tied together a network of vacant lots and underused backyards gardened by volunteers who swap their labor for produce—the collective proudly labels itself as "nonmonetized"—and the surplus is donated to a food bank. Work parties rotate through the plots, organized by an online calendar, and the collective offers workshops in plant propagation, garden design, and similar food-based topics, as well as an apprenticeship program. The collective is an example of a small idea—connecting neighbors with unused land—that generated many unexpected possibilities and grew into them. It offers a model for enhancing local food security and neighborhood cohesiveness.

In West St. Paul, Minnesota, gardener Andy Russell founded what he calls a micro-CSA, a home garden that solves its surplus problem by spinning it off to a handful of nearby families. Russell had upped both the fertility of his soil and the number of beds he gardened until fruits and vegetables were erupting from the yard in vast excess. Two neighbor families agreed to subscribe, for $500, to a share of the bounty, which amounted to a large box of produce weekly from June through September, supplemented by soup stock and bread, which Andy, a professional chef, prepares. By adding a few more raised beds and cooking a bit more, Andy foresees expanding the CSA to four or five shares. The micro-CSA model transforms the problem of an overflowing garden into a networked solution that absorbs the surplus, brings income, creates community, and helps feed the neighborhood.

**Community Gardens**

Community gardens are parcels of land divided into small garden plots that can be leased (often for free) by individuals or families. The land is usually owned by local governments but sometimes by schools, churches, nonprofits, or private groups. Plot size varies from about 100 to 400 square feet. In some cities their popularity means putting your name on a long waiting list. If that's the case in your city, you may want to organize your own. The American Community Garden Association (communitygarden.org) offers extensive resources for finding land, approaching sponsors, setting up bylaws and agreements, and the other essentials for creating a community garden. One bonus benefit of practicing permaculture, as opposed to straight row-crop gardening, in community gardens is the notice your plot and methods will receive from gardeners unfamiliar with permaculture. It helps spread the word.

Financing community gardens is often the biggest obstacle to starting one. Crowdfunding has stepped into this breach. One nonprofit, ioby.org (deriving their name from the opposite of NIMBY), has assembled microdonors who give an average of $35 each to fund over 2,000 projects across New York City, including numerous community gardens.

Even apartment complexes can host community gardens. One of many examples is Whitefield Commons, a sixty-two-unit building in Arlington, Virginia, operated by a corporate developer. The owners originally allowed a few private plots, but these proved so popular that a local pro-tenant nonprofit, Buyers and Renters Arlington Voice (BRAVO), helped organize an expansion and found grant money for purchasing tools, seeds, and supplies. Along with the usual hurdles to clear for establishing community gardens, apartment dwellers also need to persuade their landlords, who are most likely focused on the bottom line. The key for success here is to show that the garden will save the owners money on landscape maintenance and to assuage any fears about liability via a well-worded written agreement. With those in hand, landlords often become staunch allies.

Sometimes vacant lots can be gardened. One of my neighbors tracked down the owner of an empty lot across the street and got permission for five households to grow food there for two years until the lot was developed. The quid pro quo that we arranged was to clean up the overgrown lot and keep it tidy, and we paid the water bill of the family next door in exchange for irrigating our plots from their spigot.

A related option is guerrilla gardening, or growing food on unused land that you don't have formal permission to use, such as vacant lots, around highway overpasses and median strips, and obscure corners of public or private land. I'm sure my publisher's lawyers don't want me to endorse these not-really-legal methods, and to be honest guerrilla gardens rarely last long. I've seen many guerrilla gardeners upset when their garden suddenly disappears. Not only are these plots open to harvest or vandalism by anyone, but annoyed landowners will raze them on sight. Guerrilla gardens are often planted by the anarchically minded in a "screw the Man" attitude, but my observation is that thus far the Man remains unscrewed. However, guerrilla gardens are an option for those aware of the risks.

**School Gardens**

A full exploration of these marvelous programs must wait for a later chapter, but if you are a parent, a teacher, a volunteer, or otherwise affiliated with a school garden, these can be treasure troves of fresh produce. Edible schoolyards have swept the nation over the last decade; California's legislature has called for a garden in every school, and other state and local agencies are implementing them in large numbers. Although snacking children and their teachers are the main beneficiaries of school garden produce, regulations often prohibit eating or cooking that produce in the school cafeteria. This means that these gardens' yields are often food in search of an eater, a sad situation that you should consider your civic duty to alleviate. If you're not a teacher or parent, volunteers in school gardens are often entitled to a share of the bounty, and overworked school staff will be grateful for the offer of help.

### Churches, Businesses, Nonprofits, and Local Government

Putting our permaculture goggles on helps us refine and broaden our search for gardening space. Using the strategy of moving to the highest generalization tells us that a fruitful question is, "Who controls land in our neighborhood?" The obvious answers are homeowners, city government, and schools, but that's not the end of the list. Another source is churches, which are often surrounded by grassy swards that are begging for a better use. Also, a core part of the mission of most churches is the fellowship and support of their congregation, their community, and those in need. Gardens fit that bill. Once the possibility of a garden is on the radar screen of church staff, they frequently become enthusiastic proponents. The garden at Gathering Grounds Ministry in Richmond, Virginia, provides produce each week to over sixty-five families and seniors, most of whom have been denied social service benefits because they earn just a bit more than the very low income threshold for those benefits. At the half-acre community garden at Trinity United Methodist Church in Grand Rapids, Michigan, which has been producing for twenty-five years, gardeners harvest some food for themselves, and the rest is distributed to local food pantries for the poor. Similar examples of church garden programs abound.

Businesses are increasingly hopping on the garden bandwagon. From behemoths such as Google and PepsiCo to small local firms, office gardens have become another part of employee benefit programs. One job title at Google is "manager of culinary horticulture," tasked with overseeing the corporate garden program. Most office gardens are maintained less formally by volunteers in what is called an employee-supported agriculture (ESA). Although office gardens are usually tidy arrays of annual row crops and flowers, more innovative firms are using permaculture design to incorporate perennials, water harvesting, and even graywater. On a busy city street, Portland architect and visionary Mark Lakeman built a food forest surrounding the office of his design firm, Communitecture, that offers snack fruit and salad greens to coworkers and passersby.

I'm including office gardens in this zone 2 foodshed category because although your workplace may not be in your neighborhood, you go there often, which places it, for this purpose, in an inner zone. Other nearby businesses that are candidates for community gardening are restaurants, locally owned food stores, and socially oriented nonprofits. Yet another landowner category to approach is membership associations such as granges, Masonic lodges, Odd Fellows, and similar groups. Any piece of land with a building on it is worth considering in this category.

Another option is shared-harvest programs run by service organizations. Low-income people can participate in food-pantry services that are supplied by gardens, but an avenue open to all is the increasing number of fruit-tree harvest programs in urban areas. Homeowners and others who have more fruit than they need—which is usually anyone with a mature fruit tree—can register their trees with these programs; come harvest time, volunteers will pick the fruit and share it among the owners and the needy. By 2013 the Vancouver (Canada) Fruit Tree Project had harvested over 55,000 pounds of fruit to share. A similar project in New Orleans collected 10,000 pounds in 2012, their second year. The Portland Fruit Tree Project, founded in 2006 by permaculture design course graduates Katy Kolker, Bob Hatfield, and others, had registered over 4,000 fruit trees in the greater Portland, Oregon, area by 2012 and that year harvested over 66,000 pounds of fruit to be shared among project participants and local food banks. Many other cities have similar fruit-tree sharing projects.

Now that urban agriculture has become a buzz phrase, many cities are inventorying land suitable for food growing and tearing down the regulatory barriers to urban farming. Seattle, Detroit, Milwaukee, Los Angeles, and numerous other municipalities have radically reformed their zoning and permitting processes to make urban farming possible. To encourage farming in cities, graduate students at Portland State University joined forces with the city government to create the Diggable City Project, which identified several hundred city-owned parcels suited for agriculture, listed the merits of each one, and proposed ways to remove barriers to using them. Available on the web, the Diggable City documents are an inspiring model for identifying food-growing land in cities and tailoring city policies to be farm- and local-food-friendly. Unused city land may be prime territory for food growing.

### Urban Foraging

In the United States and other melting-pot nations, towns are gathering baskets for the favorite species of every culture and ethnic group living there. Plant enthusiasts and landscapers have added their own selections to this cultural cornucopia. This can push raw plant biodiversity—the sheer number of species—well above that of the native ecosystem that was once there. In addition to well-known food plants and edible weeds growing in our cities, many so-called ornamental plants have edible parts. It's rare in cities to collect more than a meal's worth or so at any given spot except on mature fruit and nut trees, but foraging can supplement or broaden a city dweller's diet. I've foraged with New Yorker friends in Manhattan's Inwood Hill Park, where we gathered enough garlic mustard, pigweed amaranth, and nettles for a tasty dish of braised greens. We were careful, however, to forage far enough from paths to pick greens that hadn't been seasoned by passing dogs.

As with all foraging, be certain to properly identify the gathered plants before eating, avoid areas contaminated with pesticides, and if on private land, ask the owner first. Table 5-1 lists common urban forage plants. Good resources for more information on urban foraging are the websites eattheweeds.com and firstways.com.

With all these foodshed zone 2 options available, the urban landless and those who need someplace other than their own yard to garden have many places to turn. Besides simply increasing the square footage available to grow food, we can use foodshed zone 2

**TABLE 5-1.** Urban Foraging Plants

| COMMON NAME | EDIBLE PART |
|---|---|
| Burdock | Root |
| Calendula | Flowers |
| Chestnut | Nut |
| Chickweed | Leaves |
| Cleavers | Leaves |
| Dandelion | Leaves, root |
| Daylily | Tuber, young leaves, flowers |
| Dianthus | Flowers |
| Dogwood | Fruit |
| Hawthorn | Flowers, berries |
| Honesty | Leaves, flowers |
| Indian plum | Berries |
| Lemon balm | Flowers, leaves for tea |
| Miner's lettuce | Leaves |
| Mountain ash | Berries |
| Oak, white | Acorns (after leaching) |
| Oregon grape | Berries |
| Oxalis | Leaves (sparingly; contains oxalic acid) |
| Passionflower | Flowers, fruit |
| Pigweed | Leaves, seeds |
| Pineapple weed | Flowers (also for tea) |
| Plantain | Leaves, seeds |
| Primrose | Flowers, young leaves |
| Purple dead nettle | Flowers, leaves |
| Purslane | Leaves |
| Rose (wild) | Flowers, seedpods |
| Sheep sorrel | Leaves (sparingly; contains oxalic acid) |
| Shepherd's purse | Leaves, seedpods |
| Sow thistle | Young leaves |
| Stinging nettle | Leaves (steamed) |
| Strawberry tree | Fruit |
| Violet | Leaves, flowers |
| Wild carrot | Root (late fall, early spring) |
| Yellow dock | Young leaves |

gardens to go beyond what is possible in our own yards. At other sites we can grow plants that don't thrive in our garden for any of a host of factors, such as unsuitable soil, shade, pollution, microclimate, or pets. Neighbors and other gardeners may have skills or tools that we don't. For example, my grains were grown by a local gardener more adept than I at raising and processing those specialty crops. He grew wheat, oats, and quinoa on a large property at the south edge of Portland and had smaller plots on several urban lots, all loaned to him in exchange for produce shares.

Many urban yards have room for only one or two fruit trees, so in zone 2 gardens we can trade our surplus fruit for other varieties. Foodshed zone 2 helps us easily move beyond the limits of our own yards for rounding out the food palette. Once we've done what we can in our own yard to meet our food needs, we use foodshed zone 2—nearby gardens of many types—to fill as many remaining food gaps as we can.

## Foodshed Zone 3

Many factors influence how much of your diet comes from foodshed zones 1 and 2, such as your own passion for gardening, your connection to your neighbors and how much food they grow, and the growing season length. If the inner zones can't completely fill your food basket, extend the harvest to foodshed zone 3, which contains farmers' markets and community-supported agriculture. With the leap to this zone, we've entered the commercial realm and the money economy, but much of the money can stay local. When you first start shrinking your foodshed, you are liable to rely heavily on the outer zones, but over time, as you tune into local resources, your efforts and resources cluster more in the inner zones. This reduces the need to earn, and what money you spend supports people, practices, and infrastructure in the community. It's a benevolent reinforcing feedback loop that builds local resilience. If farmers' markets and other zone 3 resources are scarce or poorly developed, supporting them will help them grow, lessening dependence on outer zones that don't build local resources as well.

### Farmers' Markets

Farmers' markets have existed since commerce began. But specialization in the industrial era, first in the form of trading posts and later through general and grocery stores, increasingly distanced farmers from eaters via intermediate links in the food-system chain. The ascendance of cheap, petroleum-based food pushed farmers' markets into a near-fatal decline after World War II. However, a backlash to the lack of nutrition, flavor, freshness, and variety in industrial produce spurred a resurgence in farmers' markets that accelerated in the 1990s. According to the US Department of Agriculture (USDA), farmers' market numbers have skyrocketed from 1,755 in 1994 to 8,268 in 2015.[3]

Farmers markets fall into two main categories: producer-only markets, in which vendors grow all the food sold, and those that allow resale. Most markets favor local goods, and even resale markets often stipulate that goods must be grown within a defined distance. Farmers' markets are decidedly win-win: Buyers have a community-knitting experience while they get fresher, healthier food that several studies say is usually cheaper than that in stores; farmers retain more profit and gain exposure; and nearby businesses benefit from the influx of shoppers.[4] The tight feedback loop between seller and buyer lets farmers learn exactly what buyers want, and buyers can support their favorite sellers directly. Although farmers' markets focus primarily on produce, many also offer value-added farm products, such as jams, pickles, honey, and baked goods, as well as meat and dairy products and even grains.

Farmers' markets are a potent social and political leverage point, and their multiple functions are worth exploring here from a permacultural viewpoint, as inspiration, as a model to follow, and to show how they reach far beyond being simply places to buy food.

Because farmers' markets and produce stands are usually cheaper and less bureaucratically onerous to set up than storefronts, they are increasingly showing up in urban (and rural) food deserts, which are neighborhoods lacking places to buy fresh, healthy, affordable food and where, if food is available, it's from fast-food and convenience stores selling what are more properly called "food products" rather than food. Shining examples of food-desert oases and innovative farmers' markets include People's Grocery in West Oakland, California, which opened in 2002 to serve a low-income, largely African-American and Latino community. The founders didn't have the expertise or capital to open a storefront and instead created the Mobile Market, a distinctive red produce truck that parks near senior centers, parks, and schools three days each week. They soon developed their own garden for growing about 30 percent of their produce and contracted with fourteen local farmers and gardeners for the rest. They also run a summer camp and training programs that teach nutrition, cooking, and business skills to local youth. People's Grocery hopes to open a brick-and-mortar store in 2015. Meanwhile, the Mobile Market model has been copied in dozens of other cities.

On the other side of the country, the East New York Farmers Market, in a low-income, high-crime section of Brooklyn, runs two markets and manages two urban farms. Many of the neighborhood's more than sixty community gardens sell produce at the markets, as do several upstate farms and local fishermen. The East New York Farms Project offers a paid internship program for local youth to learn urban and rural farming methods. The community's roots are largely Caribbean and African-American, and the markets offer locals a place to buy ingredients for traditional dishes such as callaloo and gumbo. The market is also a place for young residents to help and learn business and farming skills from the many elderly gardeners who sell produce from their garden plots.

From these descriptions it's obvious that farmers' markets do far more than sell produce. They can become neighborhood hubs where locals gather; where young people learn from elders and elders are in turn aided by youth; where locals learn about health, food justice, and business; and where a strong local economic network grows its roots. Human societies have always gathered and organized around food, and farmers' markets operate at the perfect scale for linking to the many opportunities that food offers.

### Community-Supported Agriculture (CSA)

A CSA is a multifunctional tool for filling your food needs. Members in a CSA usually pay in advance for subscription shares for a season, for which they receive weekly boxes of produce. The payment guarantees income and some security for the farm, which frees the farmers to focus on growing food instead of hustling up buyers. Many CSA groups explicitly or implicitly share risk. In exchange for superb produce at a good price and the chance to talk to their farmer about what they want, members accept that sometimes harvests are small, variable, or imperfect—and when bumper crops come in, they share in the bounty. My own experience with CSA farms has been that knowing we're all in it together is a great community builder between members and farmers. My jaunt to our terrific CSA, Laguna Farm, is a high point of my week both socially and gastronomically.

The first CSA farms in the United States, the CSA Garden in Great Barrington, Massachusetts, and the Temple-Wilton Community Farm in southern New Hampshire, were formed in 1985 by farmers influenced by the Austrian philosopher Rudolf Steiner.[5] These farms melded Steiner's concept of biodynamic agriculture with his development of cooperative economics and new forms of landownership through community land trusts. While CSA farms retain these original tenets to varying degrees, most are also commercially oriented subscription farms. Close to 99 percent of them use some or all organic practices, whether they are organically certified or not, according to a survey of CSA producers.[6] The USDA tallied over 12,500 farms marketing products through CSA programs in 2007.[7] To find CSA farms in your area, visit www.localharvest.org/csa.

As mentioned, foodshed zone 3 is where money enters the food-getting equation. Spending your money at farmers' markets and CSA farms not only keeps your income cycling within the community but also lessens the chances that your money will inadvertently support practices you don't agree with. Directly connecting people to their food growers also eliminates the friction losses of middlemen, so more of your dollar goes to food and farmer and less to adding layers to an unwieldy food-system hierarchy.

## Foodshed Zone 4

In this zone we leave open-air gardens and farms and enter the brick-and-mortar store, but the groceries in this zone are locally owned and support regional farms and food businesses. Foodshed zone 4 can range from independent natural-food stores and the hippie-esque food cooperatives that first appeared en masse in the 1960s to locally owned chain stores such as PCC Naturals in Seattle; New Seasons Market in Portland, Oregon; and Yes! Organic Market in Washington, DC.

At these stores a lesser proportion of your dollar goes to the farmer than in zone 3, and it's harder to know what practices your money is supporting, but more money stays in the community than when shopping at a national or global chain. The smaller stores in this category are likely to have ties to local farms, although the chain varieties, even the local ones, often need more volume than small farms can provide. The industrial-scale organic farming supported by some local chains, while less harmful in many ways than industrial pesticide-based methods, still promotes enormous monocultures, vast acreages of tilled soil, and standardized crops that, when grown at large scale, may not be more nutritious than conventional food.[8] Now that organic farming has been industrialized and scaled up, the real difference, both in nutrition and ecofriendliness, may be between large- and small-farm practices.

This is also the point where, if you're not careful, your foodshed can inflate to near-global size. Even small natural-food stores carry packaged foods made by multinationals. A surprising graphic by Philip Howard, a professor at Michigan State University, shows how organic food businesses have been bought and consolidated by a handful of multinationals.[9] Cascadian Farms, a once-small organic producer in Washington State, is owned by General Mills, and Cascadian Farms in turn controls Muir Glen Organics. Santa Cruz Organic has been bought by J. M. Smucker and juice-maker Odwalla by Coca-Cola. A seeming artisanal chocolatier, Dagoba, is a subsidiary of Hershey. It's almost impossible to know where packaged products came from. Any food from a box or bottle is likely to be tied to the global food web. To avoid those entanglements, do as much shopping as you can along the outer walls of the store, in the produce, dairy, bulk, and meat sections, where more food is locally grown. Stay out of the aisles and their packaged food when possible.

In our present food system, within an economy that favors volume, predictability, and convenience over nutrition and local production, it's challenging to find locally grown food year-round, and our busy, maxed-out lifestyles make it tempting to buy packaged food. Those are real and substantial obstacles to a healthy, resilient food system, so don't beat yourself up if you occasionally succumb to convenience or simply can't find the ideal choice. Just do the best you can, and try to find ways to shrink your foodshed a little more next time.

One way to do that is to bear in mind that foodshed zone 4, like any outer zone, should be rarely visited compared to the inner zones. Zones 1 to 3 can supply a large share of your seasonal produce and, in some areas, meat, dairy, and grains. Focusing your diet on those locally grown foods and processing them at home via canning, drying, pickling, and baking will reduce visits to zone 4. That keeps your foodshed footprint small. If your life is not yet arranged to stay mostly in the inner zones, choose zone 4 purchases that support practices you agree with.

How do you do that? This is another reason to minimize zone 4 visits, because assessing the impact of, say, a tub of organic hummus or a loaf of whole-grain bread requires some research and must be combined with a little guesswork and personal preference about the relative weight to give to carbon footprint, supporting the local community, labor and land-use practices, and whatever other factors are important to you. That process can be daunting and the results arbitrary and arguable. Many lists are available that rank the carbon footprint of different foods and diets, but the practices used to grow the food are at least as important. Beef and lamb get terrible ratings on carbon emissions, but most of the data used for those calculations are for industrial meat. A grain-fed steer at a confined-animal feedlot has a vastly larger carbon footprint and overall environmental and social impact than a steer that is rotationally grazed in a pasture and butchered on the ranch, let alone an animal carefully integrated into a perennial agroforestry farm. However, simple physics and ecology dictate that regularly eating higher on the food chain—meat instead of vegetables—will inflate the carbon and resource footprint of any diet. This is shown in the fact that consuming industrially raised lamb produces 39 kilograms of $CO_2$ per kilogram versus 0.9 kilograms per kilogram for industrial lentils.[10] That's a fortyfold difference. Many local food stores display information on where their products come from and how they are produced, and some even offer carbon footprint data. I recommend patronizing those groceries and encouraging that kind of transparency.

## Foodshed Zone 5

In permaculture, zone 5 usually comprises those wild places best left untouched, and that applies here, since foodshed zone 5 contains corporate chain groceries and big-box stores. These stores rarely offer products whose more sustainably produced counterparts can't be found in locally owned stores, unless your only aim is to buy cheap regardless of ecological and social consequences. I suspect that kind of reader has long ago put down this book. It's true that many corporate chains are instituting various "green" practices; Walmart claims to have saved 3.1 billion plastic bags in 2011 by using alternative packaging, and some Costco stores buy produce from local farms. Let's encourage them to do more of this, as these stores aren't going away soon, and their impact is tremendous. But spending money and time in the inner foodshed zones instead of zone 5 will reduce your foodshed footprint and is a better leverage point for building a more sustainable food system. Encouraging a manager at a locally owned store to buy locally or reduce waste has a better chance of creating change than at a global chain, because there are fewer bureaucratic layers between the local manager's ears and those of the decision makers.

## Keeping Your Foodshed Footprint Small

I'll sum up the foodshed zone concept. The way to keep your foodshed footprint small—and to receive all the benefits that brings—is to get as much of your food as possible from zones 1 to 3. Those, depending on your growing season and how well developed your local food system is, can provide most of your fruits, vegetables, meat, dairy, juices, perhaps grains, and many of the processed foods derived from them. The leap into a nation- or planet-sized foodshed occurs at zones 4 and 5, with industrial-processed and out-of-season foods bought at retail stores.

Avoiding these can be challenging, especially if your local food scene is not robust. Are you willing to forego zucchini and peppers in December? Can you make the time to can, pickle, bake, and otherwise avoid shopping in the aisles by processing local, fresh ingredients to put by for later use, or can you find friends or local businesses that do this? It's forgivable if we fall short of creating the smallest possible foodshed, since we live in a world of constraints and multiple demands on our time. We just do the best we can and try to do better in the future.

Is the foodshed zone concept useful only for the affluent? After all, yards big enough for gardens and access to CSAs, farmers' markets, and locally oriented boutique groceries are rare in low-income neighborhoods and entirely missing from food deserts. So it's a valid question. But at the same time, this zone analysis highlights a potent leverage point. Impoverished and even middle-class communities, if they have food sources at all, are often served only by the chain and big-box stores of foodshed zones 4 and 5. What's missing are the food sources of zones 1, 2, and 3. This suggests that rather than attracting more chain stores into underserved neighborhoods, cities should be developing programs to strengthen the inner foodshed zones. These will far more effectively help eliminate food deserts, bring healthy food to the communities that most need them, help keep hard-won dollars in those communities, and build a resilient food network. This is one more case of permaculture's toolkit helping to identify powerful places for action.

The foodshed zone approach, like any good permaculture concept, casts a net large enough to wrap around almost all of our food needs, helps organize its components, and suggests ways to make choices, set priorities, and find solutions systematically and ecologically. It places you, your food, the environment, and your community in a set of healthy, mutually rewarding relationships.

## ➡ POLICIES FOR ⬅ PERMACULTURE

Farmer/philosopher Joel Salatin has written a book called *Everything I Want to Do Is Illegal*, detailing his battles with authorities over his on-farm practices that he believes are crucial for sustainability but that government has made illegal. Many permaculturists slam into the same problem. City codes often prohibit or limit chickens and other small livestock, growing vegetables in the front yard or on the parking strip, selling produce, harvesting rainwater, reusing graywater, using a composting toilet, and myriad other methods that shrink our ecological impact. Much of this is a legacy from the early-twentieth-century shift away from the farm and into the cities. When upward-bound country dwellers have saved for years to flee their rural homesteads and move to town, the last thing they want is roosters waking them at 4 a.m. and the smell of manure wafting in the window. And some of these rules are grounded in legitimate health and safety concerns. But the recent movement toward urban agriculture and reconnecting with our food—and the rest of the natural world—is forcing policy shifts in cities and towns, sometimes graceful, peaceable ones; sometimes ones hard-fought, acrimonious, and deeply compromised.

Before implementing an urban permaculture design for your home or working to set up a community garden or farmers' market, it's essential to review your local government's policies on food, water, and related issues. I'll give a few examples here and also offer some suggestions for those who want to help shift policies to a more permaculture-friendly form.

Seattle is a city that prides itself on having some of the most unfettered policies toward urban farms of any in the United States. The 2010 Seattle Farm Bill allows homeowners to keep up to eight fowl (no roosters), three pygmy or other small goats, and four beehives. No permit is needed except that hives must be registered with the state's Department of Agriculture. More animals can be kept in larger yards. An urban farmer can plant up to 4,000 square feet and sell up to $12,000 of produce per year from it without a license, as long as only household-scale equipment is used. This exemplary ordinance removes two of the biggest impediments to urban farms: limiting livestock to hobbyist rather than small-farm numbers and zoning and other restrictions against raising food for sale in a city.

A less sweeping and more typical policy is Baltimore's. In 2012 small livestock were allowed in the city, but only with permits. The limits are four fowl, two beehives, and two dwarf or pygmy goats. A major victory for gardeners in this cold-winter city was the allowing of hoop houses and other temporary greenhouse structures without a permit. Baltimore's urban farmers can sell their produce, but they need business licenses, must submit business plans, and must go through similar red tape that impedes the flourishing of urban food sources. But the city has made a step in the right direction, and let's hope officials continue further on that path.

Often the common ingredient for galvanizing a city into modernizing its farm policy is a showy test case or cause célèbre. Sacramento, California, banned vegetables from the front yard back in 1941, and in 2004 local resident Karen Baumann was stung by that obscure ruling when she was fined $750 for planting tomatoes and a fruit tree out front. This prompted food activists to flood the city government with petitions and phone calls to revise the law, and in 2007 the city wrote what became their Front Yard Ordinance, allowing vegetables and fruits. Salt Lake City, San Francisco, and other towns large and small have followed a similar pattern, wherein the city leveled an egregious fine for a veggie plot that triggered a heated public outcry, soon followed by a speedy change in the ordinance as officials quickly realized the nonsensical nature of the law. For activists who seek more open policies toward food and urban farming, a provocative test case can be an effective leverage point for rationalizing the official views toward food and land use.

## THE CITY FOOD FOREST

We can pry open our towns' policies toward food using another tool: urban food forests on public land. Until recently, many municipalities forbade fruit trees in parks or on public easements such as parking strips, fearing that rotting fruit would be messy and slippery and would attract rats. While that's a reasonable worry, I marvel at a culture so rich that we can forbid the planting of food and let our fruit rot. Meanwhile, citizens are pressing forward with their own initiatives, spawning edible forest gardens in public parks in cities from Seattle and Helena, Montana, to Bloomington, Indiana, and many others.

The first large public food forest in the United States was hatched in a 2009 permaculture design course taught by Jenny Pell in Seattle, as a proposal for the class's final design project. The site of the design was a 7-acre parcel owned by Seattle Public Utilities that had lain vacant for nearly a century. The project, presented on the final day of class by a team of students, struck three of the design group's members, Glenn Herlihy, Daniel Johnson, and Jacquie Cramer, as more than just a classroom exercise; it could be a way to manifest permaculture on the ground. So they made a few phone calls, and the Beacon Hill Food Forest was born. The project's progress is worth exploring because it shows that permaculture design is not just about laying out plant species and swales. Understanding the social and political dynamics of a project is crucial in bringing permaculture's ideas to fruition.

First, Herlihy and Cramer met informally with residents of Seattle's Beacon Hill neighborhood, an ethnically diverse community near downtown, to talk about the possibility, and the concept took fire. Some of the attendees formed a small steering committee that began heroic outreach efforts to secure community support for transforming seven acres of a public park into a blend of fruit and nut trees, berry bushes, perennial vegetables, and other edible and useful plantings such as medicinal herbs and basketry willows. The plantings were to surround a classroom, playgrounds, native species, a gazebo, and barbecue grills. In keeping with the low income level of the neighborhood, they planned to lease lots to gardeners for only $10 per year.

Outreach has been a critical component for the project's success, because city officials insisted on proof that the community wanted the food forest—and that someone would maintain it—before approving the project. Too many ideas like this wither for lack of follow-through or are blocked by envisioned impediments— the most common here was that people might steal the fruit. Savvy to social dynamics, the committee knew that a few loud naysayers could outweigh the praise of hundreds of supporters. The team sent out 6,000 postcards in five languages for comment and thoroughly canvassed nearby neighborhoods. To get suggestions for plantings that would appeal to everyone, they hired a translator to speak with elderly Chinese neighbors and met with a Samoan cricket-playing group that used the park. The result was a list of nearly 1,000 proposed plant varieties from over a dozen cultures. Much of the design was generated at public meetings, where participants from varied backgrounds sketched their ideas on large sheets of paper.

When articles about the food forest appeared in online journals and blogs, supporters replied to critical and fearful comments with polite, cogent, and well-researched rebuttals. One critic posted that a food forest would reduce the amount of open space at the park. A supporter's rejoinder carefully pointed out the expansive acreage of open space both in the food forest plan and at an adjoining park. Another critic argued that homeless people might set up camp there, harvest all the fruit, and perhaps sell it. The reply agreed that homelessness was a problem but that all parks, including the current vacant site, had the same potential for illegal activity, and that community gardens, equally open to vandalism and theft, suffer little of either and have never had "a fruit stand set up on their corner run by homeless individuals peddling plums." Permaculture course leader Jenny Pell noted that the most common worry was that people would take all the fruit, and her reply was joyful: "That would be great! Usually there's so much that it rots on the ground. And that would prove that we need more food forests."[11]

By showing this kind of pattern literacy—foreseeing objections, listening to critics' fears, and showing that they had been anticipated and provided for—the food forest's proponents smoothed the path to acceptance. A collection of hundreds of neighbors and volunteers broke ground in summer of 2012, and the city and other groups are funding the project. As of this writing, the infant food forest has been partially planted, and plots of annual vegetables are yielding produce. And in other cities and towns, public food forests are on the agenda and underway.

Food security in urban places depends on having a wide array of food sources, from individual gardens to urban farms, as well as restoring our cities to hubs of support for the surrounding rural agriculture. All this is in line with Jane Jacobs's observation that healthy "city regions" are those in which vibrant metropolises support and are made prosperous by rich natural and designed ecosystems around them. As this and the previous two chapters should show, a functional food system is a blend of personal efforts in our own gardens plus collective action at larger levels, such as farmers' markets and local business, as well as smart planning in food policy and city codes. But food is only part of the story of urban permaculture and meeting our needs in an ecologically sound way. We'll now move to some of the other petals of the permaculture flower.

Göbekli Tepe, a monumental site built 11,800 years ago by hunter-gatherers. It's possible that feeding the large numbers of people needed to build this and other religious sites was a driving force behind the development of agriculture, and eventually cities. Photograph by Teomancimit, Wikimedia Commons.

A sector map showing the unique forces and variables of a high traffic corridor at Naropa University in Boulder, Colorado, was created by students in Jason Gerhardt's permaculture design class. Sectors can be difficult to depict, because there are many kinds, from both point and nonpoint sources, and these students have met the challenge well. Map courtesy of Santiago Giraldo Anduaga.

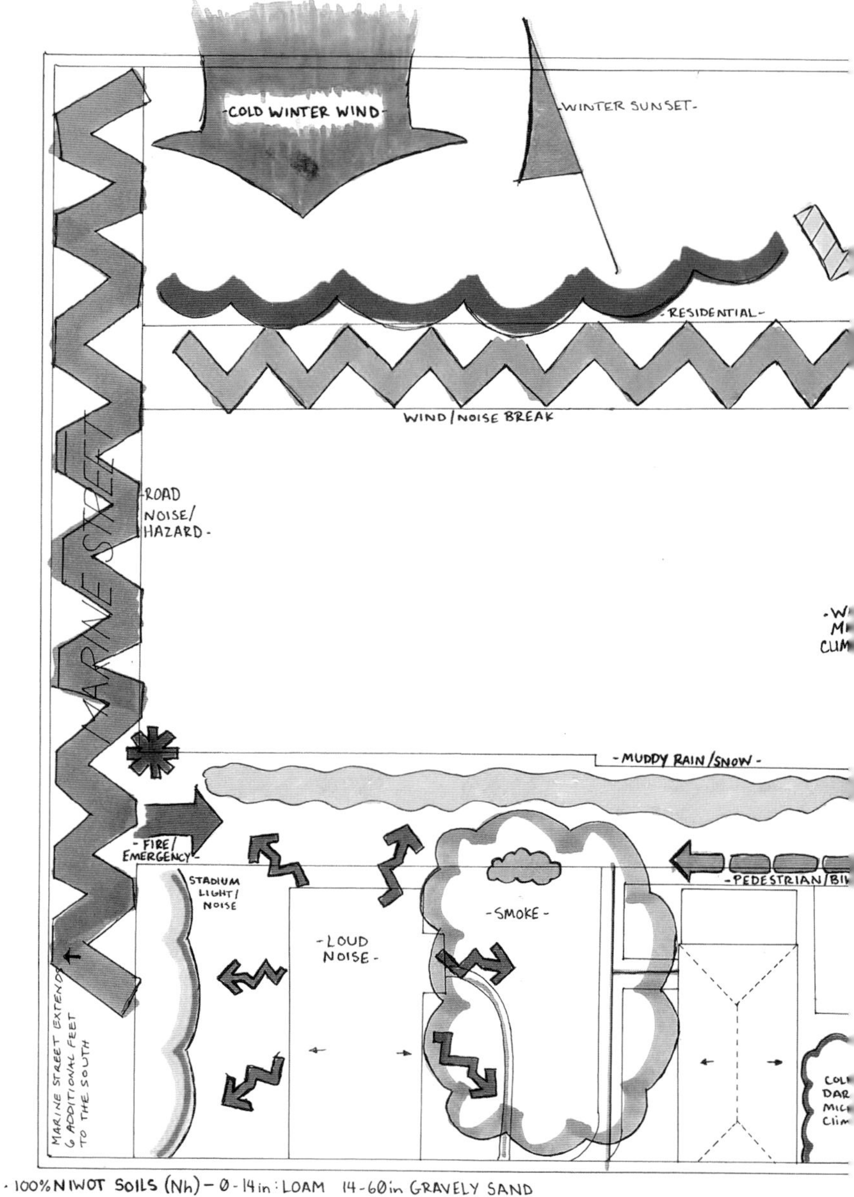


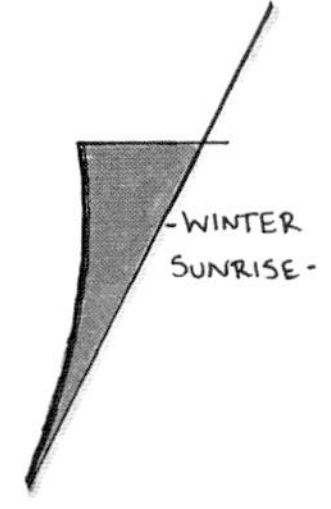

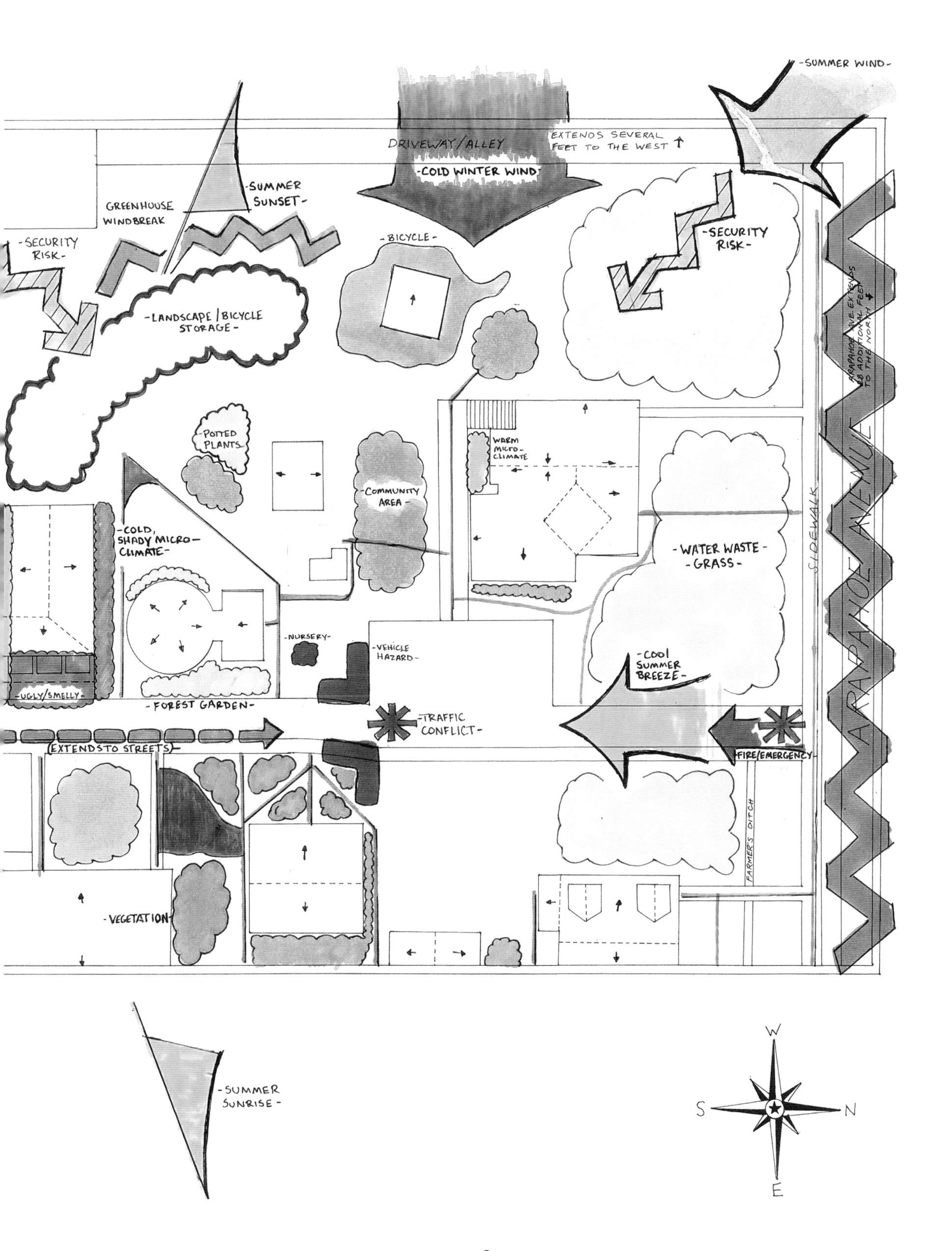
-SUMMER WIND-
DRIVEWAY/ALLEY
EXTENDS SEVERAL FEET TO THE WEST
-COLD WINTER WIND-
-SUMMER SUNSET-
GREENHOUSE WINDBREAK
-SECURITY RISK-
-BICYCLE-
-SECURITY RISK-
-LANDSCAPE / BICYCLE STORAGE-
ARAPAHOE AVE. EXTENDS 18 ADDITIONAL FEET TO THE NORTH
-POTTED PLANTS-
WARM MICRO-CLIMATE
-COMMUNITY AREA-
-COLD, SHADY MICRO-CLIMATE-
-WATER WASTE -GRASS-
SIDEWALK
ARAPAHOE AVENUE
-NURSERY-
-VEHICLE HAZARD-
-COOL SUMMER BREEZE-
-UGLY/SMELLY-
-FOREST GARDEN-
-TRAFFIC CONFLICT-
EXTENDS TO STREETS
FIRE/EMERGENCY
FARMER'S DITCH
-VEGETATION-
-SUMMER SUNRISE-
W
S
N
E

*Above,* good design makes unpleasant sectors disappear. In many cities, roofs larger than roughly 100 square feet require expensive, time-consuming permits. In Portland, Oregon, architect Mark Lakeman designed the roof over a cob bench for the Pacific Crest School as a series of small overlapping sections. Not only did this bypass the need for a permit, making that sector disappear, it avoided the overly massive look that a large single roof might have created. Photograph courtesy of Mark Lakeman.

*Left,* a guild built around a 'Reliance' peach tree. The understory includes comfrey and a number of insect-attracting plants native to the guild's Wisconsin location. Photograph courtesy of Robert Frost.

These views depict the 16th annual "barn raising" at Share-it Square in Portland, Oregon, at the intersection of Southeast 9th Avenue and Sherrett Street. Each year a new design is conceived and installed by the immediately local community for its own benefit. As the first-ever Intersection Repair project, it was created to inspire ecological place making in numerous other communities. This 2011 version of the project was designed by a 10-year-old young woman, depicting a giant dandelion with seeds floating away toward endless other intersections that await reclamation in cities and towns over the horizon. Photograph courtesy of Mark Lakeman.

An indoor/outdoor worm bin. Mary Zemach, of Los Alamos, New Mexico, has built a worm bin on a sliding shelf, accessed from outside via a small door, or from inside the house, where the bin sits inside a kitchen cabinet, handy for adding food scraps. This arrangement keeps both her and the worms warm in the high desert's cold winters but makes it easy to empty the bin from outside. Photograph by Toby Hemenway.

An elevated rabbit hutch. Connie van Dyke, of Portland, Oregon, uses a two-tiered rabbit hutch with a slanted manure-collecting trough beneath it that channels droppings to a trench under the hutch. Photograph courtesy of Larry Rogers.

Rabbit tractor. Similar in function to a chicken tractor, Connie van Dyke's rabbit tractor is a movable, bottomless pen that lets the bunnies graze and manure the soil while protecting them from predators and preventing them from munching garden veggies. Photograph courtesy of Larry Rogers.

Green roof rabbit hutch. In Jacksonville, Florida, living roof advocate Kevin Songer's rabbit hutch has a living roof planted in greens and a cover crop mix for the rabbits to graze on. The hutch is inside his chicken yard and is raised to give the birds room to forage beneath it. Photograph courtesy of Kevin Songer of metroverde.com.

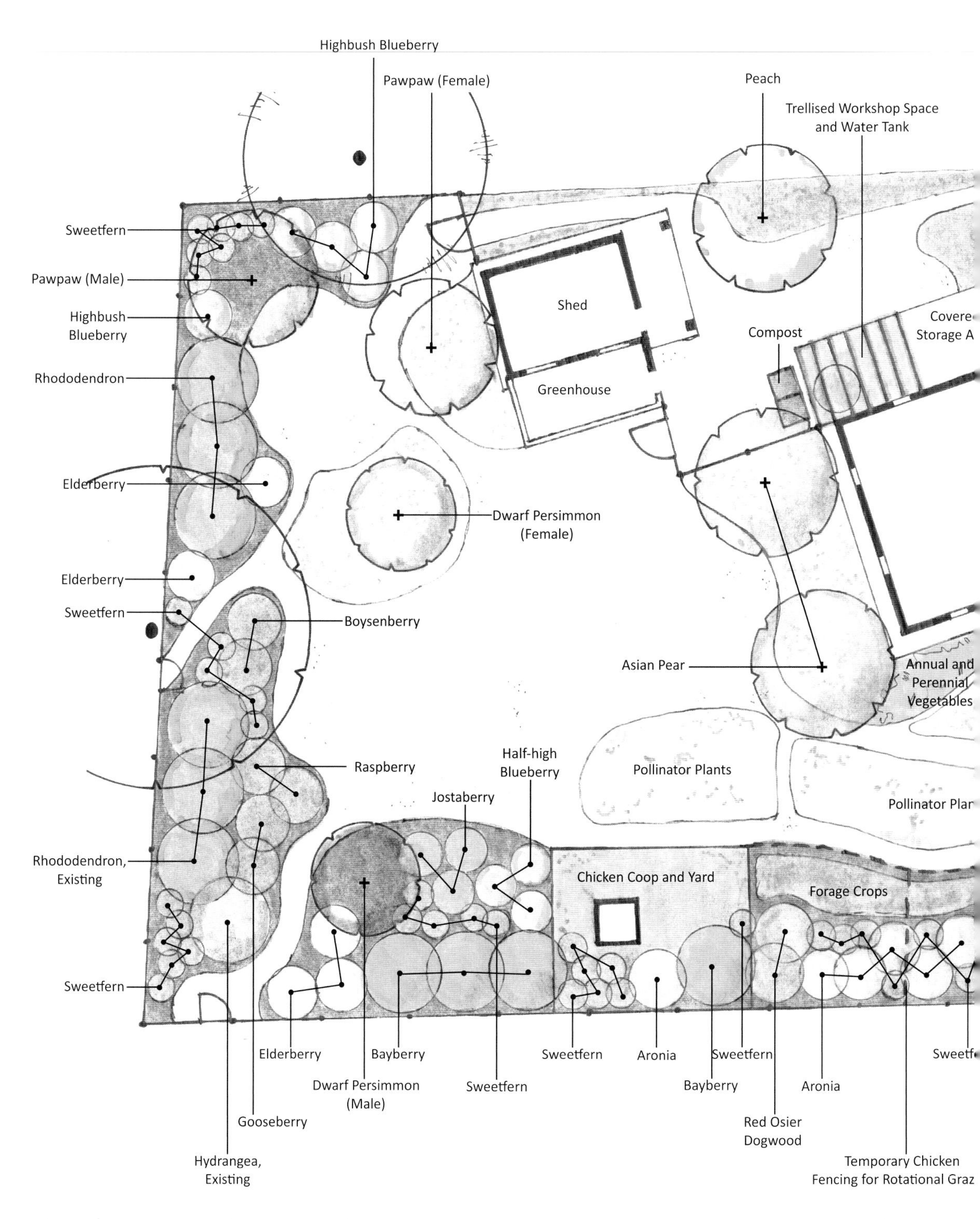


A design for a productive urban yard around a net zero energy home in Massachusetts by Jono Neiger and Regenerative Design Group.

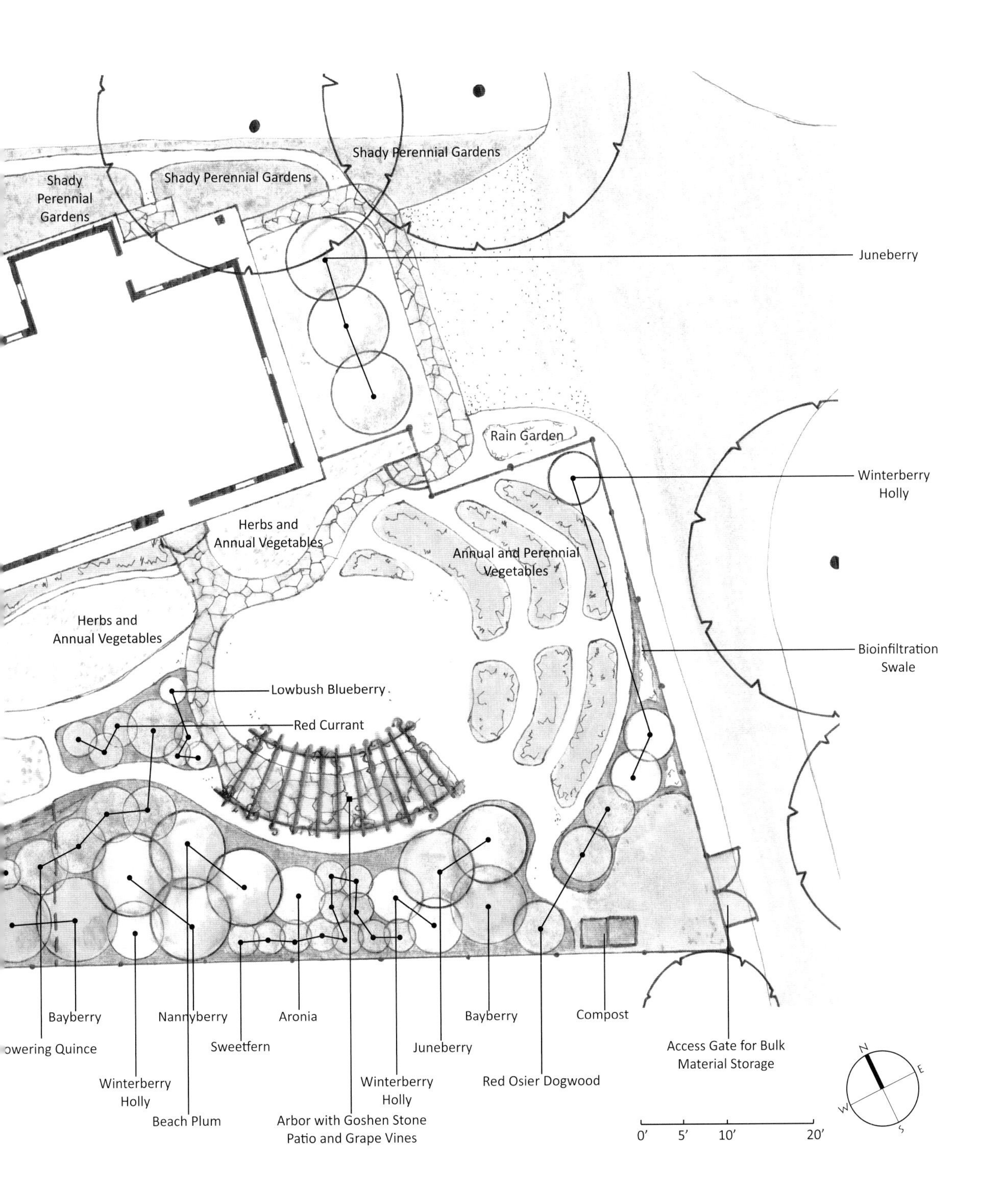

Shady Perennial Gardens
Shady Perennial Gardens
Shady Perennial Gardens
Juneberry
Rain Garden
Winterberry Holly
Herbs and Annual Vegetables
Annual and Perennial Vegetables
Herbs and Annual Vegetables
Bioinfiltration Swale
Lowbush Blueberry
Red Currant
Bayberry
owering Quince
Winterberry Holly
Nannyberry
Beach Plum
Sweetfern
Aronia
Arbor with Goshen Stone Patio and Grape Vines
Winterberry Holly
Juneberry
Bayberry
Red Osier Dogwood
Compost
Access Gate for Bulk Material Storage
0'
5'
10'
20'
N
E
W
S

*Top left,* Langstroth bee hive. Photograph by Toby Hemenway.

*Top right,* a Warré hive in an exploded view to show the interior parts. Photograph courtesy of David Heaf of www.bee-friendly.co.uk.

*Bottom left,* a mason bee house and trays with removable paper tubes. Photograph courtesy of Dave Hunter of Crown Bees, www.crownbees.com.

A Kenyan-style top-bar hive. Beekeeper Les Crowder of www.fortheloveofbees.com is checking one of the easily removed combs to see if it holds enough honey to harvest. Photograph courtesy of John Denne.

Andrew Millison's chicken paddock rotation system for his suburban yard in Corvallis, Oregon. *Left*, the coop at the center of the back yard, with chickens foraging in an established vegetable bed. Other paddocks separated by fences and not presently being grazed are to the left and above the paddock in use. *Below*, the chickens grazing in a fava bean cover crop. The birds rotate through the paddocks as explained in the text. Photograph courtesy of Andrew Millison of www.permaculturerising.com.

A private rain garden in Stelle, Illinois, that gathers rainwater off the roof of the home of Bill and Becky Wilson, owners of the design and education firm Midwest Permaculture. *Right*, the swale and two drainage basins are doing what they should after the first big rain. *Below*, after six years the swale and basins are invisible in the mature rain garden. Photographs courtesy of Bill Wilson.

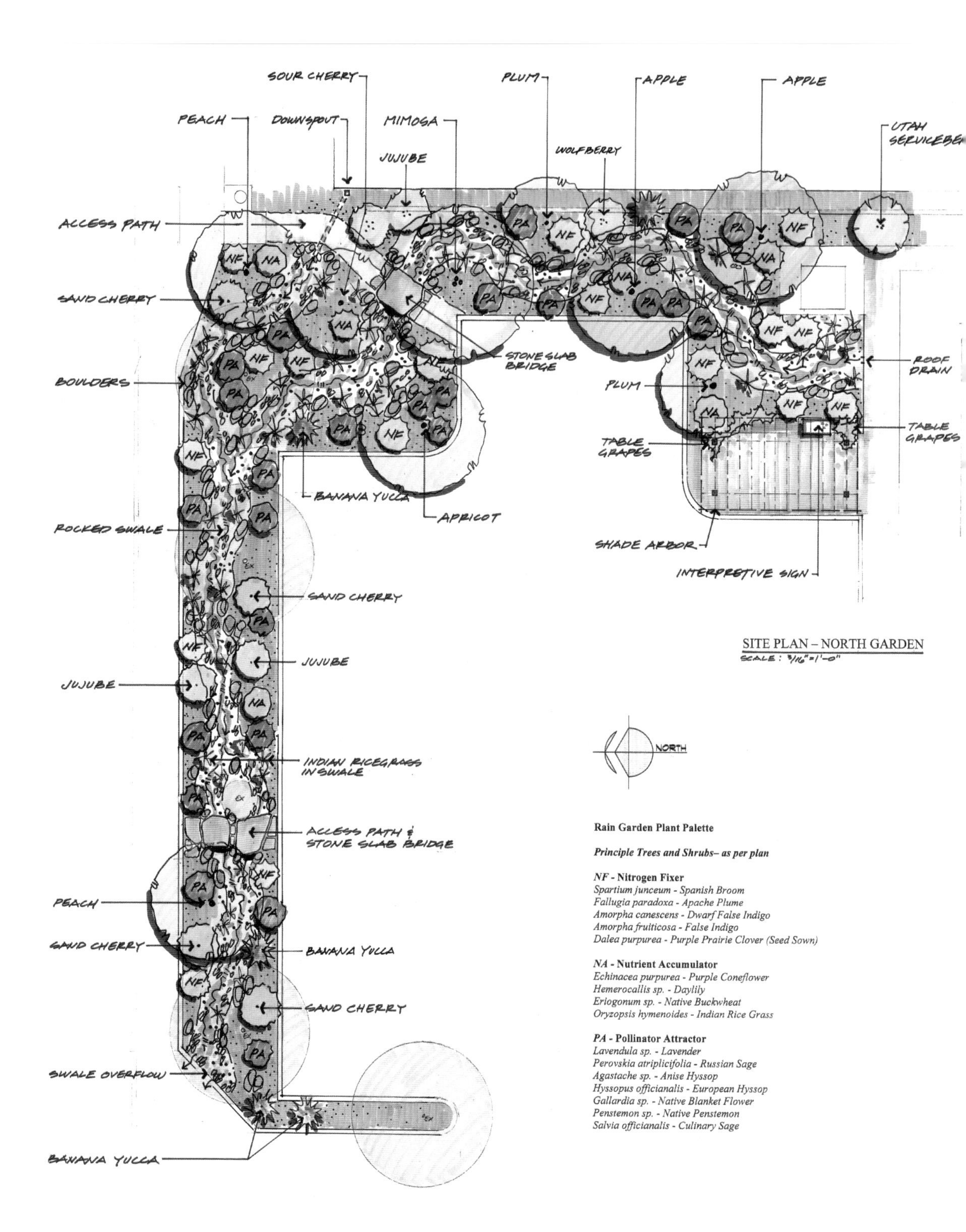

SOUR CHERRY
PLUM
APPLE
APPLE
PEACH
DOWNSPOUT
MIMOGA
JUJUBE
WOLFBERRY
UTAH
ACCESS PATH
SAND CHERRY
STONE SLAB BRIDGE
ROOF DRAIN
BOULDERS
PLUM
TABLE GRAPES
TABLE GRAPES
BANANA YUCCA
APRICOT
SHADE ARBOR
ROCKED SWALE
INTERPRETIVE SIGN
SAND CHERRY
JUJUBE
JUJUBE
INDIAN RICEGRASS IN SWALE
ACCESS PATH & STONE SLAB BRIDGE
PEACH
SAND CHERRY
BANANA YUCCA
SAND CHERRY
SWALE OVERFLOW
BANANA YUCCA
NF
NA
PA
EX
NORTH
SITE PLAN – NORTH GARDEN
SCALE: 3/16"=1'-0"
Rain Garden Plant Palette
Principle Trees and Shrubs– as per plan
NF - Nitrogen Fixer
Spartium junceum - Spanish Broom
Fallugia paradoxa - Apache Plume
Amorpha canescens - Dwarf False Indigo
Amorpha fruiticosa - False Indigo
Dalea purpurea - Purple Prairie Clover (Seed Sown)
NA - Nutrient Accumulator
Echinacea purpurea - Purple Coneflower
Hemerocallis sp. - Daylily
Eriogonum sp. - Native Buckwheat
Oryzopsis hymenoides - Indian Rice Grass
PA - Pollinator Attractor
Lavendula sp. - Lavender
Perovskia atriplicifolia - Russian Sage
Agastache sp. - Anise Hyssop
Hyssopus officianalis - European Hyssop
Gallardia sp. - Native Blanket Flower
Penstemon sp. - Native Penstemon
Salvia officianalis - Culinary Sage

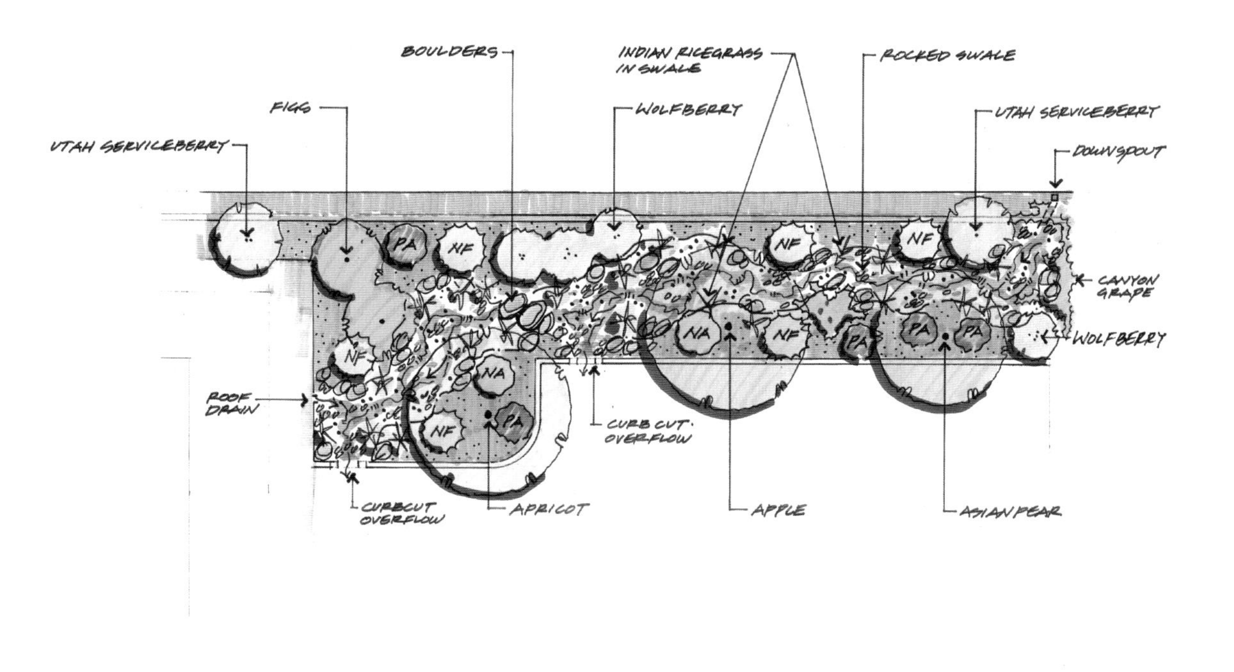


SITE PLAN – SOUTH GARDEN
SCALE: 3/16"=1'-0"

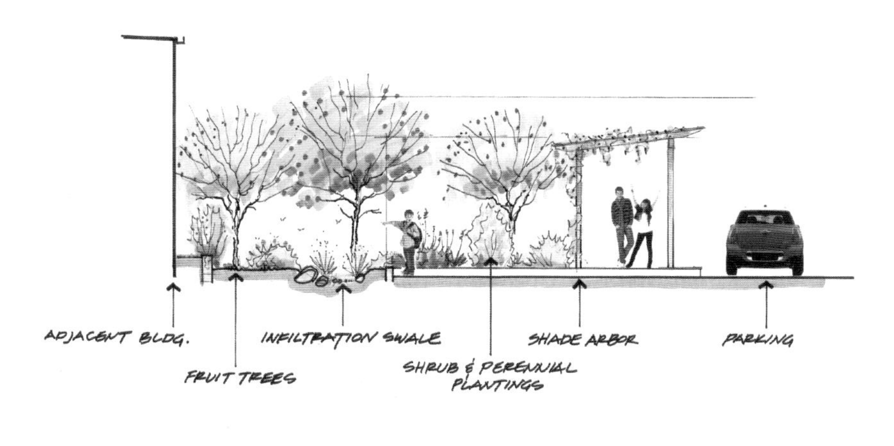


CROSS SECTION
SCALE: 1/4"=1'-0"

The Utah State University campus in downtown Moab, Utah, was a mostly bare expanse of desert-sun-baked asphalt prior to the design of water-harvesting tree islands and food forest to create greater inhabitability. Shade, educational gardens, and visual interest were created through making use of over 75,000 gallons of annual urban runoff from the buildings and parking lot themselves. Designed by Jason Gerhardt and Barnabas Kane of Real Earth Design.

A narrow-profile tank and first-flush diverter at the home of Erik Ohlsen in Sebastopol, California. Photograph by Toby Hemenway.

A rocket mass heater built by Ernie Wissner and Erica Ritter. This clean-burning rocket stove releases some of its heat into the room via the heat exchanger, the black barrel in the photo. Most of the heat from the stove goes into a long run of stovepipe through the bench. This transfers heat into the cob's thermal mass. The exhaust leaving the stovepipe is nearly all carbon dioxide and water and is barely warm. Photograph from the DVD "How to Build Rocket Mass Heaters with Ernie and Erica" by Calen Kennett.

## CHAPTER 6

# Water Wisdom, Metropolitan Style

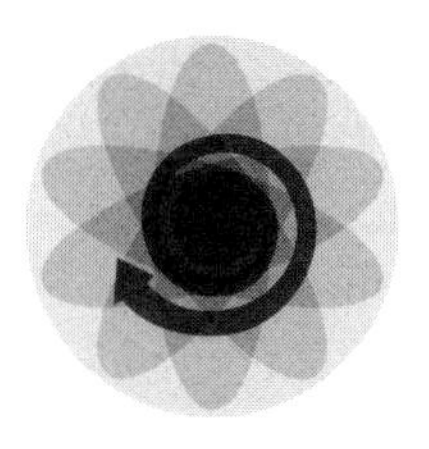

When it comes to cities and water, Los Angeles instantly springs to mind. The titanic scale and landscape-obliterating impact of that city's water system has made its way into American mythology. Immortalized by the film *Chinatown* for the corruption of its politics and on screen in countless chase scenes screaming down the concrete trough of the Los Angeles River, Southern California's water-gathering complex is one of the largest engineering feats on earth. One chunk of LA's water system, the Los Angeles Aqueduct, ferries water 338 miles from the Mono Lake Basin on the far side of the Sierra Nevada. It uses mammoth pumps—able to crank at 80,000 horsepower and fill an Olympic-sized pool in six seconds—to loft this water 2,000 feet over the mountains.

Another piece of Southern California's waterworks, the Colorado River project, runs 242 miles from a huge dam stanching the river at Lake Havasu, Arizona. The project's water sluices through 92 miles of tunnels, 63 miles of canals, 55 miles of conduit, and 144 underground siphons. A third massive system pipes and pumps water 440 miles from the snowfields in the north of the state.[1] All this is to water the farms and homes in the onetime desert of Southern California. Water-related energy use swallows about 20 percent of all California's electricity. The cost of the entire water system is almost impossible to calculate, but to get an idea of its scale, consider that an upcoming overhaul of one miniscule portion of it, a set of tunnels to send water under the Sacramento–San Joaquin Delta, will cost $25 billion to build. The history of water in Southern California proves that water does flow uphill—toward money.

Although Hollywood and the dreams drifting out of La-La Land have spotlighted LA's water system, other American cities have put efforts nearly as prodigious into their own water supplies. New York City's water system pulls 1.2 billion gallons a day into the city from as far away as 120 miles, captured in a series of upstate reservoir complexes that span 2,000 square miles. Three enormous tunnel systems pour water into the city. One of them, the Rondout–West Branch Tunnel, which is 45 miles long, 13.5 feet wide, and buried up to 1,200 feet deep, is leaking 20 million gallons a day, prompting a $250 million repair job. The tunnels feed into 6,500 miles of water mains, but before the city's water is released, it is treated with chlorine, fluoride, sodium hydroxide to raise the pH, and orthophosphate to coat archaic pipes that might otherwise leach lead into the supply.[2]

Chicago draws its water from the adjoining Lake Michigan. In the city's early days, the Chicago River poured the city's sewage, industrial waste, and slaughterhouse offal in a fetid stream into the lake. Unhappy with fouling their own nest this way, the city fathers in 1900 engineered a series of locks that *reversed the river's flow*. Yes, you read that right—the Chicago River now flows backwards. To keep the river draining the city's runoff toward the Mississippi instead of the lake, Chicago now sucks 3,200 cubic feet per second into a pipe 2 miles offshore. About half of this water, 1 billion gallons per day, sluggishly drains from the lake into the reversed river, and the rest pours into two purification plants to scrub out the lake's pollutants before being split among twelve pumping stations.[3] The cleaned-up water is then boosted into elevated reservoirs and pressurized into the city water mains.

In water-rich Florida, Miami is fed by the Biscayne Aquifer, a 4,000-square-mile mantle of spongy limestone and sandstone that is saturated with water. Wells, some gushing at a Niagaran 5,000 gallons per minute, pump the aquifer's water into 2,000 miles of canals and 2,800 miles of levees and berms. The Biscayne is being drained at the rate of over 800 million gallons a day.[4] Much, but not all, of this is recharged by rainfall, and this deficit siphons the saltwater that underlies the aquifer upward into the freshwater, posing serious problems for this fast-growing region.

A glance at any city's water system shows the costs and risks we accept to keep our faucets running. Eighty-five percent of San Francisco's residents drink water that flows 165 miles from the Hetch Hetchy Reservoir, via a dam built ninety years ago and pipes that cross at least two active fault lines.[5] In Tucson 42 percent of the electricity consumed in the city is used to pump water.[6] The list goes on. This litany should make it obvious that watering our cities is an enormous job, consuming resources, energy, and land in staggering quantities. At those scales municipal water systems are vulnerable to energy shortages, financial problems, labor breakdowns, wear and tear, and natural disasters ranging from drought to earthquakes and storms. City water supplies are a potential weak point in meeting basic needs, and the savvy permaculturist will have spotted this. How do we strengthen our relationship with and access to this critical resource?

## ➡ LOOKING AT ⬅ WATER SYSTEMS: THE OBSERVATION STEP

Putting our permaculture lenses on, we can first look at the sectors that influence a typical city water system before moving on to designing personal and community water security. A sector analysis will tell us what influences and forces, both physical and social, shape and control the flow of water to where we use it. It will also help us see the strengths and weaknesses of our water system, since we will be able to see what factors we can affect, which ones we can't, and which ones are the least resilient and most vulnerable.

Potential sector influences include:

- Extent, age, and state of the physical water infrastructure.
- Type and amount of energy required to run the system. (Some city supplies are mostly gravity fed, such as New York and Portland, Oregon, while others require extensive pumping, such as Los Angeles, which means that cheap energy is needed to keep the system running affordably and smoothly.)
- Development and condition of the watershed supplying the water.
- Policies and planning of the agencies in the entire water district.
- Annual rainfall in the collection area.
- Rate of use at the consumption end.
- Factors influencing rate of use, such as the amount of industrial and agricultural use and their growth rates. These fluctuate with the economy, while residential use remains fairly steady per capita.

As you can see, these factors, like all other sectors, are beyond the direct control of individual users. Some influences, such as rainfall and whether the system is gravity fed, can't be altered. Most others are so extensive that affecting or working with them takes large-scale efforts, huge budgets, political maneuvering, and serious organizing at the community and city level. Because water is such a critical resource, it's wise to be engaged in water policy, but we also know that the gears of urban water departments move slowly. However, there are some useful leverage points if you want to have a voice in your region's water policy. Two of the best are:

**Water district boards.** Nearly every community falls within a water district that oversees its water supply, and each water district has a board of directors, usually elected, that shapes policy. While in some cities, seats on the water board are recognized as power centers and thus are hotly contested, in many districts these spots don't attract much interest, and a dedicated activist can get elected via a simple grassroots campaign.

**Watershed councils.** The term "watershed," in its ecological sense, was a specialized term until the 1980s. The word entered popular usage when the abysmal water quality of almost every lake and river on the continent became a public concern. Environmentalists seized on "watershed" as a word that would help people break down a seemingly overwhelming problem to a size—that of the land area draining into a local water supply—that was comprehensible and manageable. Watershed councils by the hundreds appeared, staffed by volunteers who took on simple tasks such as monitoring the health of local lakes and streams, going to water district meetings, and making water quality an issue that was constantly in front of local officials. It was a movement based around a single evocative word that helped communities see that they could boost water quality and, with it, the overall ecological health of their region, simply by paying attention and letting the right people know. Joining the local watershed council is a leverage point available to everyone.

The scale of systems that move water into cities is matched by the task of getting it out. In many cities maintaining aging and overtaxed storm and sewer systems is prompting creative responses in an era when simply raising taxes or floating bonds for billion-dollar drainpipes isn't an option. Pavement and roofs in cities are vast acreages of impermeable surfaces that translate to equally vast volumes of water that must be dealt with during rains. Most cities use combined sewer systems in which rain runoff swirls into the same pipes that drain sewage to treatment plants. Heavy rains sluice enough water down the drain to overwhelm those plants, and the excess is shunted to emergency channels called combined sewer overflow (CSO) that dump rain-diluted raw sewage straight into a river, lake, or ocean. CSOs are a major source of water pollution. Since it's getting mighty expensive to build bigger pipes and treatment plants, one answer—and a smart place to intervene—is to shrink rain runoff by reducing impermeable surfaces and building rain gardens, curb cuts, living roofs, and bioswales. These are neighborhood-scale solutions that harvest runoff and not so incidentally also create beauty and biodiversity. Bioswales are simply permaculture-style swales engineered to handle street runoff. They run beside streets and are often fed via a cut-out section of curb. Bioswales are usually much bigger than backyard swales(2 to 4 feet deep), are heavily planted with wetland species, and are usually tied to the sewer system with raised drains in case they, too, should overflow.

Progressive cities are sinking bioswales into parking strips, often widening the water-absorbing rights-of-way, narrowing streets, and tearing up pavement to do so. In Portland and Seattle, some neighborhoods feature bioswales on every block. New York City began installing bioswales in 2010, and the pilot program has been expanded with nearly $3 million in grants for "green infrastructure" on private property and a total budget of nearly $100 million per year for public projects to reduce CSO dumping.

Rain gardens serve the same purpose as bioswales—to absorb rainfall and keep it in the soil—but are larger. They can be built at the home scale to soak up water from a roof, as the accompanying photos show. A home rain garden offers the mutually beneficial payoffs of reducing a house's runoff burden on city infrastructure while giving the residents an excuse to occupy more growing space. (See photographs in the color insert, page 13.)

More common are public rain gardens, usually installed to gather runoff from parking lots, cul-de-sacs, and other expanses of commercial or city-owned pavement large enough to overwhelm a simple bioswale. They feature constructed wetlands and well-drained excavations that hold and soak up enormous volumes of water. Public rain gardens feel like parks, even to the point of offering benches and tables for users to enjoy the wetland plantings and the birds and beneficial insects that they attract. In the long run, these projects will reduce taxes and infrastructure replacement while building soil health, urban greenery, and social connections. If those benefits appeal to you, check with your city's water and sewer department to see if these water-harvesting, value-enhancing features can be added to your neighborhood.

Not quite ready to engage on the policy level? You can always install a rain garden in your own yard or, as Brad Lancaster did, chip out a section of curb to divert street runoff. Although that latter may get you engaged at the policy level—or the citation level—with whatever local authority maintains your streets, it can be, as it was for Lancaster, an opening for a conversation with local agencies about smarter stormwater policy.

Being active in the policies and planning at the regional level of the local watershed is a powerful intervention point, but the glacial pace of reform at that scale can be discouraging. To avoid being discouraged, and also to be able to spot and work at other points of maximum leverage, each of us can act at the neighborhood and personal level, where we can get results quickly. That's the focus of the rest of this chapter.

## ➡ THE BEST LEVERAGE ⬅ POINTS FOR SAVING WATER

To begin our personal-level watershed observations, let's look at how the typical household uses water. That will help us see the best places to intervene. A study using data from the US Geological Survey found that typical household water use breaks down like this:[7]

- 29 percent outdoor
- 19 percent toilet
- 15 percent clothes washer
- 13 percent baths and showers
- 11 percent faucets
- 10 percent leaks
- 1 percent dishwasher
- 2 percent other use

A few points leap out from these numbers. The first is that outdoor use, which usually means irrigation, is the big water hog, especially in the arid West, where outdoor use can make up 70 percent of a home's water use, so finding ways to reduce that number or shift the source from piped water to rainfall and other less infrastructure- and energy-intensive sources are smart strategies. Those could entail large DIY projects that involve new cisterns, channeling downspouts, and installing a drip irrigation system. Those are not small jobs, but they should go on our list of things to do. If those seem daunting for now, the water-use statistics point us toward other high-impact ways to cut water use in small steps without spending too much time or money. Some of these involve replacing old fixtures with newer, more efficient ones (and, yes, I'm aware that we're not going to consume our way to sustainability by buying more stuff). In many cases, the life-cycle savings of super-efficient fixtures can outweigh the energy and ecosystem costs of making them.

### Water-Saving Projects

Following is a list of water-saving projects, ordered by my own informal cost-benefit analysis, which is a combination of water saved, cost, and amount and complexity of labor, to help you decide what to do first based on your own budget of time, money, and potential water savings.

#### Replace Your Toilet

Because the toilet is the biggest indoor water consumer, it makes sense to start there. Replacing a high-volume toilet with a new low-flow unit, if you do it yourself—and it's a simple plumbing job—costs about $200. This can pay itself back in one to three years in water-bill reductions. Whether your toilet is new or old, make sure that the flap valve seals without leaking. You can test this by putting a few drops of food coloring into the tank. Don't flush, and check the bowl a few times over the next thirty minutes to see if color is seeping into it from the tank. Clean or replace the flap if it leaks. Also, if the toilet runs after flushing, replace the float and flap with good-quality hardware. But really, if we're starting from first principles, using water to dispose of human waste is a bad idea. It converts pristine drinking-quality water directly into sewage in one step, pollutes groundwater, disperses a once-concentrated pollutant across miles of pipe and waterway, and turns potential fertilizer into poisonous garbage. You might consider buying or building a composting toilet, which uses no water at all and safely turns your food back into nutrients.

### Install Low-Flow Showerheads

Install low-flow showerheads and a stop valve on the showerhead stem. The latter lets you easily shut off the water flow while lathering up without needing to reset the water temperature. Reducing the odds of being frozen or scalded when the shower fires up again will increase the chances that you'll shut off the water between rinses.

### Repair or Replace Leaking Faucets and Pipes

Once you've fixed the obvious leaks, use your water meter to find any others. Older meters will need to be read over a period of time to see if the numbers change—two to three hours should be enough—while all fixtures in the house and yard are turned off. Newer meters have a flow indicator, usually a red light that indicates whether water is flowing. If water is running when it shouldn't be, try to isolate the leaking area by shutting off any valves that serve individual parts of the house or yard, such as valves under sinks, at laundry and toilets, or an irrigation network. Then check those sections of pipe for leaks.

### Replace Your Washing Machine

Those above are all modest-cost, quick, high-payback steps. This next option is more expensive but will save plenty of water: Replace your clothes washer with a low-volume unit. New, these machines are expensive, but not only do they save 20 to 30 gallons per load, most have a high-speed spin cycle that wrings your clothes dry enough to save energy in the dryer as well.

### Store Water in the Soil

As noted, the biggest water user is the yard. Although there are parts of the continent where rainfall is regular enough to rarely warrant irrigation, our increasingly erratic weather makes it likely that no matter where you live, there will be times when plants will need irrigation. Although some of the tasks for creating a yard that rarely needs municipal water are simple, a full program can be a big project. Again, we can start small and work up.

In keeping with this book's theme of looking for smart strategies and high-value leverage points, I'll offer one of permaculture's most-heard mantras here: *The best place to store water is in the soil.* Because the biggest consumer of household water is the yard, it makes sense to deliver and store water right where most of it is used: in the soil. Inexpensive, low-tech, passive ways to do this, such as catching runoff and increasing the water-holding properties of the soil are where we start, before moving onto spendy technological solutions such as drip irrigation and storage tanks. Soil can store an enormous amount of water: A foot of organic-rich topsoil can hold the equivalent of 3 inches of water, so you can think of your yard as a lake, but one that won't evaporate quickly because the water isn't out in the open. It's in the ground, protected from the sun and kept cool by mulch and shady plants. Thus permaculturists work on ways to get water into the ground quickly and to help that ground store as much water as possible until plants can use it.

Water wizard Brock Dolman, who works in Northern California at the Occidental Arts and

Ecology Center, has another mantra that helps guide us in getting water into the ground effectively: Slow it, spread it, sink it.

*Slowing water* shrinks, in a big way, its ability to erode or carry off soil. Reducing the speed of running water by half drops its capacity to hold sediment by seven-eighths, making this a brilliant leverage point. Small berms, tube-like straw wattles, brush or logs laid out on contour, plantings, and mulch can all be set in the path of runoff (whether overland or out of downspouts) to trim its speed and temper its erosive power.

*Spreading water* increases its friction with the ground (another way of slowing it) and boosts the area it reaches (which increases edge), an elegant stacking of functions that both reduces erosion and irrigates a larger area. Water that is spreading out along a level contour is water that is slowing down, dropping sediment, and not running downhill and away.

*Sinking water* into the ground is the result of slowing and spreading it, and we need to give it places to sink in via berms, basins, and dish-shaped lawns and gardens and by fluffing up soil with organic matter and soil-busting plant roots to turn it into an absorbent sponge.

## Ways to Store Water in the Soil

Any drought-proofing program has two main aspects: saving and sparingly using the water that's available and employing ecologically sound means to increase how much water is available. How you tackle the task of drought-proofing your yard will depend on your time and money budgets. There are many facets to reducing water use in the landscape, and once again I've broken them out and ordered them by effectiveness and project cost and labor.

### Mulch Everywhere You Can

Cover up bare earth. Even a light dusting of mulch a half-inch deep will reduce evaporation dramatically. Deeper mulches save more water, but they also can prevent light rains from reaching the soil, especially if the mulch has dried out. In small yards my strategy is to pull mulch back in the spring to let the soil warm and make sure that rains can soak in. Then when summer heat and drought starts, I rake the mulch back over the soil to lock in that precious moisture. In places with dry summers, apply several inches of mulch over drip-irrigation emitters. Then it doesn't matter if the mulch dries out, as the water arrives from underneath. One caveat: be careful not to slice the drip lines by carelessly punching a shovel or spade through the mulch.

### Add Organic Matter to Your Soil

Organic matter holds two to four times its weight in water, so beefing up your soil's humus content will help hold water right where it is most needed: in the soil around your plant roots.

### Contour the Soil to Hold Water

Ways to do this include:

- Swales, in permaculture parlance, are long, thin basins built level, following contour lines. The soil from these modest excavations is piled on the downslope side to make a berm, and the berms and basins can be planted or used as paths. These will catch water running downslope (or from gutters or other sources), slow it down, and sink it into the ground where plant roots can get it and where it won't evaporate. *Gaia's Garden* and many other permaculture resources explain swales in detail.

- Small semicircular berms, 2 to 6 inches high, can be built 3 to 6 feet downslope from trees or other major plantings. These "boomerangs" create basins that catch runoff and help soak it into the soil.
- Shallow trenches dug with a slight slope can ferry water from downspouts and wet spots to dry areas.
- In drylands and warm Mediterranean climates, paths should be raised above the average grade so that water runs off them onto plantings. Garden beds in dry climates should be sunk, not raised, so water flows into them. Install major plants such as new trees and shrubs below the average soil level in basins that collect water. If you are installing a new lawn area, grade it so that the center is a bit lower than the edges, mimicking a dish that will collect water.

### Shrink the Lawn

Shrink that notorious resource hog. While some lifestyles allow homeowners to ditch the lawn entirely, many people enjoy a grassy spot for sitting,

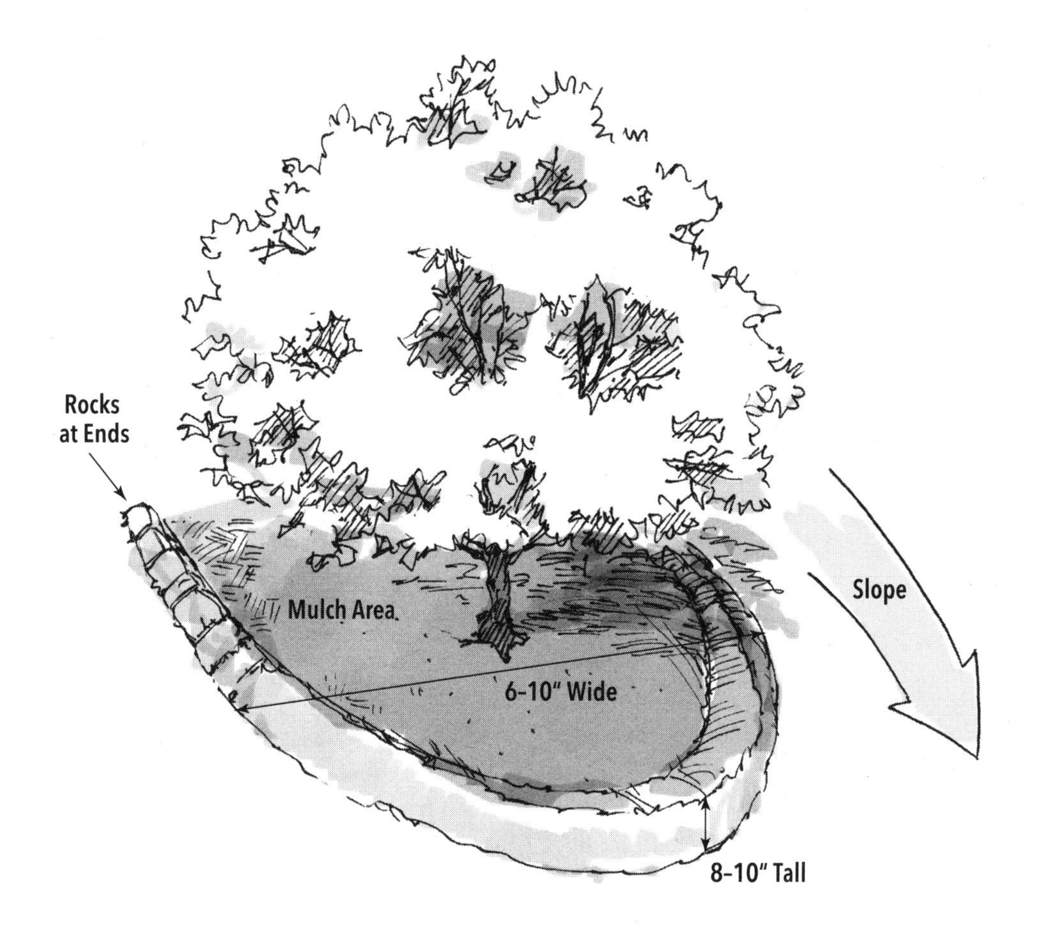


**FIGURE 6-1.** Boomerang water catchment for a small yard. Illustration by Elara Tanguy

for kids' activities, and to give the dog somewhere to romp outside of the garden beds. But we can assess the area truly needed for lawn-only functions, then trim the grassed zones to the minimal size and install less thirsty plantings in their stead.

### Create and Enhance Water-saving Microclimates

Applying the tips in the section "Cooling a Hot Yard" in chapter three will not only keep you more comfortable but will check evaporation of precious soil moisture.

### Replace Water-Guzzling Nonnative Ornamental Plants

I'm willing to use water on food plants and other working species such as nitrogen-fixers and specialized habitat plants, but species chosen mainly for looks or privacy screens don't warrant the added water use when there are so many suitable natives adapted to your climate and rainfall that can do the same job with less water. The best choices, from a permacultural viewpoint, are multifunctional natives: those that feature edible parts, attract birds or beneficial insects, harbor nitrogen-fixing microbes, act as mulch generators, or have other uses. What are these species? As much as I'd like right here to insert a big table listing multifunctional native plants and their uses, a partial tally of North American native plants runs to over 10,000 species, and the list varies for every county and habitat on the continent. There are good online databases and books about native plants for every region, as well as chapters of native plant societies nearly everywhere, and I recommend that you explore them.

### Install a Drip Irrigation System

Install a drip system in planting beds and for any irrigation-requiring plants in your yard. Drip systems are high in embodied energy; the plastic parts are made from petroleum. But the HDPE used in drip tubing is one of the least-toxic plastics around, and plastic that saves water for many years seems one of the best uses for petroleum—surely better than burning it in a gas tank. An extensive drip system can cost a few hundred dollars and take many hours to install, but the project can be done in phases. Start with the highest water-use areas, which are often vegetable beds. Expand to fruit trees and berry bushes, where irrigation will increase growth and yields, and then to other water-needy areas. Many drip-system retail websites offer useful tips, but my top resource for comprehensive coverage of designing, installing, and using drip systems is Robert Kourik's book *Drip Irrigation for Every Landscape and All Climates.*

### Reuse Your Home's Graywater

Graywater systems range from a simple basin in the sink or tub to complex, constructed wetland-pond-swale affairs and many systems in between. The subject is large enough to warrant its own section on page 143.

### Harvest and Store Rainwater from the Roof for Irrigation

Rainwater catchment projects range from modest to large scale and are covered in a separate section on page 143.

## Using Graywater Indoors and Out

Most household wastewater is only slightly used by the time it splashes into the drain. Guided and even compelled by the design method of highest use, as good permaculturists we are inspired to pull further yields from that still-useful water before it leaves our system in unusable shape as groundwater, as vapor, or in a sewer pipe. Reusing the most contaminated water in the house, blackwater from the toilet, is technically daunting, has a high yuck factor, and is usually illegal, but about 75 percent of household wastewater is graywater—from shower, tub, laundry, and sink—which is eminently and easily reusable. Graywater not only holds all the wondrous properties of earth's most precious liquid but also contains organic matter in the form of dirt, soaps, and effluvia sloughed off the living and once-living things it has contacted—us, our food, and our clothes. Microbes in soil will happily clean graywater by transforming the "gray" into fertilizer for plants, providing another yield besides water.

Graywater can turn a dry landscape into a green paradise without the guilt of further drawing down municipal reservoirs. And instead of racing away into a sewer, graywater can recharge local groundwater and aquifers. Just as we now separate precious compost and recyclables from spent waste, it makes sense to split almost-clean, reusable water from true sewage. We'd be crazy not to harvest all that multifunctional bounty.

Before we look at how to use graywater, there are some guidelines to consider. After all, it is wastewater and thus a potential source of disease and general ickiness if not dealt with properly—although graywater guru Art Ludwig asserts that there has never been a case of illness traceable to a graywater system. But let's preserve that record by using it sensibly.

### Basic Precautions for Graywater Use

Graywater is new for most of us. We're all used to potable water systems; the stuff comes out of a faucet, and we use it with little thought. And if we're lucky, we've never had to become intimate with sewer plumbing. Graywater is neither drinking water nor sewage. It's a different substance that has its own personality, and as with any new acquaintance, there's a getting-to-know-you period before using graywater becomes second nature. Graywater use can be simple, sanitary, and trouble-free, but to ensure that it remains so, there are a few basic rules to follow. Here they are.

1. **Never store graywater for more than twenty-four hours.** Fresh graywater contains bacteria, which, if left standing, will multiply and convert harmless graywater into smelly sewage.
2. **Do not create open pools of graywater;** let it soak below ground level immediately. Get it on the ground, into mulch, or into a constructed wetland quickly. Those places are rife with beneficial bacteria that will rapidly outcompete the sewage-producing microbes in the graywater. Soil critters also speedily break down soaps and other chemicals, converting them into nutrients.
3. **Apply graywater to the soil directly,** not via overhead sprinklers, which will create graywater aerosols. Don't apply graywater to edible parts of plants or to crops you will eat raw.
4. **Graywater does not include** washwater from dirty diapers or a wash load that contains bleach or borax, which are toxic to soil organisms and plants. Send these to a sewer. And obviously, toxic chemicals and solvents don't belong in graywater or any part of our water system.
5. **Design your setup to divert graywater** to the sewer or septic system when needed (e.g., for diapers), via a valve or some other system that ties to the sewer.

6. **Keep it simple.** Make your graywater system easy to maintain and reach, with a minimum of pumps, filters, and moving parts. Be realistic about how often you are willing to deal with it. It's tempting to install lots of bells and whistles, but cleaning and repairing all those bells will take time.
7. **Water from the kitchen sink is the most challenging to use,** since it contains oils, concentrated soaps, and food scraps. Carrying that water in a basin to dump outside is fine, but if sink water is incorporated into the plumbing of a graywater system, it may require special treatment, explained below.
8. **Learn the rules for graywater reuse** in your state and county. Legal graywater use is on the rise, and some areas have permissive rules, allowing graywater systems without inspection or permit. Graywater systems can be unobtrusive, and my personal view is that they are beneficial in so many ways that a well-designed system is worth having, quietly if need be, in hopes that your government will catch up with the times. My prelegal graywater system in Portland was used as a model to guide the writing of a revised code making graywater use legal in Oregon, so having examples in place can speed the rule-writing process. Your own system can help inspire others, even local and state officials, to reuse graywater.

Most graywater is used outdoors, although some commercial systems and a small percentage of home setups use graywater to flush toilets. That means that the toilet is fed by an isolated, labeled plumbing system not connected to the main water supply. But since most toilet tanks don't need pressurized water, if there is a handy way to drain graywater (or rainwater) into the toilet tank, that can save up to 25 percent of household water. Unless you're willing to pour water into the toilet via a bucket, setting this up takes plumbing skills, and the toilet will need cleaning more often. One commercial toilet design cleverly incorporates a hand sink into the tank lid, and the sink drain fills the tank.

### Ways to Reuse Graywater

Graywater systems can take many forms. I'll list the principal types here, in increasing order of plumbing work that each requires to install. There is a graywater system that will suit almost everyone. Often the systems that require the least up-front work need the most maintenance and monitoring, so there is a rich payback to investing some effort in setting up your graywater program.

1. **A basin in the sink** to catch water is the simplest graywater system. That's a fairly high-maintenance setup, as the basin needs to be carried outdoors and poured onto the soil when full. A basin might be a good way to begin using graywater to get familiar with it, but being human, you'll probably soon be longing for something that doesn't require a trip outside as often. The upside of this method is that you can choose precisely which plants receive a shot of water. In drought-prone California, the basin system keeps me in tune with thirsty plants that I might otherwise miss.
2. **Removing or opening the P trap under the kitchen sink** and placing a 5-gallon bucket under it is slightly more complex. Most people who do this also remove the cabinet doors for easy access and to spot a full bucket before water comes gushing out of the cabinet. Again, this is a high-upkeep

system that requires carrying a full bucket outside for emptying.

3. **The laundry-to-landscape system** is next up the scale of complexity. This involves only some easy plumbing that doesn't require tying into the house drains, which keeps it simple. It takes advantage of the pump in every washing machine. This pump is powerful: Regulations require that it be able to lift water 10 vertical feet. That means that even if your yard slopes uphill, you may be able to run graywater to much of it. In essence, laundry-to-landscape means connecting the outlet of your washing machine to a pipe or tubing outside to deliver graywater to trees, shrubs, or

## Laundry-to-Landscape Graywater

The outlet hose from the washer (1) is connected via an appropriately sized barbed adapter (2) to a three-way diverter valve (3). One outlet of this valve is connected to the sewer for using bleach or doing other graywater-unfriendly laundry loads. The other outlet of the valve is connected to 1-inch pipe that goes to the outside, through the wall or a convenient opening.

Outside the house, this line runs to a tee (4) that ties a vacuum breaker (5) into the line. The vacuum breaker stops graywater from continuously siphoning out of the washer when emitters are lower than the washer water level.

In freezing climates an automatic bypass is installed above this tee so that if the outdoor graywater pipes freeze, water will gush out the bypass rather than onto your floor. This is done by attaching an ell-shaped length of pipe just below the vacuum breaker (not shown). Graywater systems are usually shut down in cold winters, as there is no need for irrigation then.

The other end of the tee at the vacuum breaker connects to a 1-inch PVC pipe that ends roughly 1 foot above the ground, where it terminates in a male hose connector. This connects via a female hose connector to 1-inch polyethylene tubing that runs to the yard. This way, if the line to the yard clogs, a hose can be connected to the line and used to clean out the line with full-pressure water. The tubing to the yard runs to four to six emitters that branch from this tubing. The emitters can simply be tee fittings in the pipe that have ½-inch outlets (the outlets must be smaller than the pipe size or all the water will run out the first outlet). Emitters can run into mulched basins directly or into inverted 5-gallon plastic flowerpots (emitter shields) sunk in the ground, which will allow water to penetrate over a larger area.

If the emitters are above the level of the washing machine, a backflow preventer must be attached near the beginning of the main line to prevent graywater from running back into the washer.

With this setup you can irrigate downhill any distance without harm to the pump or pump uphill to places up to 2 feet below the lid of the washer for a distance of 100 feet. It's possible to irrigate 1 to 6 feet above the top of garden beds via outlets below the soil or mulch surface. In reality, there are a few extra features that avoid burning out the washer pump or flooding the laundry area, but those extras are easy to add. Code officials like the laundry-to-landscape system because the graywater stays in a pipe until it's in the ground. Dean of graywater use Art Ludwig was one of the first to come up with this idea in 2008, and since then it's been refined, taught in hundreds of workshops, and installed in thousands of homes with few reports of trouble. I describe the laundry-to-landscape system in a sidebar.

4. **Diverting other household graywater to the yard** using gravity flow is a larger

the washer over a very short distance, but it may shorten the life span of your washer pump.

Detailed instructions and a parts list for building a laundry-to-landscape system are available at Art Ludwig's website, www.oasisdesign.net, and in his excellent book *Create an Oasis with Greywater*. And of course videos abound on YouTube.

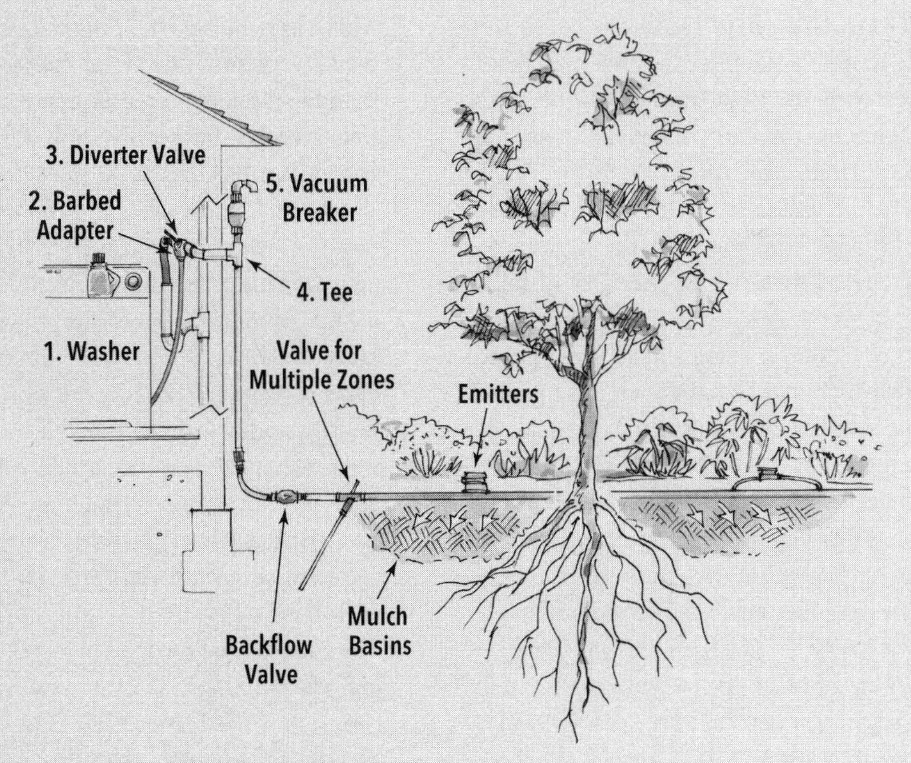


**FIGURE 6-2.** A laundry-to-landscape graywater system. Illustration by Elara Tanguy

project that you can do separately from or combined with a laundry-to-landscape system. The task here is to cut into and divert drainpipes running from the sinks, tub, or shower, so it takes plumbing skills or a professional plumber. It may also require opening up walls for access to pipes; thus, its feasibility depends on how deeply buried your home's drainpipes are. If you can't get to all of the drains, then after the washing machine, the shower or tub that is used most—and thus will deliver the most graywater—is the drain to go for. Sinks use far less water, so if their drains are hard to get to, you can skip them.

If you live in a one-story house with the pipes exposed in the basement ceiling, you can easily get to the shower, tub, and sink drains as long as they weren't combined with the toilet line up in the walls somewhere. You can tell because sink and shower drains are usually 1½- or 2-inch black pipe, and toilet drains are larger 3- to 4-inch pipe. If all you see is the larger pipe, you're faced with the tougher task of finding the graywater drains before they meet the toilet line, and that takes cutting into walls, which may be more than most people want to tackle. The most difficult cases are basement-less two-story houses with an upstairs shower or tub, because reaching waste pipes requires not just chopping into walls or floors but also running the new graywater pipes down the inside or outside of walls. (By now you can see why the relatively simple laundry-to-landscape system has become so popular as the entryway into graywater reuse.)

Another challenge is the slope of the yard. The section of the yard destined for graywater infiltration needs to be lower than the drainpipe in the house, and the farther the irrigation site is from the drainpipe, the more downward slope you need, by roughly a ¼-inch drop per foot of distance. If you can beat the challenge of finding the right drainpipes and having a downhill slope from them into the yard, congratulations! You can run graywater to your yard via gravity.

I'm not going to go into the technical details of the in-house plumbing required, as that would require a couple of chapters on basic plumbing, and it varies enormously from house to house. However, the idea is to tie into the drain lines with a similar-sized pipe using a tee and run new pipe outdoors while preserving the original drain run. Install a shutoff valve near the tee to close the graywater line so that any really nasty water can be shunted to the sewer and the system can be shut off in freezing weather if needed. Where the graywater leaves the house, pipe or tubing can be installed running downhill to emitter boxes like those in the laundry-to-landscape system. The book *Create an Oasis with Greywater* gives detailed instructions on various outdoor drain systems, as do videos and sites on the web. Once again, my role here is to give you enough information to make permaculturally based decisions on whether and when to use these systems, and yours is to take it from there.

5. **The constructed wetland** combines plumbing with biology, wherein the graywater runs into a lined, gravel-filled pit or trench to nourish microbes that clean the graywater by converting the soaps and effluvia into nutrients that feed wetland plants.

Constructed wetlands are primarily built for treating graywater for further use other than irrigation, such as a pond or other storage or open site. A backyard wetland also creates a spot for beautiful water-loving plants, lush biomass production that's boosted by the high nutrient levels, and wildlife attraction, all in a way that doesn't use additional city or well water.

But for simply irrigating with graywater, there's no need for the extra work of building a wetland. Just get your graywater onto the soil quickly and simply. Take it from me: Don't make your system more complicated than it needs to be. I built a constructed wetland as part of my demonstration graywater system in our Portland yard. While it was terrifically successful as part of my backyard ecosystem, it also was an object lesson in how projects can balloon to ridiculous size. It started because I wanted a pond, and graywater that had been cleaned via a wetland was a guilt-free way to keep a pond filled in Portland's dry summer. Guided by permaculture's earth-care ethic I knew that the soil from the pond and wetland excavations had to go somewhere other than the dump, which complicated matters. So in a community project—I couldn't have done it alone—we transformed the dug-up soil into an earthen sauna and rocket-stove-heated cob bench next to a pond that was fed by the

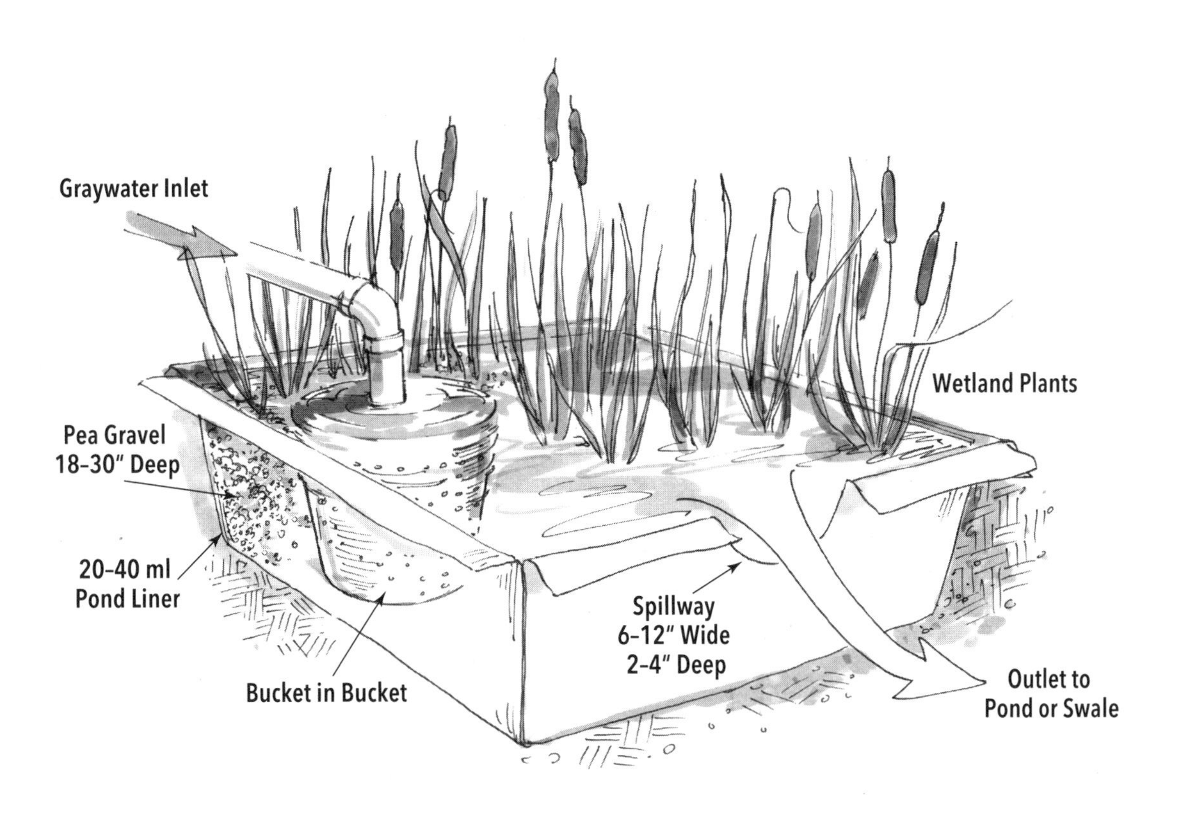


**FIGURE 6-3.** A graywater wetland with a bucket-in-bucket inlet for easy cleanout. Illustration by Elara Tanguy

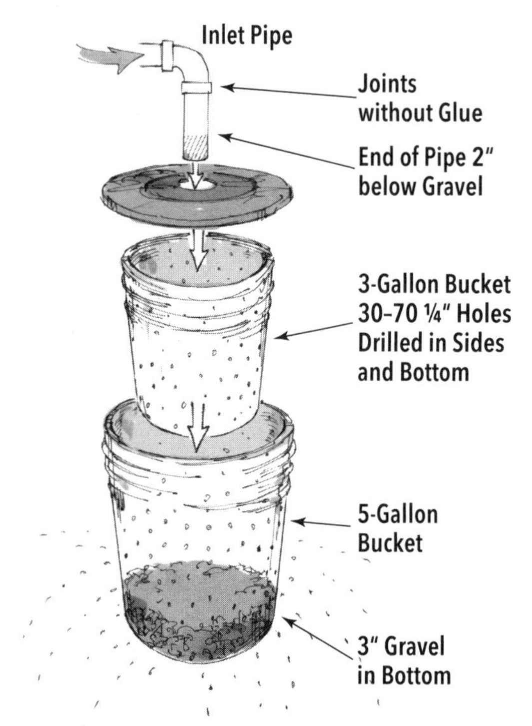


**FIGURE 6-4.** Detail of the bucket-in-bucket graywater inlet system. Illustration by Elara Tanguy

graywater wetland. It was all wonderful at the end of the year-plus that it took to finish the ever-expanding work, but be warned: One project leads to another. Keep it simple.

### Tips for Living Happily with a Backyard Wetland

Detailed instructions for designing and building a backyard graywater wetland are given in *Gaia's Garden* and in *Create an Oasis with Greywater*, but here are some hard-won lessons I've gained from years of living with backyard wetlands.

- Don't mulch the surface of the wetland. The exposed gravel may look unattractive at first, and you may be tempted to mulch it. Once the plants grow in, you won't see much of the gravel, and if you mulch, even with coarse wood chips, the hyperenergized microbes in the wetland will chew the mulch into compost in a few weeks, which will filter into the gravel, clogging it.
- Build an inlet graywater dispersal system that is easy to clean. The length of perforated pipe shown in various designs, including some editions of *Gaia's Garden*, may eventually clog. Figure 6-4 shows a setup that is easy to remove and clean.
- The outflow can be a simple spillway set at a level that keeps the water below the surface of the gravel. Although I once built the more complex adjustable-level bucket design shown in *Gaia's Garden* and *Create an Oasis with Greywater*, I never needed to adjust the level, and the design means losing several inches of elevation that may be needed to gravity-feed the yard or pond with water. It's not hard to remove or add a little soil under the liner at the spillway to adjust the water level if needed.
- If you are running the treated graywater into a pond, it will take one to three months until the microbial ecosystem matures enough to remove most of the contaminants/nutrients from the water. Using a water test kit from a pond-supply store, I found that the water running from my wetland initially had a pH of 8.2, more alkaline than is healthy for some pond plants, and it also held significant, though far from toxic, levels of ammonia, nitrate, and phosphate. After about six weeks, the pH had dropped to 7.0, and the other compounds were undetectable. The microbe food web had evolved into a happy balance.
- If the wetland is in full sun, water loss through evapotranspiration will be substantial. This is

why, if the water is only for irrigation, you don't want the intermediate step of a wetland. Friends in hot, dry Phoenix built a graywater wetland, and for much of the year no water at all came out of it; it all went up through the plants. Mine was in partial shade and yielded plenty of water, and the plants didn't just thrive but grew rampantly. I was harvesting cattails and rushes for compost in multiple wheelbarrow loads once a month, and multi-flowered canna lilies grew 9 feet tall.

Somewhere in the hierarchy of varyingly complex graywater systems is one that will match your building skills, budget, and desire for tweaking and maintenance. Water truly is the essence of life, and recycling what comes into our home ecosystem this way, harvesting the nutrients we have put in it and returning it to the soil to clean it and slow its race back to the sea, makes too much sense not to do.

## ➡ CAPTURING AND ⬅ USING RAINWATER

Weaning ourselves from the municipal water supply takes us toward a whole-systems approach to water conservation, especially when linked to graywater reuse. Doing this reduces or even eliminates our tug on public water supplies and can help shrink their immense ecological and infrastructure footprints. Remember from this chapter's first few paragraphs how big-city water supplies ensnare thousands of square miles of watersheds? If enough Angelenos harvested their own rainwater and returned the resulting graywater to the soil, some of California's 80,000 horsepower pumps could be shut off, and the commandeered watersheds could be returned to the wild. I look forward to the day when the West's aqueduct system is only a dry relic admired for its ingenuity, the way the crumbling Roman aqueducts are today. The water-savvy nonprofit TreePeople has documented that the rainfall landing on the Los Angeles Basin is enough, if caught, to meet all of LA's water needs. The same goes for Tucson and most every other city.

I've saved rainwater harvesting for last because although catching rainwater can be as simple as a barrel under a downspout, a system that weans us off city water more significantly than that can require a comprehensive design project using methods that span several technologies, depending on how deeply you want to burrow into this fascinating rabbit hole of projects.

Every dwelling comes with its own rainwater harvesting equipment: the roof. An astonishing amount of water is intercepted by a typical roof, and here's a good guideline to calculate just how much:

> For every 1,000 square feet of roof area (the amount of ground covered, not the total area of a sloping roof), 1 inch of rain will yield about 600 gallons.

This means that if a house has 1,500 square feet of roof footprint, and the annual rainfall is 40 inches, the roof will catch over 36,000 gallons that year.

This takes us to one of the most-asked questions in green design, one that I've found can chew up a great deal of Q&A time in workshops: What is the best roofing material for water collecting? Some of the research-based answers are counterintuitive, so here I'll offer some guidelines based on the sometimes surprising data assembled by the Center for

Rainwater Harvesting from academic and government studies.[8]

- Most of the contaminants in collected rainwater come from what has landed on the roof, not from the roofing materials themselves. This means that no matter what your roof is made of, rain from roofs near heavy industry, chemical agriculture, or smog is the most likely to be contaminated and should be tested for heavy metals and harmful organic compounds.
- The most harmful stuff coming off a roof is usually fecal coliforms that grew there rather than some roofing-secreted toxin. The most effective strategy for keeping roof water safe is not to change your roofing material but to keep the water clean. There are two good ways to prevent roof water from getting fouled. The first is to clean gutters regularly, because the moist, rotting goo in dirty gutters is an excellent growing medium for E. coli and other harmful organisms. The second is to build a first-flush system (described in the subsection on tanks and cisterns on page 146) to divert the first few gallons away from the storage. Those first gallons from a dry roof are the dirtiest, and water quality improves the longer it rains. An oft-cited rule of thumb has been to discard the first 10 gallons per 1,000 square feet of roof, but studies show that diverting the first 20 gallons per 1,000 square feet leaves water far cleaner. The amount of contaminants decreases dramatically between the 10- and 20-gallon data points.[9]
- Rainwater from roofs made of tile, often touted as an ideal material, contains some of the highest levels of copper, lead, and zinc of any roofing material tested, depending on the type of clay and glazing used.
- Asphalt shingles, feared in many green circles, are among the lowest contaminant releasers of any roofing and are a good choice. These shingles were once thought to release polycyclic aromatic hydrocarbons (PAHs), which are serious carcinogens, but tests show that the PAHs don't originate from the roofing but from deposited smoke and other air pollutants.
- Water from colored metal roofing consistently has the lowest numbers of contaminants and coliforms, probably because the smooth surface washes clean quickly.
- Old galvanized (unpainted) metal roofing, particularly when it's rusty, releases high levels of zinc, especially in acid rain. Humans are tolerant of high zinc levels, but soil microbes and invertebrates are not, so unpainted galvie is a poor choice for collecting irrigation water.
- Cedar shakes are not a wise choice, as they leach natural fungicides that can harm soil fungi and plants, and they are often treated with microbe-killing borate as a fire retardant.
- Green (living) roofs seem to pull metals and organic contaminants out of rainwater. Coliform levels are a bit higher than with other roofs, but so slightly as not to be a big concern for irrigation, and the low levels of coliforms can be easily filtered out of drinking water.

In general, starting with clean rain and keeping it from picking up toxins that have landed or grown on the roof and gutters is a better leverage point to invest in than replacing roofing materials.

## Ways to Catch Rainwater

Once caught, rainwater needs to be stored for a nonrainy day. Directing it into a swale or drainage basin is one option, but there might be dry parts of the yard that a swale won't deliver water to or you may need to store the water for other purposes. Some kind of container is the answer. Here are ways to store water and the best uses for each method.

### The Classic Rain Barrel

This is a simple place to start collecting water. The barrel can be a plastic or metal food-grade drum of 35 to 55 gallons or, for better looks, a wine barrel. Most barrel systems store water for gardening and rarely include additions such as first-flush disposal, which makes them easy to build, and they usually can be installed without a permit. The chief concern for city authorities is that the overflow will not flood a basement or create some other hazard. Thus in designing the setup, you need to provide excess water with a dedicated escape route via a pipe or overflow into well-drained soil that is away from a house foundation or, less ideally, into a sewer. I know people who drink rainwater caught in food-grade barrels, but urban air usually throws so many contaminants into rain that it's not safe to do in cities. Most people use the water for irrigation.

The downside of the barrel system is that 30 to 50 gallons won't irrigate much garden. Fifty gallons, at the oft-recommended watering rate of 1 inch per week, is enough to irrigate 80 square feet—a 4-by-20-foot garden bed—just once. Careful spot-watering can stretch that out considerably, but it's still not much water. You'll need multiple barrels to store irrigation water for a larger garden or to tide you over a few dry weeks.

However, even a single barrel has its virtues. I have observed, as have many colleagues, that rainwater yields far better germination rates than tap water, so an excellent use for rainwater is for starting seeds and nurturing young seedlings. And a rain barrel is a potent symbol to remind you and anyone else who sees it that water falling free from the sky can be caught and used. I am constantly surprised how few Americans realize this. When we lived in rural Oregon, eight of the twelve households on our road had no well because of a cantankerous local geology that wouldn't hold water. This left a trail of expensive dry holes across the landscape. But only two of the eight well-less households caught rainwater. The other six trucked water weekly into a cistern, when for about the amount of money they spent on water in one or two years they could have built a superb water-harvesting system. Rainwater just isn't on the radar screen for most Americans.

### Pond

A pond can store water for irrigation or other nonpotable use. And yes, pond water can be used for drinking, but proper treatment can be expensive and complex and, especially in cities, should be backed up by regular, comprehensive testing, also expensive. Since drinking water makes up such a small fraction of household water use, making pond water drinkable is usually not worth doing compared to using other drinking sources.

Ponds are not only less expensive per gallon than tanks, but they are also much prettier and provide instant habitat and other gifts to nature and people. Hence I'll usually go for a pond before a tank unless I need really clean water. Their downsides are that they attract wildlife that is not always wanted, such as raccoons; they pose a risk of drowning to small children; and they require maintenance, such as occasional cleaning, leak repair, and readjustment of edges or the liner after disturbance. But the serenity they bring to a small yard is immense, and even a small pond can hold a good bit of water.

The key to good water storage in a pond is depth. Going deep rather than wide keeps the pond's footprint small and also cuts evaporation by lessening the surface area. A 12-by-12-foot pond that is 4 feet deep will hold over 4,000 gallons of water.

There are several other important steps to follow as you design a pond:

- Give the water a place to go. In heavy rain, ponds can overflow, so direct that water via a spillway and drain (such as a faux creekbed or a swale) to where it will be useful and harmless.
- Create a low-maintenance edge. The standard arrangement of flat rocks placed on the pond-liner lip on top of soft soil always fails. Someone stands on the edging rocks when the ground is moist, a rock squishes downward, water flows out of this new spillway, the soil gets wetter, more rocks sink, and so on. To avoid this scenario, you can install a rock or used-concrete foundation under the liner edge, or build a beachlike edge. Pond-building books and videos show details like these, as does *Gaia's Garden*.
- Aerate the water with a fountain, waterfall, pump, or some other method. Ponds with low oxygen levels build up anaerobic microbes that can smell or turn the water funky colors.
- Grow plants in the pond. Ponds should be built with a shelf to hold aquatic plants in pots. Pond plants will consume excess nutrients from runoff or rainwater, and they also help aerate the pond. Also, they attract beneficial insects, including uncommon ones such as dragonflies, and are soothing to look at.

How do you get pond water to the garden soil? One way is to direct the overflow to a swale that winds through plantings. This is more useful when the pond is fed by a graywater wetland rather than only by rain. Graywater will feed the swale every day, while overflow from rain only feeds the swale when plantings are already being watered by that same rain. I've also hand-bucketed my share of pond water, and it's not that much work in a compact garden. Another handy solution is to multipurpose an aeration or waterfall pump (preferably solar) by adding a diverter valve that sends the pumped water to a hose or length of poly pipe for irrigating.

### Tanks and Cisterns

Tanks and cisterns open the option of using rainwater inside the house and out, as the water can be kept very clean. Let's set aside an indoor system for now and start simply. A tank-based water system used only for irrigation needs the following:

- A location such that the top of the tank is below the gutter level and that lets the water drain by gravity to the garden sites. Gravity flow won't provide enough water pressure to run sprinklers and probably not even drip emitters unless your lot is steep. Water pressure builds at the rate of roughly ½ pound per square inch (psi) per foot of elevation, so to run a standard 10 psi emitter would take 20 feet of head, and few small yards have that. But even a foot of elevation will push enough water out of a hose or other tubing to soak soil.
- A way to keep debris out of the tank and outlet line. This can be as simple as a screen over the inlet and an outlet located 4 to 6 inches above the tank bottom to let any solids settle below it. I've had great results using an old panty hose leg tied over the inlet pipe as a filter, and my wife found this a fitting way to

mark her transition out of corporate life and into urban homesteading.

- An inlet that doesn't let light into the tank. Algae and mosquitoes need light to thrive; a dark tank discourages them.
- An overflow line directed to a safe place.

That's about as low-tech as a rainwater irrigation tank can get. Users might not be satisfied with the water pressure, so a pump is probably the first next step in complexity. Simply connecting a modest pump to the outlet will give plenty of pressure, though the pump will run continuously when water is flowing. To save energy, add a pressure tank. This is a small tank of 30 to 70 gallons filled with air that is pressurized by the pump, and this air drives the water. It acts as a pressure storage, and the pump runs only when the pressure tank needs topping off to keep the pressure within the useful range. But study up before doing this; proper sizing of pumps, tanks, and other parts is critical.

If your roof water is dirty or if you want to use rainwater indoors, you'll want a first-flush system to divert the first ⅛ inch or so of rain—which will contain most of the roof's debris—away from the tank. A first-flush diverter goes between the gutter and the tank. The best versions of these devices are passive, with no moving parts. Here's how one works: Water leaving the gutter flows into a tee. The far arm of the tee connects to the tank, but the diversion arm of the tee, which points downward, is connected to a large-diameter pipe or 10- to 20-gallon tank (see figure 6-5). The first, dirty flush of rain runs into this diversion tank and fills it. By the time the diverted water rises to the level of the tee, most of the debris has been washed off the roof and is in the tank. Water now flows across the tee and into the main tank.

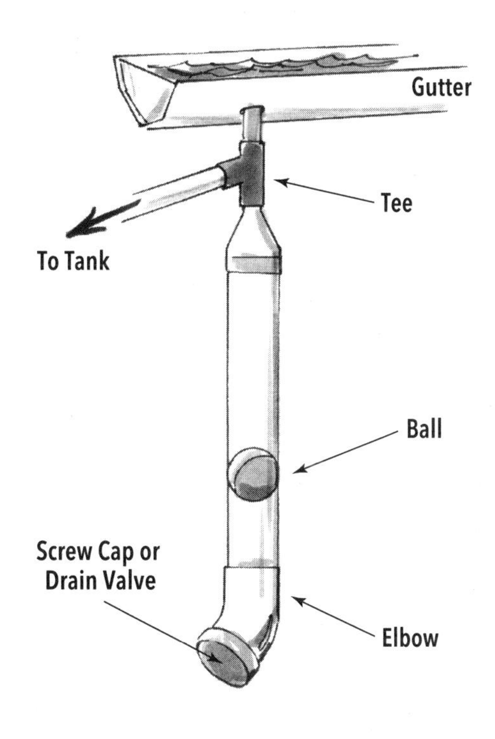


**FIGURE 6-5.** A first-flush rainwater diversion system. Illustration by Elara Tanguy

The small amount of turbulence right at the tee is not enough to suck debris or dirty water out of the diversion tank. Putting a small valve (or even drilling a ⅛- to ¼-inch hole) at the bottom of the diversion tank lets it empty between rains. If you drill a hole in this tank, be sure to check it often to make sure it doesn't clog.

Tanks come in a bountiful array of materials and shapes. Most common are cylindrical tanks made of polyethylene, but other plastics and galvanized metal are common. Plastic tanks cost on average about 50 cents per gallon of storage; metal costs more. Another option is a large bladder that inflates and deflates with use. A space-conserving design consists of tall, rectangular tanks that are wide but not deep—like a giant cereal box—that can be set against a wall or run along a property line to serve as a fence. These go by various names, such as slimline and waterwall. It's a nicely multifunctional product, but they cost at least twice as much conventional tanks. Do-it-yourselfers can build their own tanks from rammed earth (with a liner), ferro-cement, cinder blocks, or wood. Disguise these utilitarian objects with trellises or wire for climbing plants or with paint. (See photograph in the color insert, page 16.)

Although a tank with a pump and pressure tank can also supply household water, rainwater from this arrangement is not considered safe to drink without further treatment via filters, UV light, or reverse osmosis. An indoor rainwater supply needs to be kept separate from municipal water by shutoff valves and backflow preventers, so there are some complexities, but I'm always looking for ways to reduce the drain on large-scale infrastructure, and this is one of them.

## Designing Your Water-Harvesting System

I'll limit the scope of this section to catching rainwater for irrigation, not indoor use, since a system for the latter is much more complex. A rainwater catchment project beyond a water barrel or two can be a large and pricey undertaking, so it's important that it not be built too large, which wastes time, money, and materials, nor too small, which renders the whole project more of a toy than a water supply. Five factors interact to guide water catchment design toward the right size, location, and relationship among the parts:

1. How much water is used, when, and where
2. How much rain falls and when
3. The area of roof available
4. What size storage can be built
5. Where to place the storage relative to the roof

Let's explore these in detail and see how they affect one another.

### How Much Water Is Used, When, and Where

It can be difficult to calculate how much water is used for irrigation, but it is usually about half of the total yearly water consumption, more in dry climates, a bit less in places such as eastern and midwestern North America that get regular rain or don't have a pronounced dry season. Check your past water bills to get annual and monthly or quarterly use. Most water bills are tallied in hundreds of cubic feet, or CCFs, and there are 748 gallons in one CCF. You can make a reasonable guesstimate of outdoor use because most of us irrigate in the summer, while indoor use usually stays constant year-round. The difference between summer and winter water use can give you an idea of how much irrigation your landscape uses.

### How Much Rain Falls and What Its Pattern Is

If you live in a region where rain falls every few weeks throughout the year, you will need much less storage than in a dry-summer region, only enough to tide you over between garden-soaking rains. The National Weather Service provides monthly and annual rainfall averages for most US cities, although their stations are often at airports and may not match your own microclimate. A rain gauge (or a neighbor who's kept track of rain for years) is a more accurate source of information.

Here are two ways to estimate how much storage you will need:

1. Determine the longest typical dry period or, if you can't get that data, the amount of rain that falls during the driest month. Then determine the number of gallons of irrigation you would use during that time period. For example, if your average summer irrigation use is 600 gallons per week, and the longest typical dry spell is one month, then four to five weeks of storage, or 2,500 to 3,000 gallons, will carry you through most rainless bouts. It's up to you and your time and money budget to decide if you want to catch the full amount of rain, a bit more (which is always comforting), or less (and supplement it with tap water).
2. Check your water meter before and after a typical irrigation session (when no other water is being used), calculate the number of gallons used, estimate how many times you would irrigate like that between garden-soaking rains, and multiply those two numbers to arrive at water use between major rains. That number is probably close to the minimum storage you'll need to irrigate without using tap water.

### The Area of Roof or Other Catchment Available

Another factor that influences tank size is how much rain your roof can catch between dry spells (which are the times when you'll be using rainwater). This is dependent on the roof area and how much rain falls between droughts. If you use 3,000 gallons during a dry spell, and rainfall before that dry spell will put only 1,500 gallons onto your roof, there's not much point in installing a 3,000-gallon tank unless you can store rain that fell in a wetter or nonirrigating part of the year. (You can now see why designing a catchment system is as much art as science. There are complex interactions among the amount of rain, how often it falls, how much roof you have, how much irrigation you use, and your budget for building storage, and at some point you either fabricate a many-factorial spreadsheet or make a best guess.)

The amount of roof area available for catchment can depend on the pattern of gutters and downspouts, because water from some sections of roof may not be catchable without elaborate plumbing schemes. Only the usable parts of your roof should be included in calculating the potential harvest.

Pavement and other hard surfaces can also be called into service, even if that water won't go into a tank. Some folks toss a sandbag into the street gutter at the downslope lip of their driveway when it rains, diverting an impressive river of water up the driveway if the slope is right. This water can run into a swale, or down a ditch or pipe to fill a pond or irrigate plants. It's best to wait to divert the gutter until the street has been washed of surface petroleum by a ¼ inch or so of rain.

### What Size Storage Can Be Built

In areas with long, dry summers, it may not be possible to fit (or afford) enough storage to do all the irrigation needed. Here, budget, space, and aesthetics are all factors. Tanks are more expensive than ponds but take up less space. Ponds are much nicer to look at than tanks, but tanks can be hidden underground or even built in the cellar.

### Where to Place Storage Relative to Roofs or Other Catchment

If the storage can be located higher than the garden, then gravity, rather than a pump, could power the irrigation system. Aesthetics and ease of construction will also play a role in placement decisions. Most important, the top of the stored water needs to be lower than the bottom of the lowest gutter supplying it.

Note that all of these factors except the amount of rainfall are controllable. This gives the gardener a lot of leverage. The four other factors can be tinkered with to design the best system for your region and needs.

## ➡ GETTING WATER ⬅ TO THE GARDEN

In keeping with one of the major themes of this book—that permaculture helps us select, from a plethora of possibilities, the best technique for each situation—I'll show how permaculture's zone system can guide the choice of what method to use to water the garden. At first this may seem silly—um, don't we just water our plants?—but watering requires labor, water, and equipment, and how it's done can save or waste plenty of each.

If we consider criteria such as reliance on municipal infrastructure, total home labor involved, project scale, and so forth, we can, once again, rank potential methods from lowest to highest work involved. Obviously, the easiest way to water your yard is to turn on the spigot, but we're trying to avoid dependence on the Roman aqueduct scale of most city water systems. Given that, possible methods to deliver water are as follows.

### Bucketing from a Pond, Tub, or Sink

A bucket or watering can to transfer stored water works well in a small yard or garden or in one that holds a high proportion of natives or other low-irrigation species compared to the more thirsty varieties. A practical limit to size here is 100 to 300 square feet of plants needing regular watering, ideally clustered, and located close to the source—a sink, tub, or pond, which we can think of as zone 1 for this task (remember that zones can move based on where the user or center of activity is). For very small yards, this may be the only watering method needed, and it relieves the strain on city water systems while avoiding an investment in tanks and plumbing. If the bucket-water source runs low, supplementary squirts from the hose will be so infrequent that the permaculture police will probably look the other way. And not only does this method require only a few minutes per watering, but it follows the old adage that the gardener's shadow is the best fertilizer, because the regular attention enforced by hand-watering attunes us to each plant's health.

### Downspouts Directed to a Garden Bed or Tree Basin

If the yard's slope permits, water from downspouts can be used, via a shallow trench or a pipe, to soak soil or mulch. This system has two components:

1. A way to move water from the downspout to where it is needed in the soil. Home-supply stores carry a variety of connectors that attach a downspout to 4-inch corrugated drainpipe or 3-inch and 2-inch plastic pipe. This pipe is laid on or in the ground to carry rainwater to its destination. The water can also travel overland. If the soil has a high clay content, a shallow drain trench will deliver the water without much of it soaking into the ground on the way, but in

well-drained soil the drain trench will need a liner or else water loss will be substantial.

2. The destination where water soaks in. This can be a slightly sunken area of lawn, a mulched basin around a tree, or an on-contour swale or level length of perforated pipe (with perforations facing down, contacting the soil) that waters a garden bed.

Each downspout can water a different zone of plants, which means that a fairly large area of garden can be irrigated this way. Also, if neighbors are not using their roof water, especially if a portion of their roof is close to your yard, they may be agreeable to letting you use it. Their roof may help irrigate distant or upslope parts of the yard.

Obviously, these systems water the garden only when it rains, so the soil or mulch at the destination needs to be able to hold the surplus water and soak it in for storage. But in rich soil or deep mulch, the extra water will reduce the need for watering for days or even weeks after the rain stops.

These passive systems save labor; just remember to check the plants regularly between rains to make sure they're getting enough water.

## Overflow from a Pond or Graywater Source

Using a similar delivery and soak-in system as the setup for downspouts, gravity-flow graywater or the overflow from ponds fed by graywater can run to a mulched basin or into a garden bed via a mulched swale or perforated pipe adjacent to plantings. Remember that graywater should not be open to the air, hence the mulched rather than open swale. Small trickles of graywater won't travel very far in a mulched swale, so a level length of perforated pipe with the holes facing down may be a better choice for getting the graywater into the soil over a large area. Be sure to size the system to be able to deal with the largest expected volumes of graywater to prevent backup or overflow. Depending on how much graywater is generated, this system can have multiple destinations for the graywater, selected by a diverter valve or by blocking alternative swales or pipes to direct water flow where it is wanted. In Tucson water guru Brad Lancaster's washing machine has several different drainpipes next to it, each labeled according to which fruit tree it runs to. He shifts the washer's outlet hose to a different drain (and tree) with each load of laundry.

## Cistern or Pond to Passive Watering Systems

Adding a cistern or storage pond increases the complexity and cost of a passive watering system but lets you water during dry spells. The critical variable here again is where the storage is located relative to the beds to be watered: The top of the stored water needs to be uphill from the destination, or you'll need a pump. Water storage also means that instead of being limited to watering only where swales and pipes have been built, you can use a hose. The gravity flow from the hose will be slow but usable, though it usually won't be enough to power a drip irrigation system.

## Active, Pumped Watering from Pond or Cistern

Adding a pump increases the cost and work substantially, but the benefit is a big one: water pressure. This allows long hose lengths, sprinklers, and drip irrigation systems. A drip system can save enormous amounts of water, 50 percent or more compared to overhead sprinklers, and it offers the option of using the house water supply when the home storage runs low. Pumps, pressure tanks, and the other components of these systems are best installed by someone with substantial plumbing and wiring experience.

## ➡ WATER SECURITY ⬅ IN A NUTSHELL

To sum up: Reducing municipal water use is a powerful leverage point. Most water bills get you coming and going, literally, in that you are charged for how much water comes into your home, and in many cases that number is also the basis for a sewage charge. So shrinking incoming water use drops the sewer charge as well. And, municipal water supplies are often aging, overtaxed, treated with chemicals such as chlorine and fluoride, and increasingly vulnerable to disruption. The least expensive and most potent place to reduce water use is in irrigation, where methods abound for holding water in the soil and elsewhere, reusing graywater, and reducing your yard's need for water via plant choice, placement, and related strategies. Indoors the biggest water users are the toilet and laundry, so finding water-conserving options for these is another strong leverage point. Addressing just those few systems can reduce your city water use by half or more.

With all these options for saving, harvesting, and conservatively using water, we should be able to reduce the burden on resource-intensive public water supplies and sewers, have more control over the quality of our water and the impact of its use, and be more water secure.

# CHAPTER 7

# Energy Solutions for Homes and Communities

In one year the average American uses an amount of power equal to the work of 100 people doing hard labor round the clock, seven days a week, for that entire year.[1] Energy experts call this small army of phantom workers "energy slaves," and it's easy to forgot how much they do for us. Access to this vast pool of power has made Americans rich beyond the dreams of the most profligate royalty of a few hundred years ago. It's also brought us face-to-face with two linked challenges: the end of the era of cheap fossil fuels and the human-caused climate disruption that using these fuels has spawned. (I know that anthropogenic climate change is still disputed in a few quarters, but I'm going to side with over 97 percent of the world's climate scientists, every major scientific organization, and most oil companies in saying that it is real. And the reality of climate change doesn't change the wisdom of using energy wisely, which is this chapter's focus.)

Energy is a slippery concept to define, but one way to think of it is the ability to do work. One key to humanity's becoming a dominant force on this planet has been our ingenuity at lowering the cost of getting work done—the cost of energy—steadily for the last 10,000 years. Look, for example, at how the effort involved of moving people and goods has steadily shrunk. In 1630 the first of my ancestors to emigrate to America paid for his crossing from England by indenturing himself as a laborer for four years. Four years' wages is a steep fare for an ocean cruise, but it was in line with the resources it took then to make the voyage. Today, less than a week's work at the median American wage buys that same journey, and it's done in a soft airliner seat instead of a damp, rat-filled boat, taking hours rather than weeks.

The consequences of this trend—the shrinking human-labor cost of obtaining useful work—have been rippling across earth and society since the dawn of agriculture. About three centuries ago our ability to grab more power for less effort, and with it the human impact on the rest of the planet, boomed with the discovery that the energy in coal, and later oil, could be substituted for human labor. Productivity soared, and when economists talk about increasing productivity, what they nearly always mean is finding more ways to get a machine to do work previously done by a person. Until the recent skyrocketing of fuel costs, which made efficiency increases more cost-effective than boosting brute power, this was done by trading the energy stored in fuel of one kind or another for the work of human muscles.

## THE MIRACULOUS POWER OF FOSSIL FUELS

Fossil fuels store an astounding amount of energy. One gallon of gasoline can release 33,700 watt-hours of energy. Let me put that sterile number into perspective. A person working vigorously at a sustainable pace, say, mowing lawns or framing a house, can put out about 250 watts per hour. But look what she could trade for that physical work: If she takes home $20 per hour, one hour of work will buy 5 gallons of gasoline at a price of $4 per gallon. Thus she is trading 250 watts of her energy for 168,500 watts of fuel energy, a phenomenal bargain. Translated into typical workweeks, 1 gallon of gasoline yields the same amount of energy as three months of one person's physical labor.[2] Oil geologist Euan Mearns has calculated that the work available in one barrel of oil, as human labor at $20 per hour, is worth $933,000.[3] Compare that to oil's recent price of between $50 to $100 per barrel. No wonder we live better than kings did before the oil age.

Liquid fuel is also energy dense and gets energy to us quickly. To generate the energy in a gallon of gasoline, a standard 120-watt solar electric panel needs to run for about sixty days in average US sunshine (five hours per day). In fossil energy we have been given an incredible bargain, even at today's prices.

That bargain has been the key factor in human prosperity for centuries. I know that folks on the green side of the political spectrum are appalled at humanity's insatiable appetite for energy and resources and the problems that has caused. However, humans aren't unique in this trait. All organisms tend to gobble up available resources as fast as they can until they hit some limit. That's part of the instruction set for life: Find and exploit the resources that help you survive and breed, and do it as fast as possible. Humans have done that, too, and now we're hitting the limits. So we don't need to feel guilty about having done what all creatures do. But we do need to be conscious of the consequences of having outpaced our resources and do something about it, because the signs of passing that peak tend to appear in the form of starvation, plague, mass killing, population die-off, and sometimes extinction.

One major effect of our centuries-long resource binge has been explosive population growth. This, too, is perfectly typical of an organism that's found a lush energy source. A rich resource base triggers biological signals to breed. With this in mind, it's worth briefly examining the story of how our own species discovered the trick of turning energy into people. This can set the stage for a permaculture approach to thinking about and using energy.

The human population took many thousands of years, stretching from dimmest prehistory to 1800, to slowly rise to just under one billion, most of it living rurally. Over the next century, from 1800 to 1900, we added 600 million more people, and the move toward cities accelerated.

Then came the real explosion. From 1900 to 2000, population increased by 5 billion, about an eightfold rise in the growth rate over the previous century, and cities soared in size.[4] What had happened?

Mechanization did some of it by shifting the work of food growing off people and onto machines and fuel and by creating the factories in cities that drew the now-unemployed rural dwellers into new jobs in town. But the big push came from fertilizers and what we did with them.

By the late 1800s, the productivity of much of the world's farmland had begun to stall. The last great prairies in North America, Russia, and much of South America had been scalped by the plow and seeded with grain, leaving little room for expansion of cropland. The cubic miles of guano mined from a cluster of islands off Chile, which had fertilized the farms of Europe and Asia for decades, were running out. Other natural sources of fertilizer were being quickly depleted. In 1898 the president of the British Royal Academy of Sciences, Sir William Crookes, gave a speech predicting worldwide famine within a few decades, and his careful enumeration of undeniable facts and figures left his audience stunned. But as has happened before and since, the doomsayers weren't considering human ingenuity at designing ever more effective resource pumps to convert planetary wealth into people.

In 1909 the German chemist Fritz Haber invented a way to synthesize ammonia-based fertilizers by combining hydrogen with gaseous nitrogen. The latter, because it makes up 78 percent of the atmosphere, is fantastically abundant but not in a form usable as plant food. Haber's invention let humanity tap into a vast new resource pool. An engineer, Carl Bosch, scaled up the process, eventually building a single ammonia-making machine that sprawled over 3 square miles. The Haber-Bosch process is terrifically energy- and fuel-intensive because the prodigious quantities of hydrogen needed are stripped from natural gas, and the reaction mix must be compressed and baked four times through 3,000 psi and almost 1,000°F. Today almost 2 percent of humanity's energy use goes to powering the Haber-Bosch and related processes, as ammonia-making has become one of the single largest consumers of earth's fuel.

What Haber and Bosch did was to rip away nature's cap on fertilizer production and thus on food growing. For every organism, an abundant food source is the trigger for biological shifts that stimulate breeding. And so our numbers exploded. The Haber-Bosch process has boosted the human population by at least one-third. Eighty percent of the nitrogen in the typical human body now comes from artificial ammonia.[5]

But by the 1960s, new predictions of global famine appeared, most famously by Paul Ehrlich in his 1968 book, *The Population Bomb*.[6] Ehrlich crunched the numbers much the way Sir William Crookes had and predicted that by the late 1970s hundreds of millions of people would be dying of starvation. Ehrlich was wrong. Again, cheap energy bought us more time, here in the form of the green revolution, which added about 2 billion more mouths. This feat had three key steps.

The first was the breeding of new seed varieties that responded to enormous doses of fertilizer. In traditional crops, adding more fertilizer yields more food, but only up to a point. After enough fertilizer is added to roughly double yields, pouring on more fertilizer doesn't hike yields further; the plant's metabolism is maxed out. Green revolution scientists bred new crop varieties containing what have been called high fertilizer-response genes. These new strains continue to ramp up when given fertilizer doses much larger than older varieties responded to. With green revolution crops, the point of diminishing returns on fertilizer was shoved much higher.

That meant we needed a lot more fertilizer. So the second component of the green revolution was a Haber-Bosch building boom and vastly more fuel use for fertilizer, along with refinement and greater efficiency of the process. Fertilizer factories multiplied around the globe. Cheap energy made it easy to expand fertilizer production to feed the nutrient-hungry new strains.

The third factor was the addition of potent pesticides. The new crop strains were particularly susceptible to insects and disease. One of several reasons for this is that the mountains of nitrogen fertilizer applied made the plants nitrogen rich. Insects crave nitrogen because bugs are mostly protein, which has a high nitrogen content. With natural selection relentlessly humming along, insects have evolved sensitive mechanisms for sensing nitrogen, just as we can smell our favorite foods. Bugs zeroed in on the nitrogen-pumped crops of the green revolution. But we had an easy solution: Spray the crops with multiple passes of insecticide. And herbicides could kill the giant weeds stimulated by all that fertilizer. Because many pesticides are made from petroleum, these, too, were cheap because of low oil prices. World pesticide use has exploded to fifty times the 1950 levels, to 5.2 billion pounds applied in 2007 to crops and soil.[7]

The staggering increase in agriculture's use of cheap petroleum and natural gas has rewarded us with vast increases in food production, and this is what drove the past few decades of population growth and expansion of cities. World grain yields shot up from about 600 million metric tons in 1950 to 2,000 million metric tons in 2012. Grain production tracks crude-oil production nearly perfectly over those years, as does population. Grain yields have been roughly flat since 2008, in part because of a leveling off of conventional petroleum production and in part because in many nations farmers are abandoning green revolution methods because of the rising price of fertilizers and pesticides, salting of soils by excess fertilizer, and the loss of the large subsidies that made these resource-intense methods affordable in the less developed world.

The point of this historical overview is that cheap energy launched humanity into a population and consumption overshoot. To keep up current levels of both, we'll need to continue finding cheap energy. Imagine if 250 watts of your work bought only a fraction of the 168,500 watts that it buys today. We'd lose whole platoons of our phantom energy slaves, which would mean paying a far higher price for getting useful work done of any kind: raising food, making goods, getting around, keeping the lights on, staying warm—all of it. Besides, even if we develop new cheap energy sources, greenhouse gases and other toxic by-products build up in parallel with growth in resource use, and we then must deal with those. This all suggests—or rather shouts—that we need new approaches to energy use.

By now you won't be surprised to learn that I think permaculture design can guide us toward those needed new approaches. As in other chapters, let's first explore some thinking tools, and then we'll look at specific strategies and techniques for using energy.

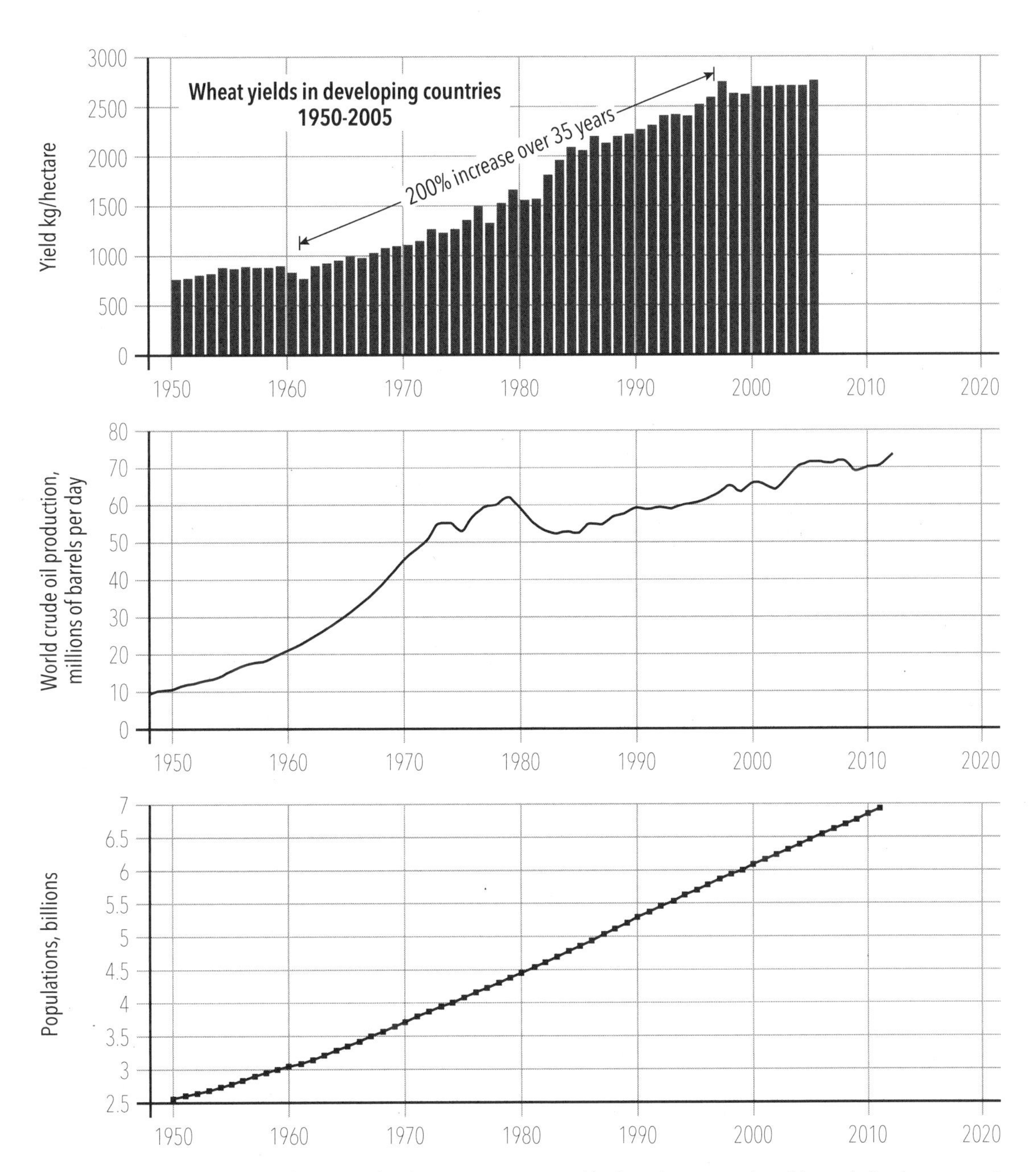


**FIGURE 7-1.** World wheat production in developing countries, world oil production, and world population have moved almost in lockstep since 1950. What happens to food and population when fossil-fuel production begins to drop? *Sources:* Wheat: R. A.Fischer, Derek Byerlee, and G. O. Edmeades, "Can Technology Deliver on the Yield Challenge to 2050?" Expert Meeting on How to Feed the World in 2050 (Food and Agriculture Organization of the United Nations): 12, ftp://ftp.fao.org/docrep/fao/012/ak977e/ak977e00.pdf; Oil: US Energy Information Administration, http://www.eia.gov/totalenergy/data/annual/pdf/sec5_8.pdf; Population: US Census Bureau International Database, http://www.census.gov/population/international/data/worldpop/table_population.php.

## ➡ FIVE TOOLS FOR ⬅ THINKING ABOUT ENERGY

In addition to the usual permaculture principles and design methods, I use five thinking tools for making decisions about using energy. These are:

**Efficiency**, or how much energy is lost during the processes of making or using a product or process

**Emergy,** embodied energy, or the total amount of energy used to build and deliver a product or service

**Life-cycle assessment,** or how much energy is consumed by a device over the entire course of its use and disposal or recycling

**Transformity,** a measure of the quality of energy used to do or make something

**Energy return on (energy) investment (EROI or EROEI),** the ratio of energy out to energy in: how much usable energy was produced by an energy source compared to the amount of energy used in acquiring that amount of the energy source

Let's look at these and at how each is useful in designing and using energy devices and solutions.

### Efficiency

One of the sad truths of energy use is that nothing can be done with perfect efficiency. Losses always manage to sneak in. Scientists in the energy-study field of thermodynamics have teased from nature a few inviolable rules for how this is so, and two of these laws are relevant to permaculture design. The *first law of thermodynamics* says that energy can be neither created nor destroyed, merely converted from one form to others. The *second law of thermodynamics* tells us that when energy is used, or converted from one form to another, its ability to do useful work always declines. Another way to say this is that energy can't be turned into work with 100 percent efficiency because there are always losses. An example is burning gasoline to power a car. The car takes you where you want to go—you get useful work from the fuel's energy—but in the process those watt-packed, complex molecules of gasoline are burnt up and degraded to carbon dioxide, water, smoky pollutants, and heat, all of which are less capable of doing work than the gasoline from which they came. You can't run a car by pumping engine exhaust back into the tank. (The first and second laws have been ironically stated, respectively, as "you can't win" and "you can't break even, either.")

These laws help us think about using energy wisely because they reveal that there are two kinds of efficiency, named after the laws that describe them: first-law and second-law efficiency. *First-law efficiency* tells us how much work is done when one form of energy is converted to another. For example, converting the stored energy in gasoline to the dynamic work of spinning the chain of a chain saw has an overall efficiency of about 30 percent, typical for internal combustion engines. The other 70 percent of the energy in the fuel is dissipated as heat, friction, and exhaust. So from a first-law point of view, this chain saw is 30 percent efficient.

But that just tells us how efficient the saw is at turning stored fuel energy into the work of spinning its chain. Chain spinning is not our objective when we fire up a saw. We want to cut stuff. That first-law-efficiency number doesn't tell us how much of the fuel's energy truly goes into the useful work of cutting something with the saw. For that, we need to know the *second-law efficiency*. That's defined as the ratio of the minimum amount of energy needed to do a task to the amount of energy actually used.

In other words, what is the efficiency of the tool we're using compared to that of the ideal, most efficient tool for the job? Environmental scientist Amory Lovins describes second-law efficiency by talking about using a chain saw to cut butter.[8] A chain saw is about 30 percent efficient in its conversion of fuel to cutting power. But in Lovins's example, nearly all of the fuel's energy is spent whirling the chain and splattering butter around the room. Only a tiny fraction goes to cleaving the butter. The second-law efficiency of chainsawing butter is far less than 1 percent. As should be obvious without resorting to thermodynamics, a knife is a better choice for that job. The same work done in the actual cutting of butter is achieved with either tool: the energy used by your arm gently pressing on a butter knife or by the full-body workout of swinging a chain saw plus the energy in the burning fuel. Although it's obvious that chainsawing butter is stupid, many cases of unwise energy use are more subtle. Second-law efficiency guides us in matching tools to the work we want done. This difference between types of efficiency is important to grasp in order to guide us toward wisely using energy, because second-law efficiency doesn't stop at fuel conversion; it tells us how much energy we are delivering to the real task at hand. I'll give some examples to make it clearer.

Here's a typical case of how we could be saving energy by thinking about second-law efficiency. About 30 percent of North America's energy use is for low-temperature heating, mostly for keeping buildings at a comfy room temperature. But much of that heating is done by furnaces blazing inside at 1,000° to 1,800°F. How efficient is it, really, to use an 1,800° flame to warm a roomful of air by a handful of degrees? Even if your furnace is a model rated at 95 percent efficiency, that's first-law efficiency, calculated as the ratio of how much of the fuel's energy becomes heat inside the burner versus what goes up the chimney. But this heat then needs to warm the house's air. The thousand-degree difference between the burner and the small air-temperature change needed is monstrously inefficient in second-law terms.[9] That fiery furnace is, in this case, the screaming chain saw compared to the butter-knife suitability of, say, the sun's warmth or a heat pump or a geothermal system of buried pipe, which all use the slight difference between cooler air and a mildly warmer heat source to warm air by a few degrees. Those other methods have much better second-law efficiency. This not only saves energy but preserves very high-quality, nonrenewable energy sources such as fossil fuels for the jobs they can do more efficiently or conveniently than any other energy source.

In countless cases, we've made a drastic mismatch between the high quality of a fuel or energy type and the low quality of work done at the consumer end. In many cases all we want is heat, and from a thermodynamics point of view, heat is low-quality, quite disorganized energy compared to the refined, complex structure and processes of gasoline or the electric grid. The result is needless waste. A classic example of this is electric heating. We go to a lot of work to refine a fossil or nuclear fuel into a pure, highly organized form, then burn it at over 1,000°F to vaporize steam to spin a turbine to whirl a magnet in an armature to generate electricity. That's a complex process. Electricity is a supremely high-quality type of energy that can power delicate electronics or LED lights and do other tasks that only electricity can do. But instead, these precious amperes travel many miles at huge transmission losses until the few remaining electrons are finally pumped into a resistant coil of wire to jostle each other until they give off low-quality, low-temperature heat—heat that could have been created by any number of simple fuels or heat sources. That heat is then radiated into the air, probably never to do useful work again. This violates just about every permaculture principle there is. These basic rules about energy use tell us that electric heat for general home use should be avoided in all but the most specialized or desperate cases, when nothing else is available.

The take-home message on efficiency is that although the efficiency of a device or process in converting fuel to power is important, we also need to pay attention to how much of that power actually does the job that we are using it for and whether some other energy source would be a smarter choice. So let's look at how we can assess how much energy we need to do the work we want done.

## Emergy

Emergy is the total energy used—all the work done and fuel spent—to create an item or a process. The word comes from the term *embodied* or *embedded energy*. Because it's useful to compare apples to apples by using a common unit and because the energy that drives most of the processes of life and makes nearly all our goods can be traced back to the power of the sun, emergy is expressed as the amount of solar energy it took to do something. For example, it takes about 67,000 units of solar energy to make one unit of coal energy.[10] (The official unit of emergy is the solar emjoule, but defining that takes us into technicalities that we don't need to bother with here. We can get away with just thinking of an emjoule as a small amount of solar energy.) The emergy in coal climbs that high because there are energy losses at each step in the winding construction path from sun to carbonaceous lumps of fuel. The first energy conversion step, photosynthesis, is only 3 to 6 percent efficient at converting sunlight to biomass, and other cycles of metabolism that assemble plants and algae aren't much better.[11] Then more losses occur as that ancient greenery is squeezed and fossilized into coal over many millions of years, and we also need to add the energy to mine and process the coal, nearly all of which comes from the stored solar energy in fossil fuels.

Not all scientists in the relevant fields are happy with the concept of emergy. Calculating the various contributions of energy from many sources and over vast distances of time is bound to involve uncertainty, possible fudging, and assumptions. However, the concept is an important and useful one for becoming energy literate, because however we calculate it, making different goods and services uses differing amounts of energy, and almost always that energy originates with the sun. Converting back to solar-energy equivalents gives us a common currency to evaluate energy and resource choices with. Getting a sense of which products are serious energy hogs will help us make smart choices.

Emergy is not the only factor to keep in mind when choosing materials and methods. Some substances have low emergy but deplete these initial savings by wasting energy during use, and that fact carries us to the next thinking tool.

## Life-Cycle Assessment

Life-cycle assessment (LCA, also known as cradle--to-grave analysis, ecobalance, and life-cycle analysis) gauges the environmental effects of a product over its entire existence. It measures the energy used and the pollution left in a product's wake during the processes of extracting its raw materials, manufacturing and distributing it, powering it over its lifetime, and recycling or landfilling it. Just as with emergy, there are many ways to calculate life-cycle assessment and plenty of places for biases and fudging to creep in, but proponents of the concept are hard at work to create uniform standards. Besides helping compare alternative products, one of the strengths of life-cycle assessment is in spotting wasteful or high-impact stages in a product's life. Sometimes the high-impact phases are not the ones we assume they are, and that helps avoid wasting conservation work.

For example, recently the concept of food miles has been getting a lot of press. This is the idea that we're wasting a lot of energy by shipping food long distances from farm to table, on average about 1,500 miles in the United States. Food miles is a charismatic concept that grabs attention: Shipping food long distances has got to be bad for the environment—we just know it! To be sure, there are excellent reasons to go local on food. But life-cycle assessment revealed that only 4 percent of the energy burned in food's farm-to-table path is used in transport. Most of it, 84 percent, is consumed in growing and processing.[12] It seemed like shipping just *had* to be a major chunk of food's energy footprint, but it's not. A campaign to shrink food miles wouldn't have nearly the impact on emissions and energy use that focusing on farm and processing conservation would. In this way, life-cycle assessment helps spot powerful leverage points, both in the segments of the energy trail of an individual product's life and also when we compare different products or methods for getting a job done. A database of life-cycle assessments for many products resides at the US National Renewable Energy Laboratory website, www.nrel.gov/lci/assessments.html.

## Transformity

Transformity is a measure of energy quality based on how many steps were needed to convert a diffuse and indirect source of energy, such as sunlight, into a concentrated or more flexible form, such as gasoline; that is, it's a way of putting numbers to the idea that getting energy back from some sources is easier for us than from others.

I'd like to explain transformity in a bit of detail because it's a tool that makes several energy mysteries clear. Let's compare the ease of getting energy from coal to getting it from the sun. We saw above that fossil fuels are a very dense form of energy, while sunlight is diffuse. That's one important aspect of energy quality. If I want to boil water with coal, I just set a lump of coal on the ground, light it, and hold a pan over it. To boil water with sunlight, I first need to build some kind of collector to concentrate the sunlight. That means making and gathering the various manufactured parts I need for, say, a mirrored parabolic reflector, then building the gadget over the course of an afternoon or so. That's a lot of work. In most cases it takes more technology and conversion steps to extract useful energy from sunlight than from coal; that's one reason that photovoltaically generated electricity is more expensive than that from coal-fired plants. And, the sun doesn't always shine, whereas coal will always burn. Those are all informal ways of assessing energy quality. Transformity is an attempt to standardize and measure that.

The official definition of transformity is the amount of emergy (embodied solar energy) of one type needed to make a unit of energy of another type. It's a ratio: how much of the sun's energy was used in all the processes that built a product or delivered a service compared to the amount of useful energy that was actually delivered or contained in that product or service. Transformity, put simply, is the ratio of total solar energy inputs to whatever energy comes out. It's worth getting a handle on this. It is one of the most powerful concepts in choosing what goods or services to use, which ones may be more reliable, and also which ones are liable to stick around during times of scarce energy or other crises. Another example or two will help make this clearer.

Meat, for example, is a high-quality food when we consider how many calories per pound it has, how easy it is to eat, and how complex its structure is. To build that richness takes a lot of solar energy. How much? We can run through the food chain and see. For the sake of keeping things simple, let's say that a person lives exclusively on meat and only one form: trout. How much solar energy does it take to grow a year's worth of trout for someone?

If a person were living solely on trout, he or she would need roughly 300 of them per year. In one year, those 300 trout need to eat about 90,000 frogs, or 250 per day. In turn, those frogs would be eating 27 million grasshoppers each year (74,000 per day), and over that year, that plague of grasshoppers would chomp down 1,000 tons of grass, or 2 million pounds.[13]

In this scenario, one person consumes the solar energy needed to make 1,000 tons of grass per year, because that is how much solar energy is used at the bottom of the food chain to grow enough grass to feed the whole pyramid of beings up to the human. If you could eat and digest grass directly (you can't, but let's pretend you can), avoiding all those losses in conversion to higher-quality forms, you would probably need only a few thousand pounds per year, not 2 million. The losses approach 99.8 percent.

What the concept of transformity is measuring and making visible are those conversion losses along the trail from sun to grass to bug to frog to fish to person, as well as the up-converting of diffuse light to cellulose, to protein and fat, and then to human brain- and muscle power. (Though it's not my point here, transformity also illustrates why eating lower on the food chain is usually ecologically smart: The solar energy, and thus the land, needed to grow a meal of grain for a human will be less than that needed for growing the grain to feed a steer for a meal of meat.)

Transformity helps make visible the losses that occur as thinly spread solar energy is upgraded to more useful energy sources. We can do that by calculating how many units of sunlight are needed to make one energy unit of the final product. For example, when sunlight is refined via the green biochemistry of plants into leafy tissue, roughly 6,500 joules of solar energy are needed to build one joule of leaf, so the transformity of leafy plant production is 6,500 (the units here are solar emjoules/Joule, or seJ/J). The harder work of building tough wood from sun results in a transformity of 35,000 seJ/J.[14] The transformity of fossilizing plants into coal is about 67,000 seJ/J. (Note that while emergy and transformity are similar, emergy is a total amount, such as the total emergy of making a washing machine, while transformity is a ratio, measured as the amount of sun energy needed per unit of energy available from the source being analyzed.)

One thing that the developers of the transformity concept are trying to show is that not all forms of energy are the same. For example, among its many forms, energy can show up as sunlight, gasoline, or electricity. Let's say we have each one of those ready at hand. If we're trying to warm up some water by 5°F or even 50°F, sunlight could be the least complicated energy source to use. We would just need to buy or build a pan or some tubing, fill it with water, and set it out in the sun. Gasoline and electricity would take more equipment: a pan or tubing as above, but also a burner of some kind or a heating element to warm that pan or tubing, with all the embedded energy that those entail.

If instead we want to spin a drill bit in a rotatable chuck, sunlight isn't very useful until it's been converted via a photovoltaic (PV) panel into electricity, which then needs to be wired into a motor (which we must buy or build) that we can mount to the drill chuck. PV panels are a complex, hard-to-build technology with a transformity of about 170,000 seJ/J. Alternatively, we could focus sunlight via a curved mirror into a tight beam, build a boiler to cook up some steam, and make a turbine to attach to the drill chuck. The point is, we have to build a lot of stuff to convert sunlight into rotary motion.

Gasoline is more convenient than solar for spinning a drill. We need to build an engine or turbine as with the solar steam system and use its rotation, or perhaps make a gasoline generator, but those are simpler technologies than a PV panel—we came up with engines and turbines centuries before we had the technology for photovoltaics. And if we are using electricity, we need only a motor, without the PV collector needed for solar. The point is all energy sources are not equal; all need conversion steps to do useful work, and depending on the type of work—heating, rotating, computing—some sources need fewer conversions and ancillary technology than others.

So we can see that for every energy source, there is a lengthy trail running *from* that source: all the conversions and equipment needed to make that energy source do useful work. And for every energy source except sunlight, which arrives at earth ready to use, a long energy trail leads *to* that source as well. For petroleum, the trail begins with sunlight and photosynthesis by plants, runs through a few hundred million years of compression and decomposition, followed by mining and refining, before the trail pauses at the fuel itself—which then needs conversion to be able to do work for us. For electricity, the trail leading to its creation is even longer. We need to bear these trails in mind in order to make smart choices about energy.

Transformity is a way of measuring those trails. It not only tells us that energy quality varies from source to source but also reminds us that the energy from gasoline and electricity each used quite a bit of solar energy to become so useful, just the way those 300 trout did. And in most

**TABLE 7-1.** Transformity of Common Resources

| RESOURCE | TRANSFORMITY (SEJ/J) |
|---|---|
| Nonwoody plant biomass | 6,500 |
| Woody tissue | 35,000 |
| Coal | 67,000 |
| Natural gas | 80,600 |
| Ethanol from corn | 183,000 |
| Electricity from coal | 208,000 |
| Ammonia fertilizer | 1,860,000 |
| Educated labor | 8,900,000 |
| Cement | $1.9 \times 10^9$ |
| Gold | $2.5 \times 10^{12}$ |

*Sources:* H. T. Odum, *Environmental Accounting: Emergy and Environmental Decision Making* (New York: John Wiley and Sons, 1996); www.emergydatabase.org/transformities-view/all.

cases the higher the transformity of a fuel, the more energy dense and versatile it is. A steady breeze or a sunny sky can provide energy, but a gallon of gasoline is a lot easier to carry around.

How does this help us make energy decisions? Think of transformity as the inverse of efficiency. The higher a thing's transformity, the more energy has been lost during its making. So it makes sense to choose products and processes with low transformities. Energy researchers have assembled tables of the transformities of common goods and services.[15]

Just as with emergy, calculating transformity takes some assumptions and arbitrary decisions—how do you measure the amount of energy in cement or a financial transaction?—and not all energy scientists are on board. But the methods have become standardized so that we can at least compare the relative, if not exact, magnitudes of transformity among items.

Another reason to understand transformity is that it may help predict which human activities are more likely to get very expensive or simply impossible to retain as resource costs continue to rise. Transformity is a useful tool for those of us concerned about how well a complex civilization can be kept running in an era of declining returns on energy resources.

And that—the energy we get back from the energy we invest—takes us to our final energy thinking-tool.

## Energy Return on (Energy) Investment (EROI)

At its core, EROI is another way of saying, "There's no such thing as a free lunch," or "It takes energy to make energy." EROI is the ratio of the energy produced by a resource to how much energy was used to make that resource. In other words, if producing 100 barrels of oil—the drilling, pumping, refining, transporting, and so forth—requires the energy of one barrel of oil, then the EROI for that batch of oil is 100/1, or 100. To be useful, any energy source needs to have an EROI greater than one, or all the energy in it will be consumed in getting it, which is pointless.

For much of the industrial era, the EROI for oil has been 100 or more. Once again, we see what a sweet deal we've had with fossil fuels. Imagine being able to drop a dollar into a slot machine and get $100 back every time. You'd feel awfully rich and would probably pump a lot of dollar bills into that one-armed bandit and go on a spending spree for as long as the magic lasted. That's essentially what we've been doing with oil for the last century or more. Over millions of years, the sun and the earth's greenery had built up a treasure trove of stored energy, and we humans stumbled upon it and got to use it, gratis. We didn't have to make it or pay for it—our only task was to haul this massive inheritance out of the ground and start burning it. Free energy. Lucky us!

The oil shocks of the 1970s and 1980s were in part a signal that the EROI of oil was starting to drop. By then the oil bonanza was putting out only about 30 barrels for each one invested, because much of the shallow-well and high-quality oil had been pumped and burned. Current oil projects are averaging an EROI of 16 or less, and unconventional fuels seem to be even poorer. Tar sand oil has an EROI of about 5, and biofuels clock in at roughly 2 to 5.[16] Studies of renewable energy sources vary widely, but none report a return approaching that of the oil that fueled the industrial era; most range from 2 to 20 or so, with the exception of hydroelectric power, which comes in at 40 to 80. The most comprehensive study of solar electric power to date seems to confirm the low EROI of renewables. The authors of that study reviewed Spain's massive conversion to solar electric power, looking at over 50,000 installations across that very sunny nation, and concluded that the EROI of solar energy in Spain is only 2.45.[17]

As with the other terms I've introduced in this section, EROI can be calculated in several ways and has been the subject of lively debate by energy experts and amateurs. Googling the term will plunge you into some fascinating discussions about how to measure EROI and what it means, all of which are beyond the scope of this book but well worth pursuing. One superb place to start is the archive section of the now-closed, much-missed energy website www.theoildrum.com.

The decline of EROI is rippling through our culture in many ways. A growing number of energy experts believe that running an industrial civilization requires an EROI of 8 or more.[18] Unconventional fuels and some renewables that we will increasingly rely on in the future have EROIs below that. This doesn't bode well for industrial culture. In general, a declining EROI for fossil fuels means that anything requiring those fuels—which is nearly everything our society uses or does—will cost more, take more work to get, or be less available. An analogy is to imagine that you've needed to work just five hours per week to pay for a week's food, when suddenly the cost of food skyrockets. Now you need to work twenty hours per week for the same food. You have only a few choices: you need either to work more hours per week (either at your job or growing food), consume less food, or devote more of your paycheck to food and give up some other things your paycheck once bought. The various financial and resource crises of the last decade or so suggest that our society is in the process of juggling the global equivalent of all three of those solutions. Most of us are working more, doing with less, and seeing if there are luxuries we can give up so we can afford the basics.

The whole subject of energy descent can rapidly lead to a doom-and-gloom morass if we're not mindful, so it's time for me to stop taking us in that direction and shift to looking at solutions. With these five thinking tools in our minds, we're ready to tackle a permacultural approach to getting and using energy in metropolitan areas.

## WHERE AND HOW WE USE ENERGY

To design our energy solutions, we start, as always, with observation and assessment. Here's how the average American uses energy in his or her personal life, arranged by energy source, with each type of energy converted to kilowatt-hours to give an apples-to-apples comparison.

Each year, at home or in the car, the typical American uses:[19]

- 441 gallons of gasoline at an average of 19.8 mpg (14,861 kWh)
- 4,759 cubic feet of natural gas (1,394 kWh)
- 3 gallons of airplane fuel, or one round trip (117 kWh)
- 11,040 kWh of electricity

These add up to a total per capita personal energy use of 27,412 kWh.

As I was researching these numbers, I found something that puzzled me at first. Although personal energy use averages 27,412 kWh per person per year, the Energy Information Administration (EIA) says total annual energy use in the United States is 98,418 kWh per person. Why the huge discrepancy between those numbers? A little digging showed that the missing 71,000-odd kWh is each person's share of industrial and agricultural energy use. That means that three-quarters of each American's energy is used outside the home, mostly to make consumer goods that we buy and to run the businesses and farms that make those goods.

Here's what that tells us: Though conserving energy at home is a wise move and saves money and fuel, the most powerful leverage point in shrinking energy use is to buy less stuff. Also, the EIA's figures do not include energy used in other countries to make the mountain of imports that pour into our stores, so the true per capita energy use is much higher. If we want to save some of the planet's energy, the message is clear: *Stop buying stuff.*

My perusal of energy data pulled up two notable trends. First, although household size has shrunk since 1940—back then the average house held 3.7 people, against 2.5 in 2012—the size of new houses has bloated from 1,100 square feet then to 2,300 today. So each American has triple the room today that her predecessors had in 1940. Clearly, we need our space, man.

The other trend is that heating and cooling are a much smaller share of our energy

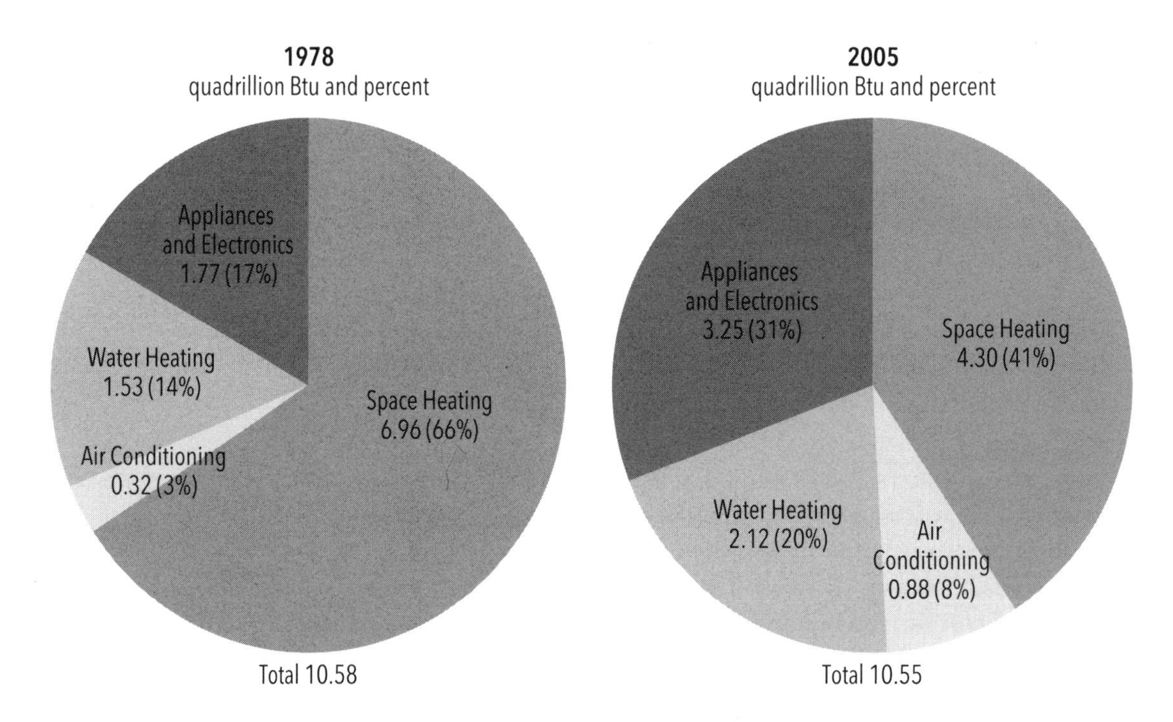


**FIGURE 7-2.** Home energy use. Although Americans are using less energy for heating than they were in 1978, they are using much more for everything else: hot water, air conditioning, appliances, and electronics. In spite of efficiency increases (or perhaps because of them), total energy consumption has not decreased since 1978. *Source:* US Energy Information Administration, http://www.eia.gov/consumption/residential/reports/2009/electronics.cfm.

use today, but—no surprise here—we're using twice as much electricity for home electronics and appliances. In 1978 we spent 69 percent of home energy use on heating and cooling and 17 percent on powering appliances. In 2009 those numbers were 48 percent and 34 percent, respectively. Because houses are better insulated today and furnaces more efficient, they use less of our total home energy budget than they did, but we've made up for this by doubling the proportion of juice drawn by our plug-in gear. We've gotten more efficient, but we own far more toys. This is living proof of Pareto's law, which says that increases in efficiency won't result in less consumption but in more use of these efficient devices. All we've done is juggle where we use energy, and the result is that total household energy use is about the same as it was in 1978.[20]

So what can a permaculturist do to be smart about home energy use? First, it makes sense to think of energy as a sector. After all, a sector in permacultural terms is indeed energy (or an influence) coming from off the design site. And, remember the three ways we can interact with sector energies. We can:

1. harvest, collect, and store them;
2. deflect or block them; or
3. let them pass by unaltered.

These three possibilities are handy guides for our home energy decisions, and we will explore how that works.

We've seen the pie chart showing how energy is used in the home, but another way to think about energy consumption is at least as useful. Instead of dividing the energy pie by what kind of equipment we use, a more permacultural approach would be to look at what jobs the energy does—what functions it performs. Then we can select efficient ways and tools for doing those forms of work.

The jobs done by energy in our homes turn out to fall into three major categories.

### Moving Heat

This is done in two ways. First, releasing and concentrating heat in furnaces, heaters, and stoves so that it can be delivered to living spaces, for household water, or for cooking food. Second, removing heat via cooling equipment such as air conditioners and refrigerators. This work can be done with a wide array of fuels and forces, including burning liquid and gas fuels, wood, or coal; by electricity; by expansion and compression of gas; via friction; and using the sun.

### Moving Things

Energy moves things: spinning fans and motor parts, compressing gases such as refrigerator coolant, pumping water, and raising and lowering objects. In most cases it does so by making a shaft spin in a motor or pump, then translating that rotational energy into linear force or a pressure change. Electricity is the most common power source for this, but fluids (liquid and gaseous) such as water, compressed air, or wind will also do the job. We can also use internal combustion engines, steam and other types of turbines (though these last are uncommon in the home), and human or animal muscles.

### Moving Electrons in Circuits

Electricity is the primary energy form here, and it's doing tasks that are almost impossible to do in other ways, such as transmitting images or sound long distances, calculating rapidly, and

## How to Enhance the Indoor Climate via Outdoor Microclimate Management

1. Use plantings and exterior structures to moderate the sun and wind striking the house. Shade the walls and windows from too-hot sun, block winds that would create chilly or hot conditions, and open the access for cool summer breezes or warm winter sun.
2. Begin improving the microclimate as far from the house as is practical. Your town may have programs for planting street trees, which are among the most powerful agents there are for improving microclimates as well as for building beauty and preserving property values. You can also help your neighbors to see the wisdom in microclimate design by showing them your examples (and their improved microclimates will benefit yours as well). A joint picnic on your cool, shady deck on an otherwise hot day will go a long way toward persuading them.
3. Work with moisture and humidity, too. In dry-winter areas, putting moisture into the soil via watering and water harvesting will reduce nighttime lows and keep tender plants from freezing. In dry-summer zones, misters and sprayers (if water is abundant), keeping moisture in the soil by mulching, layering of plants to cast shade, building organic matter, and other waterwise methods can cool the air around the house.
4. Use deciduous plantings, such as trees or vines on trellises, to create shade in summer on the south and west sides of the house. In winter the bare branches will let in the now-wanted solar light and heat.

converting one kind of information to another, such as converting a temperature reading to a number on a spreadsheet. The electrification of most of the planet, which has delivered on-demand lighting, radio, television, telephones, and the Internet to most of humanity, is in the running for the most transformative technological change since agriculture.

Now that we have a handle on the three main types of jobs done by energy and the forms of energy best suited to doing each, we can better evaluate how to better use and save energy in the home and landscape.

Let's start with heat. What work do we do with heat in the home, and how can we best collect and store it to do those jobs, block and deflect it, or let it pass by without its affecting us? How can we match our heat sources to the ways we use heat to avoid second-law inefficiencies like those of cutting butter with a chain saw?

First, we know that heating and cooling the house itself—regulating its climate—is the largest consumer of home energy. So we want to draw the conceptual box around house-climate care as large as we need to and not be tricked into thinking that the house itself is all we need to think about. In other words, we're thinking not just about how we can heat and cool the house but—moving to a higher generalization—how we can create the conditions for a pleasant home climate. How can we reduce or eliminate the need for home-climate energy use as well as make our energy use more efficient?

The smart place to begin is outside the house. After all, the reason we are heating or cooling the house is because the outside temperature is hotter or colder than we like, and we want to keep that heat or cold from getting inside. If we can change the temperature outside to a more likable one before it hits the house, we're ahead of the game. We've reduced the work and energy needed to heat or cool the indoors.

That means shaping the microclimate around the house. Apartment dwellers have an advantage here, as they are already surrounded on several sides by climate-controlled spaces. So they just need to work with their few exterior walls and any surrounding yard that they have control over.

In chapter three I described how to create and enhance benign microclimates, and I recommend reviewing those pages with an eye for improving the climate inside the home, not just out in the yard. Fortunately, many of the changes we make to help the yard's microclimate will sweeten the interior climate as well.

## USING PASSIVE STRATEGIES

Moving indoors, we'll use another permaculture approach for working with energy, and that is to use passive strategies—those that don't consume fuel—before moving to active, power-eating ones. There are several ways to do this when we're working with the home climate. They all have to do with using sector strategies of catching, storing, or blocking heat from the sun and other sources and moving it to where we want it. Broadly, they are these:

1. Keeping heat where it's wanted via insulation and leakproofing
2. Storing heat with thermal mass
3. Controlling solar gain (how much sunlight gets inside)

Let's look at each strategy in some detail.

### Keeping Heat Where It's Wanted

Every primer on energy conservation starts in the same place: The best way to save energy is not to use it in the first place. In home-climate control, insulation, weather stripping, and other forms of sealing in heat or cold is the right, if mundane, place to begin. Every dollar spent on insulation and other sealants will pay itself back manyfold in energy savings.

Where to start? If your home is not insulated or inadequately so, begin by insulating the attic floor or the roof, because, since hot air rises, that's where most heat escapes, and in summer the roof broils in the sun and transfers that heat into the house. The next big heat-leaker is the exterior floor, then the walls.

Another high-leverage move is to seal leaks. Doors and windows are the primary offenders here, and weather stripping and caulk are easy solutions. Outlets and switches on exterior walls also can be leak points, corrected by molded foam pads that mount under the faceplates. These sorts of cracks add up: Older houses often have enough leaks around doors, windows, and outlets to amount to several square feet—the same as having a fair-sized window open all the time. Like insulating, sealing leaks pays off double, because it will keep out both winter cold and summer heat. Some leaks will be easy to find, but the best method for locating lesser leaks is a blower-door test, in which a powerful fan is mounted in an exterior doorway and a smoke source is used to spot air infiltration. Home-energy audit firms may perform the test for free or very low cost under local or state energy programs.

Another major heat thief is glass. Windows can never be as well insulated as walls, but upgrading from single pane to double pane can pay for itself in five to ten years. As with any purchase, more money can buy more benefits. Energy-efficient windows can get very fancy and quite expensive, so crunch the numbers carefully to balance expense and energy savings.

If new windows are beyond your budget or enthusiasm, consider various window coverings. This can be as simple (and as ugly) as taping plastic Bubble Wrap over a window, which will trap more heat than the single-layer plastic sheets sold as window insulation but also block more light and view. Exterior storm windows are back in vogue, either homemade or custom built, and can be removed in warm weather. Insulated curtains are another effective option.

## Storing Heat with Thermal Mass

Most solar and furnace heating is done by warming air. Air, being relatively light, heats quickly but loses that heat speedily, too. To store heat more effectively and continuously, we can use denser materials with greater *thermal mass*, or ability to hold heat. We've all noticed that when a thin aluminum pan comes out of the oven, we can touch it barehanded within a few moments, while a cast-iron pan will still burn fingers several minutes later. The thicker, denser iron holds heat longer because of its greater thermal mass, and we can use that property in many materials to capture and store heat. One way to do this is to put a high thermal-mass object in direct sun. That mass will release its day-captured heat over many hours, reducing temperature swings.

In the 1970s the passive-solar homes of the day used thermal mass in the form of rows of water barrels near windows and full-height Trombe walls. These made rooms look like warehouses and dark caves, but like many other quaint hallmarks of that ugly decade, these have been superseded by less ostentatious ways of doing the same thing, such as tile floors and half-height rock or brick walls. Thermal mass also needs to be matched to the amount of heat available, or warming it will suck up so much heat that the space will stay chilly until the mass is equilibrated. Once a mass is warmed, though, it stays warm for a long time. Books and websites on passive solar design offer many methods for selecting and creating thermal mass. A good place to start online is the National Institute of Building Sciences web page on passive solar design, http://www.wbdg.org/resources/psheating.php, which also links to many comprehensive handbooks and documents.

One of the hottest trends in home heating currently is the rocket mass heater, which is a fine example of thermal mass tied to an efficient type of wood-burning heater called a rocket stove. The rocket stove, developed by combustion researchers at Aprovecho Research Center in Cottage Grove, Oregon, converts wood fuel nearly completely to water and carbon dioxide, which gives the stove both tremendous fuel efficiency and ultralow smoke emissions. One of Aprovecho's founders, Ianto Evans, has designed a rocket stove within a large earthen bench. Nearly all of the stove's heat is captured by the enormous thermal mass of the bench: If you hold a hand over the smoke-free chimney, you can barely feel any heat, and your palm dampens from water vapor. The bench takes several hours to heat but remains warm for at least as long after the stove has gone out, slowly releasing its heat into a room. The heated bench (or bed) is such a delightful and cozy place that it often becomes the focal point and gathering spot in a home. Unlike conventional woodstoves, which are usually too smoky to be ethical or legal in urban areas, rocket mass heaters are clean enough for city life. Rocket stoves for cooking and rocket mass heaters can significantly shrink pollution and also fuel needs; you can run them on twigs, another plus for urban dwellers for whom firewood is expensive. Evans's book about them is called *Rocket Mass Heaters: Super-Efficient Wood Stoves You Can Build*. (See photograph in the color insert, page 16.)

Thermal mass can also keep a house comfortable on hot days. If you live in a region with hot days but cool nights, opening windows at night can do more than make sleeping more comfortable. The material in a tile floor or a thermal mass wall or bench will chill down at night. Then, as long as that mass is kept out of direct sun, it will remain cool for hours during the day, pulling heat out of the surrounding air and cooling hot bodies in contact with it. In warm climates keeping thermal masses such as tile floors shaded by judicious use of curtains and rugs will help rooms stay cool, as once a tile floor has heated in the sun, it will hold and release heat for hours. In sunny, warm climates, consider replacing or shielding thermal mass when possible: Wood and fabric store much less heat than stone and wallboard.

## Controlling Solar Gain

One obvious rule of thumb for heating with the sun is that the sunnier your region is, the less sun needs to get inside. There are several balances to strike here. The first is the balance of sun in the warm versus cold seasons. Sun pouring in a window that is welcome on a frigid winter day won't be wanted in August. This balance can be adjusted by setting the roof eaves over sunny windows at an overhang depth calculated to let in low winter sun but block the high summer rays. The Internet and books on passive solar design abound with tools for visualizing and calculating roof overhangs for solar gain.[21] If lengthening or shortening the eaves isn't feasible, adjustable or seasonal awnings, narrow shade panels, or adjustable exterior (or less effectively, interior) window blinds will do the same job.

Although tuning roof overhangs uses the seasonal height variations of the sun to block or allow sun through the windows, one design challenge here is that the sun is at the same height in the sky on both the spring and fall equinox, but March temperatures can be cool while September is often still hot. That means that a roof overhang designed to let in cheerful March sun will also let in fierce September rays. One solution is a deciduous-plant arbor or trellis over the relevant windows. Few vine species leaf out during a chilly March, but the same plant will be casting ample shade in September. The synergy of passive roof overhang plus benevolent biology gives a solar-gain tuning that neither alone can yield.

Another balance to strike when using solar gain is the proper size of south-facing windows.

In sunny climates a little window area goes a long way toward harvesting the sun's heat, especially if the house is well insulated and has thermal mass. It might seem, then, that in cloudy places we would want to maximize solar window space to haul in as many sunbeams as possible, but those same windows that warm the house on sunny days can be heat-bleeders on cloudy days. Most passive solar experts recommend that the glazed area equal no more than 8 to 12 percent of the house's total floor area, with 70 to 85 percent of that on the south-facing wall, 10 to 15 percent on the east wall to catch morning sun, and only 5 to 10 percent on the west wall to avoid overheating on hot afternoons. Full calculations of proper window area for different regions are beyond the scope of this book but are easy to find in the references I've given and on the web.

Most readers are working with an existing house rather than new construction and will be

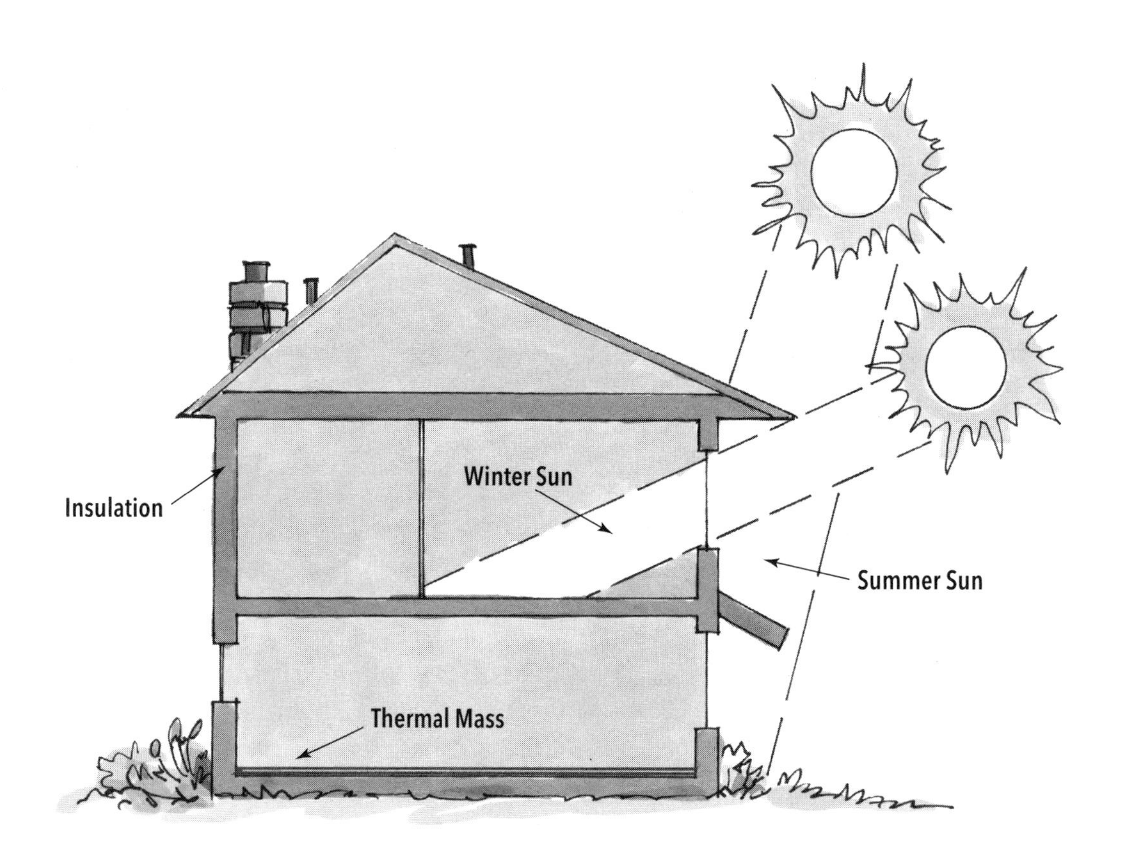


**FIGURE 7-3.** A properly sized roof overhang can block summer sun but allow winter sun to penetrate deep into the house. On two-story houses, awnings over the first-floor windows can provide the needed shade. Insulation and thermal mass will add further energy savings and comfort. Illustration by Elara Tanguy

remodeling if they want to make major changes. If sunward windows are too few to net reasonable solar gain, consider adding more. If chilly drafts are spilling off the glass on cloudy days—something you can tell by standing barefoot near the window or just by holding your hand a few inches from the glass—you can replace the windows with new, well-insulated ones.

But if you're not in a position to replace or enlarge windows, you still have options. If windows on the south or west side of the house are letting in too much broiling sun, simple curtains or translucent blinds can help, although, ideally, these belong on the outside of the window to block sun before the heat gets inside. But indoor blinds are better than nothing. If windows are too small for good solar gain, a window solar collector can be mounted below or beside windows to heat air via the sun and flow it indoors.

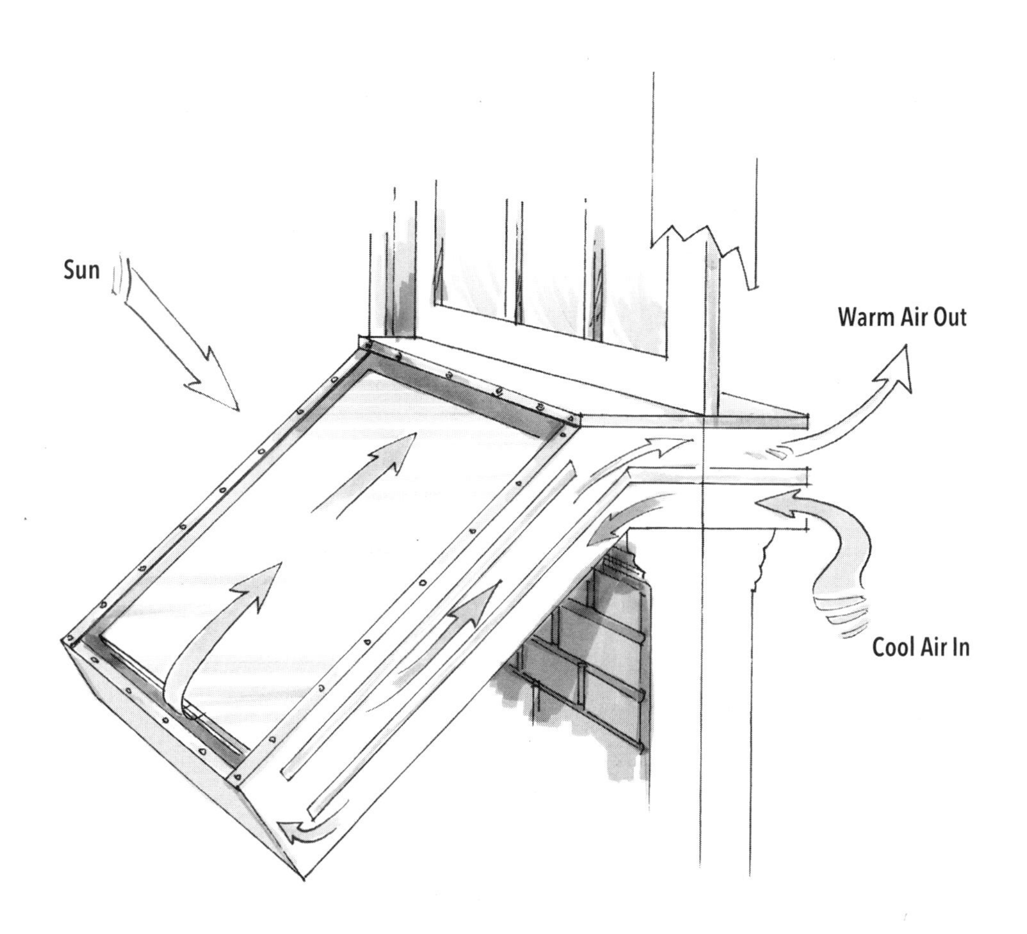


**FIGURE 7-4.** A window-mounted solar collector. Illustration by Elara Tanguy

## Gardening the Sun

The second-largest energy user in the home, after heating and cooling, is electricity. Most of us get electricity from a centralized generating plant running on coal, oil, nuclear, or hydroelectric power. But one of the primary trends of the twenty-first century is away from centralization, toward distributed networks, and electric power is following that trend. When we think of solar-generated electricity, most of us think either of giant solar-panel farms covering miles of desert or of a house roofed with a handful of photovoltaic panels giving that single household their electricity. Both paths are obvious default solutions for a society that swings between mammoth industrial projects and go-it-alone individualism, and both have their weaknesses. The drawbacks to centralized grid power are its political and physical remoteness and the helplessness of all the client households during a power failure. The weaknesses in individual home power include the high cost of a stand-alone system and the limitations of the site: If your neighbor's trees or second-floor addition shade your site, tough luck.

But a happy middle ground is emerging: community-supported solar power, sometimes called solar gardens. These are distributed solar projects collectively owned or leased by subscribers, in which the solar collectors aren't necessarily on the subscribers' properties. Solar gardens can be legally and physically structured in many ways, with panels on subscribers' roofs; on public or private buildings such as churches and businesses; over parking lots; and even, as in the case of the Solar Garden Project in Brewster, Massachusetts, over a landfill that can't be developed for other uses. This also means that subscribers don't need to cut down trees to improve their own solar access. And they can buy in bulk, saving money. In some cases solar garden groups have built their own grid rather than pay a utility to build a system that the neighborhood then pays to use. In others, any excess power is sold to the local utility, reducing every subscriber's electric bill. And, the collective political power of a large group can influence legislators and utility companies to be more renewable friendly. Depending on state laws, solar gardens can produce enough power for just a handful of houses, about 10 to 20 kilowatts, or for thousands, replacing a 20-megawatt power plant. A great clearinghouse for information on community solar is at www.solargardens.org.

As we've seen in other sections of this book, scale matters. Community power seems scaled in the "just right" Goldilocks zone, needing less money, work, and care than a single-home system but offering much more autonomy and community input and benefit than the typical region-scaled power plant. That's a permacultural solution.

## ➡ CREATING YOUR ⬅ WHOLE-SYSTEMS ENERGY PLAN

If this chapter took a conventional approach to saving energy, we would now walk through all the other energy-using devices and systems in the home and list ways to use them more efficiently, mostly by suggesting that you buy new, power-saving appliances; unplug phantom loads; and insulate the crap out of everything that makes or carries heat or cold in your house. Plenty of excellent books and websites tell you how to do that, and except for the "consume our way to sustainability" mind-set that other sources sometime demonstrate, that's all good advice.[22] Since that information is so readily available and operates at the level of technique rather than strategy, I'll assume that by now you get the idea, so I will leave the project of examining all the rest of your energy-using hardware for you to pursue on your own.

Instead, let's return to our themes of design and strategies to see how to develop an integrated energy plan for your entire home and yard. First, as in any design, we need to define our goals.

- Reduce energy use while preserving comfort.
- Replace nonrenewable energy sources with renewable ones.
- Use abundant, secure energy sources instead of scarce, uncertain ones.
- Maintain core energy needs during shortages, outages, and disaster.

To these you can add any personal goals that fit into a permaculture design plan, such as weaning your home from grid power, generating surplus energy for neighbors or for sale to the power company, or educating others about wise energy use.

### A Review of Energy Strategies

To begin developing your whole-systems energy plan, let's review the strategies and methods for saving energy in the home that we'd want to do under any conditions. These should be an integral part of any energy plan.

- **Create microclimates** around your home that will lower utility bills via shade, windbreaks, cooling breezes, reflective or absorptive surfaces, sun traps, and related methods from chapter two.
- **Insulate** ceilings, floors, and walls, in that order. Upgrade to insulated windows or wrap windows with clear coverings, and use insulated curtains.
- **Seal** leaks around doors, windows, outlets, vents, baseboards, and other openings.
- **Use the sun's free heat** for home heating, hot water, clothes drying, and cooking.
- **Shade windows** that would let in unwanted heat from the sun.
- **Increase thermal mass** where feasible to hold more daytime heat or nighttime cool as desired. In hot climates decrease thermal mass that is exposed to sun.
- **Replace or stop using energy-hogging appliances.**

### The Big Picture: Ultimate Permaculture

Most of the above fall into the category of quick-fix and retrofit items for implementing an energy plan. While making progress on these, you can be working on the big picture: ultimate permaculture solutions to create a whole-systems plan. A key step for this is to match energy resources to energy needs by understanding where your energy comes from and goes. If we use needs-and-resources analysis to help us fashion our big-picture energy plan, four questions emerge that can guide us toward our goal.

1. What energy sources are most abundant in your region?
2. What renewable energy sources make sense to use where you live?
3. How can you best match energy sources to their intended use?
4. How can you prepare for energy shortages and outages?

Answering the first and second questions will begin shifting you toward locally abundant energy sources. This increases your energy security by paring down the distance energy travels to reach you, which lessens the risk of interruption and interference. It will also probably reduce your energy bills (though in this era of irrational energy subsidies, it may not—for now). It will also preserve scarce resources for future use or, for ones that pollute or are otherwise less than desirable, help keep them from being used at all. Sometimes locally abundant sources can be unusual ones: While I was drafting this chapter, I gave a lecture in Boise, Idaho, where the houses in older neighborhoods and many buildings downtown are heated by geothermal energy. The thermally rich local geology bubbles hot water to near the surface—a major street is named Warm Springs Avenue—and Boise's denizens benefit from the cheap, plentiful heat.

Of course, not all locally abundant energy sources are environmentally friendly. Montana and West Virginia are underlain with county-sized seams of coal, but ripping them out has huge carbon, erosion, habitat, and social footprints. On balance, though, it may make more sense for the locals (not distant others, though) to use that coal for now than for them to import natural gas from Canada or oil from Saudi Arabia. All energy sources have their environmental and social downsides—photovoltaic solar uses rare-earth metals that pollute and originate in totalitarian nations, wind generators kill birds, and so forth—thus your own personal standards play a role in dictating what you can live with.

Sometimes a sensible energy source is not the most obvious one. Cloudy cities such as Seattle and Buffalo seem like poor choices for solar energy, but both cities are sunnier than Germany, which has an aggressive solar electric program that already generates more energy than Seattle uses. In Seattle cheap hydroelectric power at 5 cents a kilowatt gives stiff competition to solar electric's current 35 cents per kilowatt, but between subsidies and incentives as well as Seattle City Light's inducements—they claim that grid-tied solar systems let water build up behind hydroelectric dams for use at night—solar systems can pay for themselves there in ten years or so. The point is you may need to dig beneath first impressions about local energy sources to decide what energy sources are the most secure and sustainable for you.

To do this, learn the subsidy and rate structures for your energy sources, where your energy actually comes from, what choices your utility allows, whether or not you live in a state in which you are allowed to select your energy utility, and, if you want to generate your own power, what home or community power options are best for your region and circumstances. This will take some effort, and unfortunately there doesn't seem to be a single source for learning how to make these kinds of decisions. Most educational resources on home energy focus either on choosing a renewable energy source or on learning about your local utility provider and the related laws and price structures, so you'll need to research both.

In question three, matching energy sources to their intended use takes us back to cutting butter with a chain saw, second-law efficiency, and transformity. We can develop guidelines that will avoid needless waste. For example, if we need heat—say, for warming a house, food, or water—then the best energy source is one that is already hot, such as the sun, geothermal energy, or, less obviously, a process that's spewing waste heat, such as a power plant, engine, or even a compost pile. There's a lot of heat available for free, so use that resource before burning new fuel to make more. To grab free heat, there are two primary technology types: thermal storage and heat exchangers. Thermal storage usually uses a thermal mass that sits near something hot, warms up via conduction or radiation, then transfers its heat back out later—they move heat over time. For heat sources such as geothermal energy, engines, and compost piles, or when a solar thermal mass isn't sitting where we need the heat, we must move the heat to where we want it. That takes a heat exchanger, which is a device that passes a cooler fluid such as air, water, or a chemical solution through channels in the heat source and moves that fluid to where the heat is needed. They move heat over distance. The window solar collector described earlier in this chapter is one example of a heat exchanger. Your car's radiator is another, grabbing heat from the engine to warm the car's interior. Black poly pipe or copper tubing embedded in a large compost pile used to heat water would be another form of heat exchanger.

When there is no free heat source around, we're stuck with burning fuel. Here the best choice would be whatever available energy source can be converted into heat in the fewest number of steps and with the least pollution and equipment. Most flammable fuels fit this category—you just need to light them to release their heat—but we want to use the fuels and technologies that burn the cleanest and have the least embodied energy. Renewable fuels rank higher on this list than nonrenewables, but there is precious little renewable liquid fuel around. Biofuels are in theory renewable, but their EROI is pathetically low, so the only biofuels that make sense to me are ones made from a by-product of some other process, such as waste from growing certain food crops. The crop's primary product feeds people, fuel can be made from digesting the crop's inedible parts, and the spent fermentation mash can then be composted or fed to animals.

I don't favor using plants solely to make biofuels because of the poor second-law efficiency. Plant tissue is a highly complex, richly organized material that can power other organisms, build ornate structures, and do other sophisticated tasks that require those refined materials. If we must burn this exquisitely configured stuff, the best way, using the fewest conversions, is to pelletize it as solid fuel. To take the miraculous molecular diversity of plant matter and use fermentation or some other process—especially at industrial scale and its copious embodied energy—to simplify and degrade it into one of the least complex hydrocarbons around, ethanol, then set that on fire, instead of applying the principle of highest use and extracting benefits from it on multiple passes as it degrades, strikes me as silly as buying rare books and using them for toilet paper. There are higher, better uses for plant matter, such as building soil, feeding living beings, sequestering carbon, and constructing durable goods like shelter and clothing.

In general, liquid fuel's precious combination of high energy density and easy transportability suggests that we only use it when we absolutely require those qualities, such as in powering vehicles, not for warming a room. Keeping these needs-and-resources guidelines in mind will help save liquid fuel for when it's the smartest solution to a need.

Wood is another renewable energy source with great value. Wood fuel isn't easy to come by in urban areas, but if you have a supply of it and a fuel-efficient and clean-burning way to liberate its stored solar energy, such as a rocket stove, it's worth making it a major part of your energy mix. As I've mentioned, rocket stoves can run on twigs, so your yard's prunings—and your neighbors'—can generate a surprisingly useful amount of heat. I've made arrangements with power companies, arborists, and parks department staff to pick up branches and logs that they've trimmed around town. But consider that a better use for it might be to build soil or some other material.

Then there is biochar, which is wood or other dried plant matter that, instead of being combusted, has been carefully cooked, or pyrolyzed, to strip away everything in it but its carbon skeleton, leaving a form of charcoal. Whole books have been written about biochar, so I will not try to describe it in detail nor sum up the pros, cons, and arguments about it. Like wood, it's not abundant in cities, but I like its multiple functions: Biochar can yield heat first in the initial pyrolysis (charcoal-making) step, then again when the charcoal is burned, and it can also be a great soil supplement that builds fertility while sequestering carbon. If you need both heat and soil carbon, biochar may be for you.

The poorest choices for any energy production are sources that must be run through multiple conversions before liberating heat; that is, that have high transformity, as in the earlier example of electric heating. Another poor heat source would be wind or other mechanical forms of power such as a waterwheel, because these need to be converted first to electricity and then to heat—there's that long wasteful trail again—or at best, made to release heat via friction or compression, which are not very versatile or transportable sources of heat.

If it's mechanical energy you need, to run a pump, compressor, motor, pulley on a shaft, or something that moves back and forth such as a saw, then a wind turbine or waterwheel may be your best choice, because they create mechanical energy directly, eliminating conversion steps and their losses. As fuel becomes more expensive, we will probably see a resurgence of wind and water systems to run machinery. Whole factories were once powered by a single turning shaft mounted along its length with multiple pulleys connected to saws, presses, drills, mills, looms, and other machines.

It's best to save electricity for running motors, electronics, efficient lighting, and related tasks that can be done only by pulsing electrons. Water and wind can generate electricity efficiently, though in cities divertable streams rarely flow through our yards, and wind is erratic and buffeted into turbulent swirls by buildings. In some urban situations small-scale wind generators make sense for low-current tasks or charging battery banks that don't get heavy use.

How about question four, energy for emergencies and disasters? Temperature control and lighting are the two big energy concerns in most emergencies. One of the lessons from Hurricanes Katrina and Sandy is that in a widespread disaster you may be on your own—no power, water, heat, or visits from FEMA—for two or three days in a city and much longer in rural areas. If you're in an efficient passive solar house, you can keep warm, or at least warmer, in cold-weather power outages. Insulating and sealing leaks will pay off here. And this stacks functions, too, saving energy in the long term and being ready for disaster. It also helps to have a properly vented, efficient woodstove and an emergency stash of firewood; again, a rocket mass heater fits the bill here.

For emergency electricity, consider a small photovoltaic system. A single 80- to 240-watt solar panel, one or two deep-cycle batteries, a charge controller, and a 300- to 500-watt inverter can be had for less than $1,000 and will provide enough 110-volt and 12-volt electricity to run a few lights and a computer or other news and communication channel, periodically power a furnace fan, and charge household batteries. This last means also having a supply of rechargeable batteries and a charger as well. If you have electricity during a multiday outage, you will be very popular among your neighbors, so prepare for that, too. We have such a setup and have been very grateful when the power is out for more than a few hours. (The conventional answer to electricity in emergencies is a generator, which, while useful, usually burns nonrenewable fuel. In a long-term emergency or an era of brownouts, it will run out of gas long before the sun stops charging a PV panel.)

Barring the emergence of some new energy source, we are almost certainly in the early stages of energy descent—a period in which energy costs escalate, fuel and power are less easy to come by, and the repercussions of the end of the era of 100 energy slaves per person ripple through society. If you're ahead of that curve—in this case a downsloping one—not only will you make that transition easier and more graceful, but you'll also provide a model for others to see it as an opportunity rather than simply a possibility or even a cause for panic. And if the era of energy descent isn't on hand, it's never a bad idea to be thinking and acting permaculturally about energy.

# CHAPTER 8

# Livelihood, Real Wealth, and Becoming Valuable

It's a fact of life—a corollary to the laws of thermodynamics—that every living thing must do work to pull in the resources it needs to keep going. Bacteria wriggle their flagella to glide upstream in a sugar gradient; plants brew up photosynthetic tissue to harvest sunlight and carbon dioxide; animals browse and hunt. For modern humans, the effort we trade for life has many names: "job," "work," "earning a living," "career," "livelihood," and even "right livelihood." For permaculturists, that last term—"right livelihood"—fits well with our ethics and principles. It's important to us that we support our lives in ways that care for the earth that provides for us, care for the community of people we are embedded in, and let us limit our consumption to create a surplus that can cycle back to support the earth and people that have aided our living.

Humans have dreamed up innumerable ways of exchanging effort for life. In most of today's cultures, though, the plethora of possible ways to gain livelihood has been pruned by passage through a narrow-necked funnel; that is, what we do must be measurable in money. From a permacultural viewpoint, it is not a wise strategy to perform the critical function of livelihood via one method, moneymaking. One of the aims of this chapter will be to broaden the definitions of wealth and value beyond money. But to do that we first need to look at money itself, why it came to be, and what a few of its consequences have been.

## WHAT IS MONEY?

The formal definition calls it a medium such as coins and bills that can be exchanged for goods and services and that is used as a measure of value in the marketplace. But it is so much more than that. When we want to know how much money someone has, we ask, "What are they worth?" This language tells us that, consciously or not, we link financial status with a person's intrinsic, overall value. Money is also a form of power. Rich people wield more political influence than others, and their words often carry more weight, as if the ability to accumulate money translates to wisdom as well. Money, too, is a source of anxiety, addiction, shame, loathing, relief, and even joy when a tight spell is over and the cash starts flowing again.

We have a fairy tale about how money came to be. Once upon a time, it begins, there was no money. People bartered for everything. But if you were a cobbler and you needed an ax, and the blacksmith already had shoes, you couldn't get an ax. So we invented money to get around this inconvenience—this, in economists' parlance, lack of liquidity. Money made our exchanges more fluid.

But this world of barter never happened. Anthropologists have found no societies that have used barter other than infrequently or in special cases, just as ours does. Instead, moneyless cultures are almost always some type of gift economy. Unlike money economies, gift economies, as anthropologists Bronislaw Malinowski and Marcel Mauss showed in the early twentieth century, build reciprocal relations among the givers and receivers that never quite balance out.[1] The back-and-forth as people try to match one gift with another in return forges an ongoing community bond, or what we've come to call social capital. Here's what that looks like:

If you use money to buy an ax from a blacksmith, the transaction ends there. Neither party owes the other anything more; the smith is fully, in Mauss's language, *alienated* from the ax, meaning that he has no further claim or right to it, and you are equally alienated from your money. The exchanges cancel out, dissipate, and leave no connection behind. There is something telling in the word "alienate."

But if the blacksmith in your community *gives* you the ax, it's not completely free. You're likely to feel, if not a sense of obligation, then a feeling of reciprocity, a note to yourself that someday you ought to do something just as generous. However, that returned favor won't be an ax—after all, you're not a blacksmith—so the gift back will have a slightly different value. Thus the exchange won't perfectly balance, and this sets in motion a flow toward further exchange. Also, now you think of the smith as a good guy, so you will send business his way. That builds his esteem in the community as well as his customer network. The flow of return from this gift ripples out beyond just the two of you.

Mauss says that a gift is never completely dissociated or alienated from those who exchange it. That's not like a money transaction, where all rights are transferred to the new owner and the flow stops. (Sometimes rights are retained, such as copyright, mineral and water rights, or guarantees, but even those arrangements split and isolate the rights from the goods.)

Gift economies have become a popular topic lately. I've seen some misunderstandings about them that stem from our having been brought up in a money economy, so I'll rush to point out that a gift economy doesn't mean demanding that those who seem to have a surplus give it to you. Gifts can only be given, not taken, or else they are not gifts.

The reciprocity of gifts illustrates one important feature of a healthy economy: The exchanges work to build rather than destroy relationships. The best transactions don't quite balance. They leave a glowing residue that prompts further exchange and even a "pay it forward" attitude. This kind of economy is called "non-zero-sum." In a zero-sum economy, all transactions add up to zero: My gain will be balanced by the amount of your loss. With non-zero-sum (which is how much of the natural world works), there are outcomes in which everyone gains.

Non-zero-sum arrangements are ancient and form the basis of trade. If I make goods only for myself, creating a surplus not only does me no good but wastes time and resources; it's lose-lose. When I engage in trade with you, my surplus gets converted into items I can't make myself and gives you something you need as well. It's win-win. Well-planned non-zero-sum efforts create more than was there in the first place, as when a group collectively comes up with better ideas than any one of them had alone. How would we design an economy that has these features? And what else must we do to transition from our degenerative economy to a regenerative one? Let's explore money a bit more to see what qualities it has and lacks that we want in a regenerative economy.

A more probable creation story for money than the economist's barter myth is that once upon a time people found it inconvenient to come to market with a full herd of sheep or heavy urns of grain, so they made clay tokens in the shape of livestock, urns, and other goods and swapped them at markets as a promise and record of a future delivery. We're pretty sure this story is true, as over 8,000 of these tokens have been dug up in the Middle East. This would have been the first abstraction from goods to symbols of goods. But the receivers of the tokens could counterfeit them to get goods they hadn't earned. The solution was to seal the tokens in a clay jar, and inscribe the number of tokens and the name of the seller on the jar—and archaeologists have found such jars. But soon people realized that the tokens and the jar were superfluous. The number and signature on the outside were enough to convey what the promised goods were. This was the first money, and it points to a useful definition: Money is a claim to a good or service. To this day, money bears the ancient markers of denomination and signature of the backer. In the case of the US dollar, the secretary of the treasury and the US treasurer have scribbled their names beneath the seals on the bills as a guarantee.

Although the precise origins of money are lost in the dimness of time, the first coins and bits of precious metals seem to have represented units of livestock and grain. The Mesopotamian word "shekel" dates from 3000 BCE and means a measure of barley as well as a weight of metal. With the creation of coins, money was becoming more abstract, representing not merely one kind of good but any item with similar market value. The metal in the coin itself had significant worth, though; thus, coins were not merely symbolic carriers of value. The next step in the story appears in Tang dynasty China, when paper currency was invented around 700 CE.[2] Chinese merchants began leaving heavy, unwieldy bags of coins with trusted money-holders, exchanging them for receipts marked with the value of the coins. The receipt confirmed that the bearer possessed a claim on real goods. By 1100 CE, these private promissory notes had evolved into paper money printed by the state. With this final step, money was now fully abstract, interconvertible to anything with established value but having no physical value of its own.

This abstract quality of money brings with it some odd qualities. One is the notion that if money is a claim on an asset or a store of value like gold, then we can reckon that not everyone is liable to make their claim on the gold at the same time. Most people will keep their money with them or store it somewhere. This led clever people to realize that they could print much more money than there is gold to back it. The downside of this becomes violently clear during runs on banks, when frightened people demand, à la *It's a Wonderful Life,* to have their funds converted into whatever backs the evanescent digits on their bank statement. The upside is that it allows an economy to be far larger than the physical goods that underlie it. This system of printing more money than there are goods to back it is called fractional reserve banking, and every nation has used some form of it. The US money supply is backed only by a tiny portion of its dollar value in gold, and that fraction is stretched further by policies that allow banks to hold deposits or other forms of currency reserves totaling only 3 percent to 10 percent of the money they loan out.[3]

At this point let's overturn another myth of economics. Fractional reserve banking has morphed even further toward vapor, into an economy based on almost pure credit, where essentially nothing backs the money supply and other financial instruments. We're taught by most textbooks that when customers deposit money into a bank, the bank creates new money by making loans based on some multiple of those deposits. But it's really the other way around. In a recent article called "Money Creation in the New Economy," the redoubtable Bank of England declared, "Whenever a bank makes a loan, it simultaneously creates a matching deposit in the borrower's bank account, thereby creating new money. . . . Rather than banks receiving deposits when households save and then lending them out, bank lending creates deposits."[4] So we have it from high authority that banks just make up money. When credit is loose, they make up more of it. When credit is tight, they make up less and will often destroy money by failing to make new loans when old ones are paid off. Our money isn't in fact tied to anything tangible. Paper and other forms of money are simply made up, and their value depends almost solely on our trust in the groups that prestidigitated it into existence and on our shared agreement of, and belief in, its value.

Where I'm going with this story is here: Once we understand that our money system is something we invented and is based on our believing in it, we're empowered to create money systems that work better for us and that we have more coherent reasons for believing in. One thing we will explore in this chapter is other trustworthy stores of value besides the state, central banks, and credit card companies. And in keeping with the permaculture principle of multiple supports for important functions, we'd be wise to invest our own resources and faith in not just one but several kinds of these value stores.

## MONEY THAT GOES—AND STAYS—LOCAL

One of the most obvious and straightforward ways to diversify the way we store value is by creating money with communities of people whom we trust more than multinational bankers—a group for whom we have some evidence that they are not always looking out for our best interests. One way to diversify our value storages is via various types of local currency, also called alternative, complementary, and community currency. This is money issued by and for a particular regional group or other community. It's almost never meant to replace the national currency but to supplement it.

Conventional economists fret over local currencies, as their textbooks abound with horror stories from earlier times of money issued by small banks whose value plummeted to worthlessness when the banks failed, proved corrupt, or suffered runs. However, few local currencies are now issued by banks but instead are backed by pools of businesses, civic organizations, and citizens that are much less likely to evaporate than a single small bank. Local currencies do face obstacles, though, and some recent attempts have had brief lives. In the Wikipedia entry, "List of Community Currencies in the United States," nearly half of the more than 120 currencies listed are inactive.[5] In many cases failure stemmed from simple inertia. Once the initial thrill of launching a local currency has worn off, people find that federal dollars are simply easier to use. Also, the rise of debit cards and electronic transactions has shrunk the use of cash in general, and that includes local money. But new versions keep popping up because their advantages and appeal remain.

These currencies take two principal forms, either tied strictly to the official currency and thus viewed happily by tax authorities or tied to a unit of time or labor in a system called time banking. Time-bank currency is usually not taxable. The most venerable of modern time-bank currencies in the United States is the Ithaca Hour. Although its use, too, has declined recently, this upstate New York scrip has been around since 1991, and over 500 local businesses have accepted them. Several million dollars worth of Hours, each Hour valued at $10, have been traded; members of the Hours system can borrow them without interest for up to a year.

Another well-used local currency is BerkShares, circulating among over 400 businesses and many thousands of users in Western Massachusetts. BerkShares are also pegged to the US dollar, but local groups are considering tying its value to a basket of local goods to protect the regional economy from the uncertainty of the federal dollar. Purchasing BerkShares gets you an automatic discount, because $100 buys $105 worth of BerkShares.

We have many reasons for using local currency. Besides insulating users from the volatility of the US dollar, they encourage people to shop locally. Chain stores rarely accept local currency, so instead of having money siphoned away to corporate headquarters and far-flung shareholders, local dollars push commerce toward local businesses and keep money within the community. Classical economists object that this constitutes a restraint of consumer choice, but locally owned stores often carry the same products as the big boxes, just not in mile-high stacks, and you still have federal dollars to buy those sweatshop-made shoes if you want. This seems a small price to nurture a vibrant local economy.

Local currency is one way to support livelihoods and economic health. After all, during recessions and depressions, the problem is usually with the money system. It's not that people no longer have valuable skills or needs. It's that dollars are in short supply, and even healthy local economics can be depressed by problems that have rippled down from the macroeconomy. Local currencies are an attempt to smooth out these cycles and tighten the lags between production and purchase that occur in a national money system. If you're interested in learning more about local currencies, a good place to start is the web page on local currencies at the Schumacher Center for New Economics, at www.centerforneweconomics.org/content/local-currencies.

If there's no local currency near you and you're not ready to tackle the task of creating one, there are many other ways to build and hold value in yourself and your community that go beyond money. We'll look at some of those now.

## ➡ A JOB OR A LIVING? ⬅

We started this chapter with a list of the names people give to the way they trade their effort for sustenance, and "right livelihood" topped my list as covering the most relevant bases. My own attitudes were expanded from thinking about a job to thinking about how I wanted to live by two formative experiences when I was young. The first was when I was about sixteen, while I was wrangling my way through that period of spiritual angst that most adolescents face. I had encountered a description by Zen philosopher Alan Watts of the trajectory of most contemporary lives. It looks like this, he said: We go to nursery school to get ready for our education, then to elementary school to prepare for high school, where we strive for high grades so we can get into a good college, in which we choose a major that will prepare us for a career. In our careers we climb the ladder toward the top of our profession while we save money for and look forward to retirement, when, after all those decades of preparation have passed, we can finally enjoy life. Except that by retirement time, most of us are too worn out and sick to have much fun at living. I read this story—of a life lived in constant anticipation of something better—while at preparatory school, an irony that wasn't lost on me. And it all sounded like a bad deal, one that I wanted to avoid.

The second formative event was my first job and its aftermath. After college I delayed the next inevitable step on the trajectory I had been launched on, graduate school. In what I thought would be a brief interim, I worked for a few months in a neuroscience lab. When a grant ran out, I was laid off and took a long vacation of traveling and hiking while sorting out my next move. Still not ready for graduate school, I found another lab job and was laid off again when that grant expired. Because the economy was in such bad shape at the time, even though I'd worked only about eight months in all, I qualified for eighteen months of unemployment insurance. My habits were frugal, and those checks provided what felt like a lavish income. I thought of them as a government grant to do research and get training, so I read copiously in science books, took classes, and also learned boat carpentry and engine repair from some old-timers. I liked this pattern of working a few months in science for money, then taking a year or two to gain other skills. Not having my life centered around a job felt like the right way to live.

After a few years of this, my prep-school programming kicked in—I was too aware that my old school friends drove fancier cars, took lavish vacations, and lived in roomier houses—and I got a "real" job. Seven years later I was reporting to the chief operating officer at a 1,500-employee medical biotech company. And I was miserable. Fortunately, just as I began to be hit with panic attacks and other serious signs of stress, permaculture entered my life, and I realized that this was my path back to right livelihood.

I found my way to a centered life when I learned to stop paying attention to what I had been programmed to do and instead listen to what gave me joy in life. Today I "work" at least as many hours as I did while in corporate life, but it feels like play. Joseph Campbell's advice about following your bliss turns out to be true—joy is a good indicator of value—and it's what I wish for everyone. What I've seen is that if you make or do something that you feel is valuable, the odds are you're not the only person who will value it, and some of those other people will support your doing it. Being of value to others—being in service to something larger than yourself—is the key to right livelihood.

The distinction between "job" and "livelihood" is important. For most of human history, no one had a job in the way that we think of them. People simply went about living—gathering food, making homes, raising young—like all other animals. After the dawn of agriculture about 10,000 years ago, however, livelihood mostly took the form of farming. A few specialized livelihoods emerged to support or manage farmers, but not many. Early farmers themselves worked at varied tasks, guided by seasons, weather, day length, and the other rhythms of life. Specialization barely existed until about 5,500 years ago, when people began clustering in cities. Even then and for subsequent millennia, the shifting tasks and largely nonmonetized life of farming was the norm for nearly everyone.

Humanity's mass conversion from the rhythms of simply living to working at jobs away from home surged with the industrial revolution. To make the machinery of factories function, people had to become pieces of the machines themselves, doing one task over and over for a specified number of hours—usually long ones—rather than shifting tasks with the changing needs of plants, animals, soil, and seasons. This specialization—turning a human population of generalists with broad skill sets into workers with narrow training—produced increasingly complex technologies that made themselves essential for industrial life, and both of these phenomena shrank individual self-reliance. Access to the new necessities now depended on people's skill at serving a steeply hierarchical and often remote economic order rather than their individual ability to make what they needed. We gained a cornucopia of goods far more diverse—not to mention enticing and addictive—than what any individual could make, at the cost of much of our own independence. Machines took over as the producers in an economy that values humans most not as people but as consumers.

## WHAT SIMPLICITY LOOKS LIKE

As I've noted, the industrial era was built on the cheap, abundant energy of fossil fuels. In any dynamic system, boosting the availability of resources allows the formation of more complex patterns of interaction—a richer food web, more diverse and specialized niches—and our culture is no exception.[6] When we discovered oil, our decision to burn through the accumulated capital of millions of years of that stored sunlight in a scant three or four centuries powered our society into vast complexity. Just as one example of this, ponder the escalation of complexity from the days we traveled on horseback to that of modern air travel, replete with millions of aircraft components, the labyrinth of cockpit instruments, the air traffic control system, airports as big as Manhattan, the TSA, and full-body scanners.

Abundant resources drive up social complexity as well. Before the fossil-fuel age, people made a living in a relatively flat hierarchy of producers (mostly farmers), a few tradespeople, and a handful of rulers. Most people made or grew physical goods, and a tiny elite dictated how those goods would be allotted. The deluge of cheap fuel propelled more and more elaborate social and economic patterns. Look at the sheaf of layers we've packed into our economic hierarchy, both in the sheer number of professions we have and in the steeply stacked pecking orders that exist among them and within each of them:

the CEO at the top, then a host of executive and regular vice presidents, then even more department directors, each attended by platoons of senior managers, middle managers, line managers, and on down to the base of this teetering tower, at the increasingly empty, mechanized shop floor. Over the past few decades, our ever-expanding use of fossil energy has powered a complexity surge that has let us cram dozens more tiers of intermediaries into the cultural food chain, in a vast proliferation of white-collar jobs. Most of those employees, of course, don't produce anything other than guidance for the layer below them and reports for the layer above.

What happens when our vast pool of cheap energy starts to dry up? We're seeing it now. The middle layers vaporize, because the resource base is too small to support the immense webwork needed. This is why nearly everyone with an employer—with a "job" in that sense—feels so vulnerable. When resource flows start to slow, as they are doing in this era of energy descent, the economic system must simplify. The many tiers of white-collar jobs compress into a shorter, narrower stack. Those jobs get scarce. Production jobs, in turn, are shipped to where labor is cheaper—that is, to less byzantine cultures.

Those in the category of job-holders are now in a game of economic musical chairs, increasingly liable to be shut out the longer the game proceeds, unless they are at the apex of the social pyramid and thus insulated by their wealth. It's not a good time to be an employee. This explains in part the surge in entrepreneurs, those who are themselves producing a needed good or service and directing its disposition. Self-employment is one way to avoid the more evaporative layers of a precariously tall economic hierarchy.

Are there any clues as to which jobs are most vulnerable and which livelihoods may be the most secure in the era of energy descent? Our earlier exploration of energy itself offered a tool that I think is useful here: transformity. Remember that transformity is a measure of energy quality. It reflects how many steps are needed to convert a less directly usable source of energy, such as sunlight, into a "higher," that is, more usable, concentrated, or refined, form. Recall, also, the example in chapter seven that showed that supporting a diet of densely concentrated food such as trout means running an immense amount of energy through the entire food chain. If there is less energy in the food chain, say, due to a cloudy, cold year, we won't be eating trout as often because there will be fewer of them. We'll need to drop down to a food source with lower transformity, one that is more abundant.

The same is true in an economy that is under energy descent, and we're seeing it now. Three of the highest-transformity occupations, with transformity ratios in the billions of solar emjoules per joule, are contemporary finance, health care, and higher education. We've seen shakeouts rumble through each. The 2008 financial crisis vaporized trillions of dollars of nonphysical assets and sank several major brokerage houses, and it was prevented from cascading further only by a frenzy of money-printing by central banks in bailout programs. Medical care is unaffordable for a substantial chunk of the population, and recent corrective health plan laws in the United States also in effect simply print money to subsidize health insurance payments. The high transformities of our medical and financial systems make them tenable only as long as cheap energy can power their staggering complexity.

The same goes for education. A high school diploma was once a near guarantee of employment, but our system has become so complex that a bachelor's degree is now barely enough, and graduate school is almost mandatory for any profession. College and graduate school are attainable for many young people only via a generous program of student loans—another money-printing stopgap. Those loans will be paid back only if the economy steadily grows to provide jobs for all those graduates, and that requires an expanding pool of cheap resources and energy. These real assets are harder to come by than the swelling tide of digital money that's floating our economy presently, and the strain is showing. In other words, the cost of training for high-paying, high-complexity jobs is astronomical; competition for them is growing fierce; and the odds that they won't pay back their enormous cost are increasing. These are further symptoms of the extreme social complexity that energy descent will "cure" by simplification. As the real unaffordability of higher education becomes manifest and the need for it declines as midtier jobs disappear, student numbers will probably drop, and jobs and grants in higher education will decline as well.

As long as governments can keep printing money while clamping interest rates down, high-transformity jobs may continue to be lucrative. But barring the discovery of a new, cheap energy source, the writing is on the wall. The number and population of the economic layers above those who produce a tangible and essential good or service will shrink.

How can you not get caught in this trap? I know that for many the first response to energy descent is to draw up a plan to move to the country and try to become self-sufficient. For those with serious do-it-yourself skills, a loner's temperament, the willingness and strong back to farm from dawn to dusk, and enough money saved to survive the inevitable crop failures, varmint problems, equipment breakage, and constant presence of Murphy's Law in all its varied aspects, that plan may work. Having tried something much like that for a decade in the 1990s and having watched a lot of others try it, I can attest to its difficulty. Once that reality sinks in, plan B for many is to dream of the ideal intentional community, where you'll all be sharing the tasks and the problems together. Having also spent time in several intentional communities, I know that when they work, they can be wonderful. But well over 90 percent fail in their first few years.[7] Founding an ecovillage in the teeth of a crisis won't improve those odds, so if that's your plan, do it now.

The belief that moving somewhere else will solve our problems is so common that psychologists have a name for it: doing a geographic. But the problems often follow us, because they are generally our own problems, not those of the place.

Given this depressing scenario, what can you do? If you're already living among friends and colleagues, in a town or city that you know something about and can tolerate, where there are people who have needs you might be able to serve and skills that can complement yours, you already have many of the pieces of a successful livelihood. The grass may not be greener in a new place if you have to cultivate it from scratch. So if you decide to remain in town—and I recommend that as the place to start, since you are already there—what can you do to design and build a more reliable and meaningful livelihood? Let's look at what permaculture design tells us about that.

## DESIGNING A SECURE LIVELIHOOD

We are designing for a critical function here—livelihood—which immediately suggests a relevant permaculture principle: Important functions should be supported in multiple ways. It would be wise to have many strategies for supporting livelihood, and I will list some in very broad terms, so that we don't accidentally exclude useful possibilities by thinking too narrowly. Then we will look at ways to implement each.

Strategies for supporting livelihood include:

- Reducing or limiting consumption without unduly lowering quality of life (that is, conservation is often the easiest way to increase abundance)
- Finding nonmonetary ways of support (since money is not the only form of wealth and value)
- Developing skills that remain valued by others
- Having multiple sources of support and being able to expand or contract each as the market for them changes
- Generating and storing a surplus
- Using the stored surplus to support your living

Let's unpack these to see what each entails and to find techniques for accomplishing each one.

## Living Better while Using Less

As in any program for the wise use of a resource, be it energy, work, or money, eliminating waste is the least-effort path to creating abundance and should be done first. I've added the caveat about quality of life because if limiting consumption pinches too hard, it's like going on a severe diet, and few people can stay on a diet that leaves them hungry all the time. But preserving some quality of life, I think we can agree, isn't merely license to do whatever you want regardless of consequences. Quality of life also includes that of others.

How do we trim waste wisely? Much of it can be done simply by performing that primary permacultural activity, the observation step. Just keeping track of expenses and other consumption is often enough in itself to help spot and eliminate waste and excess. The late systems thinker Donella Meadows tells a story about a subdivision of identical houses in which one-third had much lower electric bills.[8] The savings was finally traced to the fact that in the conserving third of houses the electric meters hung in the front hall, while the meters in the other houses were down in the basement, invisible. Merely seeing those whirling wattage numbers was enough to prompt people to shut off unneeded power-sucking devices.

Mundane as it may sound, keeping tabs on monthly expenses and seeing where the biggest drains are leads almost automatically to reducing bills painlessly. Home accounting software that seamlessly imports and categorizes bank account data makes this task easy and fast, taking only minutes per month, unless you use mostly cash. It's dull and basic, but it works.

One eye-opening method that often spurs conservation is measuring the cost of purchases in terms of your own time. One of the life-changing livelihood resources for my wife and me was the book *Your Money or Your Life* by Joe Dominguez and Vicki Robin, which made clear how much time—our precious life energy—it takes, truly, to buy stuff. Three of the main facets of their program are to develop a financial goal (usually financial independence); to ask of every purchase, "Does this move me toward or away from that goal?"; and to see how much of your time it *really* took to earn a given amount of money. To do this last task, you subtract from your salary not just taxes but expenses forced on you by your job, such as clothes for work, therapeutic vacations, day care, restaurant meals because you don't have time to cook, and so forth. Then you divide that amount not just by the hours you work in a year but those plus the hours you spend commuting and in other work-related activities. I found that after I applied this logic, my seemingly substantial hourly wage dropped to about one-third of what my paycheck said. That nice restaurant meal, in truth, took four hours of work to pay for, which put a different face on it. This new arithmetic didn't force Kiel and me onto a budget, with all of its draconian baggage; instead, it simply inspired us to spend less by knowing how much of our life energy we were trading for purchases.

Keeping track of expenses acts like that electric meter in the front hall. Once you have a record of your costs (not just in money but in time and other efforts to acquire the things you use, and even in less tangible costs, such as stress on yourself and on relationships), you can spot patterns that show where your life's energy leaks away unnecessarily by asking:

- What are the big expenses?
- Which expenses are discretionary?
- Where is the fat that can easily be trimmed?
- Where are you making decisions that raise your costs, such as a house that's bigger than you need or daily restaurant lunches instead of brown-bagging it?
- What changes would make the biggest reductions in expenses?
- Which reductions would make you feel least deprived or most virtuous?

To ease the effort of implementing changes, list your potential cost-cutting moves in a few different ways: in order of the most dollars or time saved, by the easiest (least disruptive) changes to make, and by the changes you are ready to make now, soon, and not for a while yet. Then start making the changes that are high up on all these lists.

## Finding Nonmonetary Ways of Support

One of the major trends of the last few decades has been the creep of money into more aspects of our lives. For example, when people lived with or near their extended families, grandparents could look after toddlers while the parents went off to work, which also meant that the grandparents had family to care for them in their old age. No money was involved. Our shift to job-hopping nuclear families has shattered that arrangement (and divorce has split it further), so that parents today pay money for both child care and elder care. This means we need to work more at our jobs to pay for what we once did without money. Instead of child and elder care each adding to our assets as they do in a gift economy, in a money economy both deplete our assets.

Long hours at jobs means we're too busy to cook, so we buy meals instead of making them. Recreation, once mostly free, now tends to rely on expensive gear—not just snowboards, bikes, ultralight backpacks, and off-road vehicles but special clothing and accessories for each activity. All these purchases decline in value over time, whereas shared meals and do-it-yourself activities build value and skills as well as community.

Monetization has come at the loss of community. We hire help and entertainment instead of doing it for free with the people in our lives, those people who once took care of us and with whom we shared, made, and did things. This also points to a solution: Demonetizing often means simply bringing more community into your life. That can be as low-key as a potluck dinner with friends instead of a restaurant meal or as lush as a free week in a friend's vacation cottage or trading houses with them instead of spending thousands on a beach rental. Sharing resources and skills with friends and neighbors can shrink expenses. I'm constantly looking for ways to do this in my own life. For instance, we live in a rental house in expensive Sonoma County, California, but in exchange for my doing design work and land management, the property owner gives us a reduction in rent. The owner gets a permaculture paradise, Kiel and I get a financial break, and we all win.

Sharing with your face-to-face community is the most direct way to trim expenses, but the Internet is packed with resources for demonetizing, too. Most of us have seen the treasures that people give away via the free section on Craigslist and websites such as Freecycle. If you just want to borrow or share something temporarily, check out websites such as NeighborGoods.net that find people near you with whom you can share tools, appliances, and other household items. You can upgrade your wardrobe by trading unwanted clothes at sites like SwapStyle.com. Travelers can find free or very cheap housing at websites such as Airbnb, HomeExchange.com and Couchsurfing.com. It's even possible to permanently trade houses, saving the selling costs, via GoSwap.org.

Advertising and popular culture have spent decades and billions of dollars on molding us into good consumers, vacuuming up the contents of our wallets and encouraging us to max out our credit cards by playing on our fears and our desires for status. Breaking loose from that programming is particularly difficult in cities, where the temptations to spend money surround us. Fortunately, cities are where the cures—community, friendship, and innovative thinking—envelop us at high density.

Demonetizing at any meaningful scale primarily takes a change in attitude. Most of us have been taught how to create a budget, thus we've been shown tools, whether we use them or not, for not spending more money than we have. Budgeting is one path to using less money, but what I'm talking about is *using money less*. This means working on a broader level. We're not just saving money; we're searching for the aspects of our lives where we can remove money from the picture altogether. Just as child care and elder care can once again be demonetized in a way that also restores our human connections, so can all or part of many other human needs: food, shelter, health care, education, and the rest. Good design suggests that we can go through each of our needs and find ways to lessen the influence of money in each, and often—seeing the pattern here—that is best done by deeper, more direct connections to family, friends, neighbors, plants, animals, and the rest of the living world. A stronger connection to the technological world usually increases the need for money. A deeper connection to living nature can lessen it. Maybe that's why they call it livelihood.

The two strategies above, limiting consumption and demonetizing, fit into what permaculture designer Larry Santoyo calls "reducing the need to earn."[9] I like that phrase because, as with all good permaculture advice, it guides us to think at a very general level and lets us keep many options open. "Reducing the need to earn" is not demanding that you do without, and it's not even suggesting that you should earn less money. It asks that you curb those influences in your life that are forcing you to trade your time for money. You can think of those influences as one of the multiple kinds of sector energies that affect livelihood. Reducing the need to earn gently suggests we develop a set of strategies for blocking or "disappearing" those livelihood sector influences that drive us toward monetizing. But there are many kinds of sector energies that influence livelihood. The remaining strategies in our list focus on harvesting the beneficial features of the livelihood sector energies.

## Developing Skills That Retain Value

Many jobs in today's economy are not secure, and as the complexity of our economy becomes tougher to shore up under energy descent, that will only get worse. How can you make sure that you have a livelihood that won't erode or evaporate? One strategy is to be providing something that is valuable under almost any circumstances. During the recent serial bubbles in technology, real estate, and finance, jobs in each bubble-field were lucrative and abundant, but the shakeout was correspondingly vicious. These high-on-the-food-chain professions have their usefulness, but in a contracting economy only people with the strongest claims to jobs in those fields—the brokerage owners, the founding entrepreneurs—have much chance of staying in. Rank-and-file employees don't have a strong claim to their jobs—they aren't in charge—and they get pushed out in droves. Once again, an understanding of transformity tells us that the closer your work is to providing for a basic, constant human need, the less likely it will be made unnecessary.

This move toward real needs may be one of the driving forces behind the rise of urban farms and CSAs, as well as artisanal foods and crafts. The latter right now seem expensive and can't compete with heavily subsidized commodity foods whose costs in pollution, energy, and ill health have been largely externalized. For now, small producers must rely mostly on restaurants and the affluent. But I suspect that as it becomes less possible to disguise costs and quietly shift them to the commons, we'll be paying the true price for more of our goods, and that price will be nearer what the small producers need to stay afloat.

Energy-savvy writers such as Richard Heinberg and Sharon Astyk envision an America of 50 million farmers after the energy subsidies to commodity farming disappear.[10] Innovative growing methods such as small-plot intensive (SPIN) farming are helping new farmers earn upwards of $50,000 a year from urban yards. A neighbor of mine, Paul Kaiser, grosses $100,000 per acre at his Singing Frogs Farm by using no-till, high-density methods; scrupulous soil care; and greenhouse-grown transplants to speed the growing cycle. His initial investment in specialized equipment and training were high, but he says he's happy with his net income, and unlike many farmers, he is able to pay his workers well.

If food growing doesn't seem an appealing livelihood, move around the permaculture flower to other basic needs. Many people are transitioning from office jobs to develop or build energy systems, water harvesting and purifying setups, and efficient small houses. Physical systems break down, too, so repair work always has value. The intangible needs remain with us as well, such as nursing, midwifery, and low-complexity medical and dental care. And as people make this transition, we'll need teachers to train them as well as accountants and lawyers or paralegals specializing in small business. The list of possible professions is a long one, and I've just scratched the surface here. If you don't want to run your own business, then work with a partner or entrepreneur you can trust. The key quality in secure employment, I think, is to work as directly as you can with something physical that meets a basic need, as opposed to manipulating inflated, made-up assets as one of many people squeezed into a tall hierarchy of midlevel jobs.

Most of us need some incentive to make the switch to a more valuable livelihood. Obviously, losing your job is a strong motivation, but let's hope you avoid that one. If you're feeling insecure about your job, that's worth listening to, as you're probably right. For me, the impetus was increasing misery. I kept telling myself I liked my work doing genetics research, but regular minor illnesses, all the classic signs of stress, and a mysterious general anxiety were saying otherwise. My body was giving me more accurate data than my security-seeking mind. Another incentive is to study people who have already shifted to simpler livelihoods. Do they seem happier than you?

I don't claim to know precisely how the future will play out, but the laws of physics remain in effect, and a declining energy base means more people will be doing work of less complexity tied to real needs. Those niches will be where the rewarding livelihoods are. Remember, we're not talking about getting rich; we're talking about being valuable and thus needed by your community.

## Having Multiple Sources of Support

Remaining valuable and being supported becomes more certain by having several sources of income and being able to expand or contract each as the market for it changes. In the 1970s, homebrew electronics pioneer Don Lancaster coined the term "nickel generator" to describe each of several small businesses, products, and design ideas that individually would yield only a meager income but in combination would produce enough to live on.[11] And, together they would be diversified enough to respond to shifting demands and conditions. If one nickel generator went under, it was only a minor setback, while having one take off was glorious. With inflation, we'd have to call them "dollar generators" today, but Lancaster's advice on what we would now call multiple microenterprises was good whole-systems thinking. He was ahead of his time, and the idea has matured.

The less developed world has spawned hundreds of millions of microenterprises in local markets and home businesses, in part because these are a smart strategy in unstable economies where capital and credit are tough to get. Now that the developed world is tilting in the same direction, microentrepreneurship is booming there, too. At least 20 million Americans work in microenterprises, and it's one of the most effective routes out of poverty as well as an opportunity for women and minorities. Participants in one microenterprise survey said their incomes had increased anywhere from 17 to 84 percent, many had left poverty programs, and almost 90 percent were still in business after five years.[12] Many practice "income patching," combining a part-time job—another increasingly common product of our economy—with one or more home enterprises.

Cities are where microbusinesses are booming, because markets, suppliers, and workers are close by. The Internet, once again, has made microenterprise easier. You can rent out a spare room via Airbnb and related websites; use your car as a taxi with Lyft, Uber, or Sidecar; do odd jobs for others through TaskRabbit; and of course upsell used goods and thrift-store bargains on eBay and a host of other resale sites. A combination of these or other nickel generators could be a full-time livelihood, ease a transition to a new career, or offer a "start small" path to a larger business. This last was the route for Davin Vculek and Joe Blanton of Sacramento, who dreamed of opening a restaurant and, to avoid the hefty price of leasing a building, began with a food truck that served hamburgers and sweet potato fries. Success for Krush Burger was swift—it was the only food truck in town—and a second truck quickly followed. Soon Vculek and Blanton had saved enough to open not one but two restaurants. They kept the food trucks, though, to give them multiple income sources while their restaurants got off the ground, and because the trucks were by then easy to run.

Useful features for a collection of microenterprises would be businesses that:

- have synergies that support one another, as in the case of bloggers whose posts can be assembled into books and articles, while at the same time posts and publications alike build an audience for potential lectures and workshops;
- allow multiple enterprises from a single skill set, as in the food truck/restaurant example above or a beekeeper who maintains hives for others, provides pollination for orchards, sells honey and other hive products, and trains new beekeepers;
- are diverse enough to respond to shifts in the economy, such as carrying both high-end and low-end goods or offering products in both assembled and kit form.

## Generating and Storing a Surplus

Any livelihood needs to produce a surplus—a little more than is needed for subsistence—to be prepared for repairs and emergencies, for savings and investment, and to exchange for the things that the livelihood isn't producing directly. Useful activities tend to create a surplus; in fact, when resources are piling up around you, that is a sign that you are doing valuable work. Scarcity can be an indicator that you're not finding the people that value your efforts, and that may be due to poor marketing or not finding a niche that distinguishes your work from others. That niche often appears when your heart is deeply in your work.

Once you start generating a surplus, in keeping with the principle of catch-and-store energy and materials, you need to find beneficial places—catchments—for that surplus to go. One of the causes of scarcity is failure to create catchments for surplus. If there is nowhere for surplus to pool, it will drain away, the same way that water rushes off compacted soil.

Once surplus is caught and stored, it becomes analogous to a stock of capital; that is, to quote the dictionary, "an accumulation of goods devoted to the production of other goods." It's important, though, to recognize that surplus takes many forms, not just money or products. Surplus money can build financial capital, but valuable activity creates other forms of surplus that offer many benefits besides generating more money. So let's expand our definition of capital from the narrow one of economics and think of it as any surplus that we can store for the purpose of producing benefits to ourselves and to all that is around us. Permaculture consultant Ethan Roland, the owner of AppleSeed Permaculture, has looked at capital this way and has spotted at least eight different forms of capital, outlined here.

### Social Capital

This is the goodwill from others that your service to your community creates, the favors you owe and are owed, and the connections and network you have. When your social capital builds up, people are there to help in tough times, to make good times more rewarding, and to aid you in exploring and grasping new opportunities.

### Material Capital

This comprises the raw and processed nonliving resources, such as rock, lumber, buildings, tools, fuel, and so forth, as well as the stored-up goods that our activities generate. Material capital buffers you from life's vicissitudes, provides physical comfort, and makes work easier.

### Living Capital

Plants, animals, soil, and the collective properties of ecosystems, such as purifying and circulating water and storing nutrients make up our living capital. Land-based permaculture gives us good tools for enhancing these. Your own health is a form of living capital, as is ecosystem health. When living capital accumulates, we're more comfortable in our bodies and can work better, we're better nourished, and we are surrounded with beauty and a nurturing environment.

### Intellectual Capital

What we know and have learned is our intellectual capital. It is the stored knowledge and ideas that are available to us. Acquiring this reduces the mistakes we make; lets us learn and work with words, images, and concepts more easily; and helps us generate new ideas.

### Experiential Capital

Whether you call this street smarts or the school of hard knocks, experiential capital is the difference between what a novice and a master knows. It's also what we acquire (we hope) as we get older, in the form of wisdom. It makes us more certain in our actions, and it lets us take things in stride, spot opportunities, teach others, and anticipate trouble before it develops.

### Spiritual Capital

There's a reason that meditation, prayer, and yoga are called spiritual *practice*: The more we do them, the deeper our connection to the higher order grows. The ease of achieving this deep connection with and guidance from something greater than ourselves stems from the development of spiritual capital.

### Cultural Capital

This is the shared art, music, myths, stories, ideas, and worldviews of a community. This is a collective capital, not held by individuals. When we enter a new culture, it can often take time—sometimes generations—to inculcate and benefit from its cultural capital. Cultural capital is what glues communities and whole peoples together and adds depth and meaning to human life.

### Financial Capital

This consists of the money and financial assets we have acquired. Financial capital buffers us against material shortages, lets us apply it to empower businesses and other activities that we want to support, and can generate money when invested carefully. We often unconsciously measure other forms of capital, particularly material capital (e.g., a house) and living capital (the timber in a forest), in financial terms, but this discounts or just plain misses much of their real value. Each different form of capital has its own qualities, and they aren't fully convertible from one into another without grinding away the nuances and whole value of each.

While we're most familiar with capital in its financial form, this is the least stable and reliable kind, as well as the one that is often the most

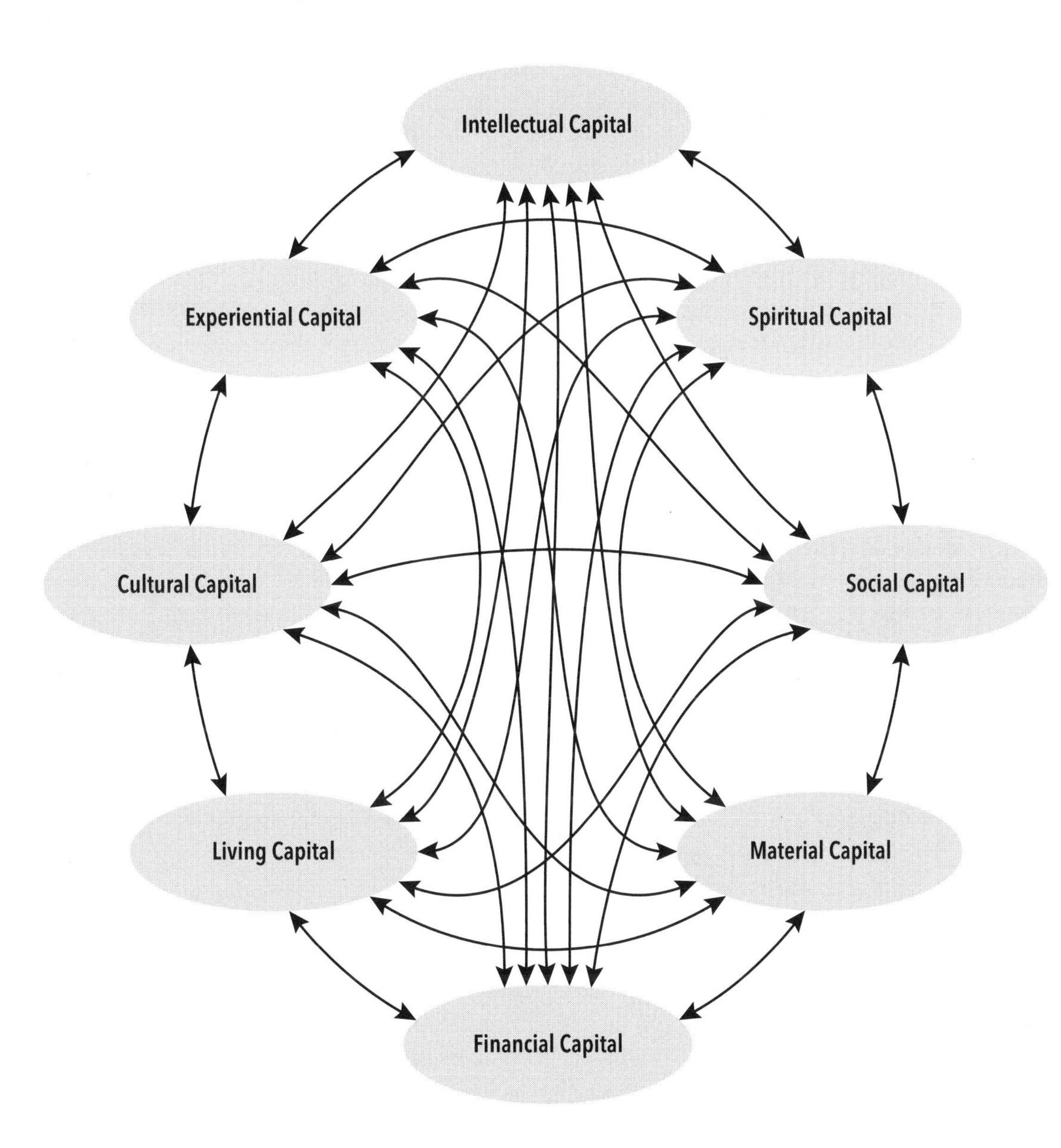


**FIGURE 8-1.** The eight forms of capital support one another and combine to reduce the need to earn. Adapted from Ethan Roland, "Eight Forms of Capital," www.8forms.org.

scarce. The value of money is eaten up over time by inflation, stocks and bonds gyrate with the market, and the value of goods when measured as money also varies and often shrinks as the goods degrade or go obsolete. But unlike money, our stores of the other forms of capital are often within our own control if we tend them well. Their usefulness rarely declines and often increases over time. We want to build capital in ways that are regenerative, both in what we do to acquire them and in the form that they take.

In a volatile money economy—especially since money is really a claim on some other kind of asset—it often makes sense to convert money into other forms of capital to preserve its value. If we spend money to build good soil, that soil will continue to grow food, even when the price of food is skyrocketing. A solar panel will produce a given amount of energy regardless of what you paid for it or what the price of energy is. When you've mastered a new skill, a stock market collapse doesn't take that skill away. If you are trusted and valued in your community, your friends will help you weather a rough patch much more cheaply than your credit card will.

Understanding how the different forms of capital work helps us develop the right methods for catching them, lets us see what ways of storing them may be the most stable and reliable given various circumstances, and aids in spotting the places where catchments and storages are missing. It's worth looking at your activities with an eye toward all the forms of capital that they might generate, because it's almost certain that you're missing some opportunities to catch and build up some of them, or at least aren't seeing where they are helping you. Being aware of the many forms of capital makes me happier in many of my doings, because I can see where they are benefiting me in ways I didn't notice before. And it's made me more generous, because I can recognize all the ways that helping others builds up my own forms of capital, and theirs, and my whole community's as well.

## Using Stored Surplus to Support Your Living

The saying goes that regular folks work for their money, but the rich make their money work for them. In conventional economics this is the goal of storing surplus to build capital, which then generates income. But again, let's broaden the focus to expand beyond money and financial capital. We can use the synergies among the many forms of capital to generate a livelihood that is richer and less effort-filled than what financial capital alone can create. This also fits our strategy of widening our lives so that our needs are satisfied in ways that don't run them all through the narrow, scarcity-minded channel of money first. Relying on the many forms of capital helps us demonetize our lives. When we are rich in many forms of capital, all those forms and all that they connect to are working for us.

To see how the many forms of capital work together, let's look at some basic needs. Most of us earn money to buy, among other things, food. An obvious way to move a bit outside of the money system is to grow some food. If you're going to grow food, the leverage point that every wise gardener starts with is to build fertile soil. By doing that, we've built living capital for the natural world, and since the garden's produce is probably more healthful than what we can buy, we're building our own living capital via better health. We'll have surplus veggies to give to our friends, and the resulting social capital triggers reciprocal gifts of food, brings them over to help us garden, and prompts invitations to shared

**TABLE 8-1.** How the Eight Forms of Capital Meet Multiple Needs

| CAPITAL FORM | NEED BEING MEET | | | | |
|---|---|---|---|---|---|
| | **Food** | **Water and Waste** | **Energy and Shelter** | **Community** | **Livelihood** |
| **Social** | Share meals, reciprocal gifts, help in the garden | Shared watershed stewardship | Community projects | Builds connections and support | Marketing, job opportunities, employees |
| **Financial** | Purchasing food and supplies | Pay water bill | Pay energy bill and buy supplies | Help in times of tight money | Income |
| **Experiential** | Cooking and gardening skills | Building catchment, restoration | Energy independence | Good people skills | Work experience, resume building |
| **Cultural** | Ethnic food and plant varieties, becoming indigenous to place | Knowledge of water use in region | Locally appropriate energy sources | Develops a local culture | Builds vibrant regions |
| **Material** | Tools, supplies | Tanks, ponds | Energy generators, buildings | Borrowing, gifting, and trading goods | Work-related goods |
| **Intellectual** | Cooking, gardening, plant breeding, etc. | Understanding water cycle | Construction skills | Learning from others | Learning a valued skill |
| **Spiritual** | Connection to a place and life | Connection to water and watershed | Connection to natural forces | Friendships | Sense of social worth |
| **Living** | Soil, biodiversity, habitat | All life needs water | Replant forests for timber, use renewables | Beautify the neighborhood | Encourages earth-friendly livelihood |

meals. We can tap into cultural capital to find varieties of plants and animals that do well in our area and perhaps contribute to it by saving seeds, and the seeds might have a good story behind them—there's more cultural capital. Experiential and intellectual capital builds as we learn how to produce food and rich soil more reliably and easily. Gradually, too, we'll get a few tools and garden supplies, and that builds material capital that makes our work easier. If we share the tools, there's more social capital. Each form of capital acts as a growing resource base that helps build capital in all the other forms. Together they reduce the need to earn money for food, along with a host of other benefits.

We often think that we can't do much without financial capital, especially if we're starting a business. But money is not the linchpin that holds our lives together; it's merely the form of capital that our culture has become obsessed with. Social capital in particular is an especially potent and flexible form, as we are adept social animals who generate it instinctively. Social capital can easily translate into help with launching our own efforts to build other forms of capital. Friends can aid not just in physical labor but in connecting us to the right people (more social as well as cultural capital), sharing a critical piece of knowledge (intellectual capital), loaning us the tools we need (material capital), and giving us support (spiritual and cultural capital). When I am able to see all the forms of capital that each action builds for me—as opposed to reducing it all to money—I notice how wealthy I am.

Many of the forms of right livelihood I've explored here depend on a certain amount of entrepreneurship, and one hurdle to starting a business has traditionally been lack of capital—but that invariably means financial capital. Seeing capital as having many forms sweeps away that hurdle. Using your own social capital can bring in not just microloans from friends and family but also help and networking connections that reduce initial expenses. Instead of expensive training programs, many entrepreneurs learn by doing small jobs for low fees for relatives and friends. They share the surplus material capital of more established businesspeople in the form of office or shop space and tools, and they benefit from experiential and cultural capital as well. When they're up and running, they're also inclined to pay back these gifts with their own accumulated capital in all its many forms.

All these interacting forms of capital support one another, and each can be converted to some extent into the others. If, say, you've used up some social capital by borrowing your friend's truck a few times, perhaps a gift to that friend from the living capital of your garden, a loan from the material capital of your tools, or a free lesson in a new skill from your experiential capital can top up the drained social capital account.

## BUILDING SYNERGIES WITH BUSINESS GUILDS

The concept of interacting forms of capital, where each strengthens the resilience of the whole, applies well beyond the personal realm. A community-scale example of this comes in the form of business guilds, a term I first heard used by designer and multiple microentrepreneur Larry Santoyo. A business guild is a set of independent enterprises that are tied together by connections among some of their needs and yields. These alliances lessen the need for cash purchases, help guarantee or attract markets, and share common and surplus resources. To illustrate, I'll describe an impromptu business guild that developed among four restaurants in Portland. Two of them were next door to each other: a breakfast-and-lunch café and a fancy dinner establishment that owned a gorgeous outdoor patio. Since the hours of the two businesses didn't overlap, the café made a deal to use the dinner spot's patio in exchange for a steady supply of the café's renowned soups. A mile or so away was a cheese shop that had space for a small lunchroom. When its owner heard of the soup-for-space deal, he offered to trade artisan cheeses with the café for sandwich supplies and soup. Then a bagel maker opened a few blocks up the street and soon delivered bagels to the café, also in exchange for food that expanded the bagel place's own lunch menu. These were four restaurants that could have seen each other as competitors, but because the owners recognized that some of their needs and yields were complementary, all were better off. They could cut costs, broaden their menus, benefit from the skills of other chefs, and share a larger customer base. Business is often seen as a zero-sum game, but collaboration will create solutions that multiply resources, develop ideas, and build synergies that are impossible to achieve alone.

Another example is the Permaculture Skills Center in Sebastopol, California. Founder Erik Ohlsen, who conceived of the center as a business incubator, runs his landscape design shop from this site at the edge of town. A large garden cranks out vegetables for a farm stand near the busy roadway, providing income for a gardener and market manager. The garden also provides produce for nearby restaurants and the center's own CSA customers. Two other designers, and sometimes more, lease office space at the site, and they shift clients among themselves depending on their workload and the specialties needed. Some of them work in a design consortium that includes outside designers, enabling them to take on projects larger than they otherwise could. The land itself showcases the work of the resident designers and brings in tours and workshops run by the staff and outside organizations, which expands everyone's client base. Several local organizations hold meetings there. Two other consultants housed there split their time between their own work and some of Ohlsen's larger projects. I teach there, hire consultants and instructors that I've met there, and use it to network. We run a nine-month ecological landscaper training program from the site that uses many of the site's visible and invisible resources. A host of other designers, teachers, local experts, nonprofits, farmers, clients, contractors, volunteers, and neighbors cycle in and out of the center, sharing extra business, filling gaps for each other, and exchanging clients, tips, contacts, and work parties. Given all the connections, every business and activity there can tie into far more opportunities, help, and knowledge than each could by itself.

Each of the strategies for livelihood that I've listed here works toward the goal of creating a resilient economy that recognizes forms of value beyond money. Money is terrifically useful, but like any metric, it can measure only certain properties, and the more of our lives we run through the simplifying filter of money, the more forms of richness are stripped out of it and reduced to those that money easily measures. A regenerative economy that offers right livelihood is based, like the eight (or more) forms of capital, on a definition of value that includes as many aspects of human and more-than-human life as we can be aware of.

# CHAPTER 9

# Placemaking and the Empowered Community

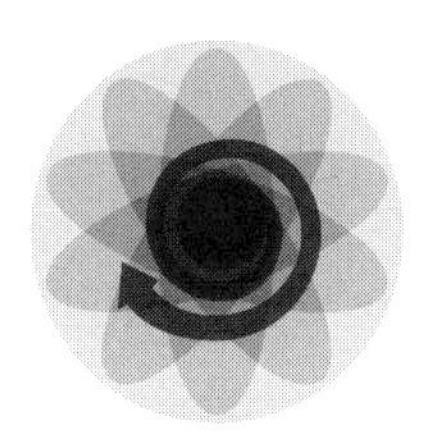

Although the people side of permaculture design has been implicit in the earlier chapters of this book, thus far we've been talking primarily about gardens, water, energy sources, and money, which are all physical things. With livelihood we began moving from the visible- to invisible-systems petals of the permaculture flower. Now it's time to focus more explicitly on people and their interactions, more deeply in the invisible-structures petals. After all, the most abundant resource in any town is people. Yet modern cities and towns are rarely designed—if they are designed at all—in ways that foster positive human interactions, spontaneous and easy gathering, and working smoothly with each other. In the past few decades, people's frustration with the overly car-centric design and inhuman scale of our cities has blossomed into a movement, variously called placemaking, the new urbanism, ecocities, and related terms, dedicated to returning cities to the people who live there. But a healthy community needs more than well-designed places. It needs well-designed relationships and social systems, too.

This chapter will begin with a look at how permaculture design tools can give us unique insights and tools for understanding and working within the human side of town life. Next we'll explore placemaking, the ways that the physical infrastructure of the city affects what people do and feel there, and how the built environment that we already have—that we're more or less stuck with—can be softened, warmed, and ameliorated to make room for those people. Then we need to look at constructive ways to behave when people *do* get together; that is, the tools that have evolved for working well with one another. As I've said, the visible, hardscape aspect of good design—the technical piece—is easy compared to the people side, the daunting and complex process of learning to cooperate with each other.

In other words, we need to be aware of and employ healthy social tools for collectively meeting our needs. If the goal is to be able to work together, permaculture design nudges us to ask, What are the components, relationships, and special sector energies acting in human communities that we need to assess and design with in order to do that? What can help us communicate well and resolve, deescalate, and transform conflict? What are useful tools for making decisions, ones that operate at the right scale for various group sizes and issues? How can we design in ways that create the circumstances, organizational structures, and management systems that seamlessly catch and store the energy of people working in groups? What are the relevant patterns to identify and nurture that will help those human energies work effectively together? Creating the physical conditions for people to come together is the beginning, but building social health is the real meat, not just of community but of living sustainably on this planet. If the problem of sustainability were simply a question of building things, we'd be living in the Garden of Eden by now. We're not there yet, because at heart sustainability is a social, psychological, and political problem. What has come to be called social permaculture is an attempt to apply ecological design methods to this challenge, and this chapter will explore some of the basic tenets and tools of social permaculture.

## GETTING PERSONAL WITH ZONES

Once again I will turn to the familiar design methods of zones, sectors, and needs-and-resources analysis to help parse the human side of urban permaculture. People and their motivations and needs are complex and changeable. Thus it's important to remember, even more than with landscapes and the built environment, that permaculture's design tools don't pigeonhole design elements into hard-and-fast categories. The design methods are heuristics; that is, thinking tools to help us understand and work with design elements. So if it helps to sometimes think of a factor in a design as an element that's located within a zone and other times consider it as a sector, that's fine. The ever-shifting qualities of human beings are a good example of this. My wife, for instance, resides deep within my personal zone 1, but she is also a powerful sector energy in my life.

How can we use permaculture design methods to help us understand and work with people? One way is to notice that we have personal zones of different kinds, and we have personal sectors; that is, influences coming from other people, groups, and organizations that affect us. What do personal zones and sectors look like? A map of personal zones could arrange the people in our lives according to how often and how intimately we interact with them, while the relevant sectors indicate in what contexts this happens and how they influence us. I use this method to get a graphic representation of the many people in my life, to show how close or distant my relationship with them is, how they affect me, and how those effects interact. It helps me see if I'm neglecting or overemphasizing certain people or groups and figure out how I can best work with their influence on me and on others.

I find this most useful when I distribute the elements (the people) into these zones not according to classes such as friends, family, and colleagues but either as individual persons or as groups that I encounter as a whole, such as "my brother David" or "my neighborhood association." In other words, although my wife and my

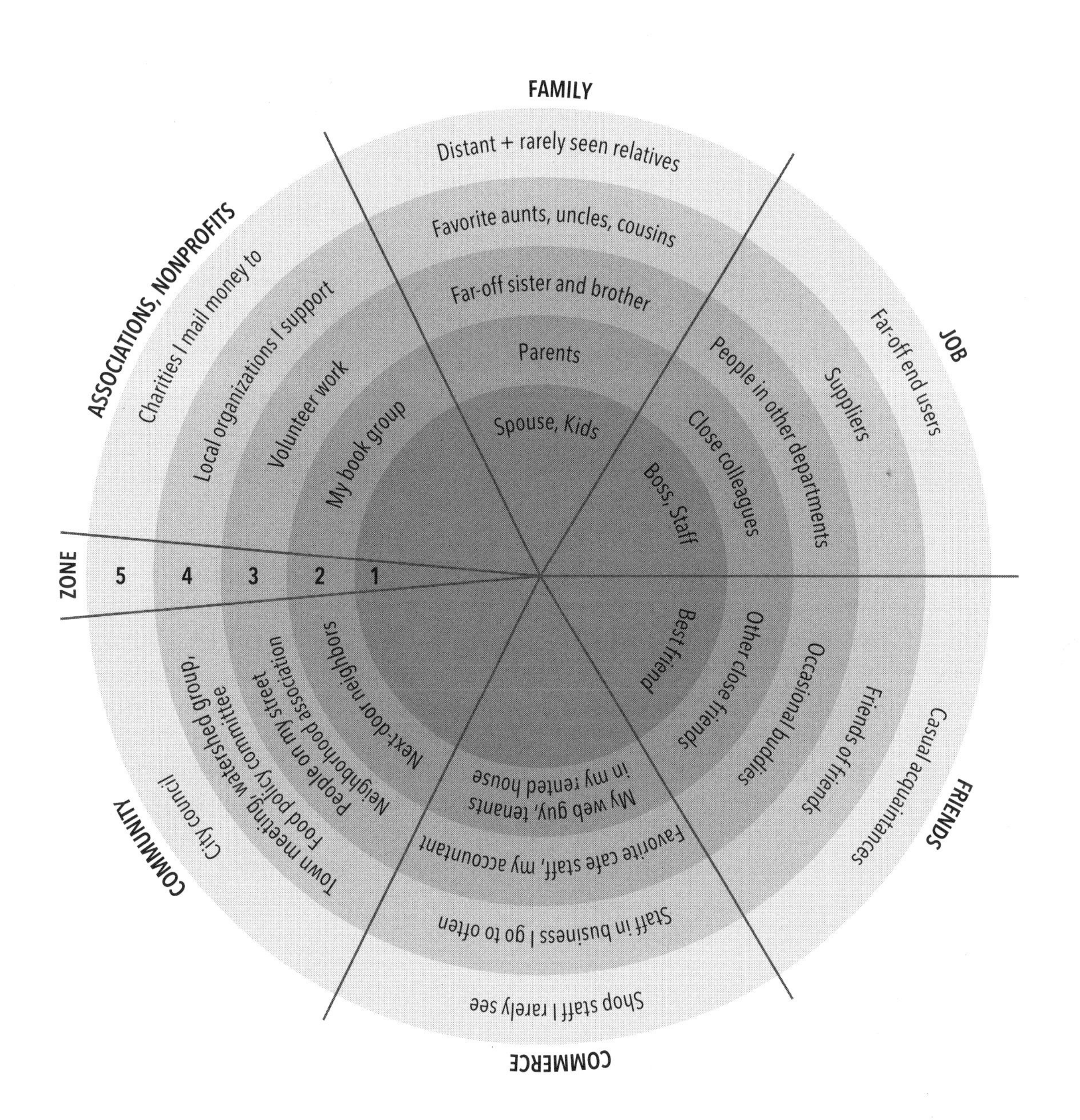


**FIGURE 9-1.** Intimacy zones and sectors, based on the major influences in a hypothetical person's life, and how directly and strongly the people in each of those sectors affects that person.

sister are both immediate family, my wife I see many times a day, but I talk to my Chicagoan sister just a handful of times a year and see her even less often. Thus even though they are close family, they are not in the same intimacy zone. The zone system would place these two close family members in different zones because I don't interact with them equally frequently. (To jump ahead, family and friends as classes are more fruitfully thought of as sector influences, not as being in a zone.)

Who is in your zone 1? When I worked a nine-to-five job, I saw my boss, my staff, and two or three colleagues about as often as I saw my wife, so they all belonged in my zone 1. They all required quite a bit of daily "care," meaning energy and attention toward maintaining a good relationship with them, just as those tender lettuces outside your door need plenty of attention for both of you to benefit from your relationship. Thus intimacy zone 1 includes the people you see the most: a spouse or life partner, children or other family living with you, your closest colleagues that you work with every day, and perhaps a very best friend with whom you spend lots of time.

Intimacy zone 2 might be your circle of good friends, regularly seen colleagues, and family you see less often but are still close to. Zone 3 might be made up of casual friends, more distant family, neighbors you interact with, and workmates you see periodically. Zone 4 would be people whose names you know—like that friendly waitress at your regular café—but whom you know little about. Zone 5 would include relative strangers whom you rarely encounter more than once.

When there are many people in your life, this method can keep track of them and help you not lose touch or wear out the welcome mat with them. I use this zone system to notice the closeness or distance of the people in my life. Sometimes I'll see that someone out in zone 3 is pretty cool and I'd like to pull them into zone 2, or I'll realize that somebody has marched into my zone 2 and I'd be better off if they were pushed out into zone 4.

Here's another way to organize personal zones, developed by Michael Becker, an award-winning public-school teacher who uses permaculture as the basis for his entire curriculum at Hood River Middle School in Oregon. Michael points out that we all have comfort zones, and he applies this system to communicating well with people in his professional and personal life. Comfort zone 1 encompasses the activities we do that we are familiar with and know well: our habits, routines, and regular doings. Zone 2 holds activities and events that we are less familiar with but that aren't stressful or anxiety-producing for us to perform. Zone 3 includes things that we push ourselves a bit to do—a new sport or travel to an unfamiliar place—that may make us a little nervous but we know will pay off. In zone 4 we're being forced by circumstances to do things we're really not comfortable with. And zone 5 covers those things that we would almost rather die than do.

As a teacher, Becker points out that for us to learn well, we can't venture much past zone 3 or we become too busy soothing ourselves—or just being freaked out—to take much in. Most learning requires that we move new endeavors from outer zones into zone 2 or 1 by tying it to activities that we are familiar with, then increasing that familiarity via practice. This system is also useful for encouraging us to try new things, break routines, and enlarge the scope of our lives. What's out there in comfort zone 4 that's a little edgy for us but would be an enriching piece to bring into our lives?

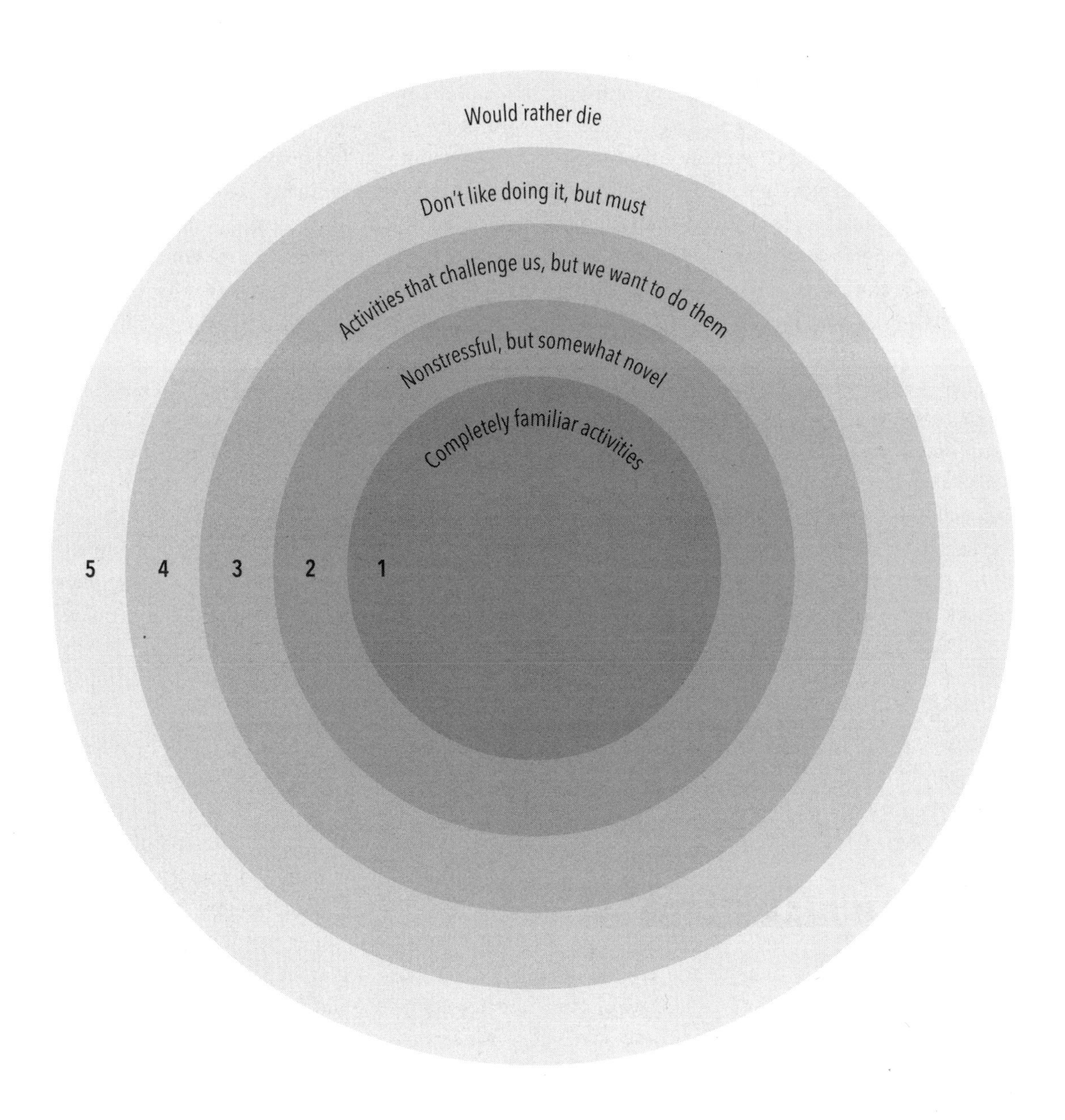


**FIGURE 9-2.** Comfort zones. Effective learning takes place in zones 2 and 3. Familiarity and mastery will move activities to the inner zones. Concept by Michael Becker.

Thus if we are trying to persuade someone to consider a new idea or endeavor, we will be more successful if we introduce it so that it links to one of the person's familiar zone 1, 2, or 3 activities. If we don't, we'll run into resistance or outright rejection. Becker relates how, years ago, he finessed the potentially momentous announcement to his mother that he was moving in with his girlfriend. He knew that his mother loathed the fact that in his living situation at the time he had been washing his clothes at a Laundromat. Putting his shirts into a machine that had held some stranger's dirty underwear was, to his mother, repulsive—out in her comfort zone 4 or worse. So he introduced the subject by saying, "Mom, Meg and I are moving in together—into a house that has a washer and dryer!" That last was such a relief to her that she barely registered the live-in girlfriend.

In many ways, using personal zones can smooth out some of the rough spots in our lives. As we've seen throughout this book, the zone concept is broadly applicable, immensely malleable, and can help us understand and work with any cluster of activities that varies in frequency of use, distance, accessibility, familiarity, or intensity or can be ranked along almost any type of gradient.

## ➡ THE HUMAN SECTOR ⬅

Let's look now at ways to apply the concept of sectors to ourselves and those around us in our homes, workplaces, neighborhoods, and communities. In the human realm, sectors are the various influences that other people and organizations have upon us. We can easily map a set of sectors onto our intimacy zones. One obvious set of sectors are the influences of easily identifiable categories of people such as friends and family. Although some of us might consider "family" as a uniform sector in itself, family often comprises several kinds of people, each with a different effect on us. Thus, to me, it pays to split family members into several sectors. One of these is the influence of a spouse or significant other. I don't know about you, but my wife influences me in a way that no one else in my life does, so I am putting Kiel into her own sector. (Note that she, as a person, is also an element in my zone 1. But she is not necessarily the only person in the "Kiel" sector. Her parents—my in-laws—are also part of the Kiel sector, just not in zone 1 the way Kiel is. This is what I mean when I say that zones and sectors are flexible heuristics.)

Kiel and I don't have children, but those who do might give them their own sector as well, and some people might sort their kids into individual sectors. The influence of an infant son is not like the influence of a teenage daughter. Parental influences are another sector. So we have several categories of "immediate family" sectors. Adjoining these we can add a category for more distant family.

Another set of sectors would be workplace influences. If work has a fairly uniform influence on you, you could label this as a single sector. But it's likely that you are affected differently by your boss, your staff, and your peers, so you may need three or more sectors for work. You may have a direct supervisor who is in zone 1 or 2 of the "boss" sector and regional and national managers who are in more distant zones of that same sector.

Then there is the friends sector, which could be split into close and casual friends. Other personal sectors could include organizations that you belong to, people or businesses you have financial relationships with, your spiritual community, clubs and membership organizations, and even

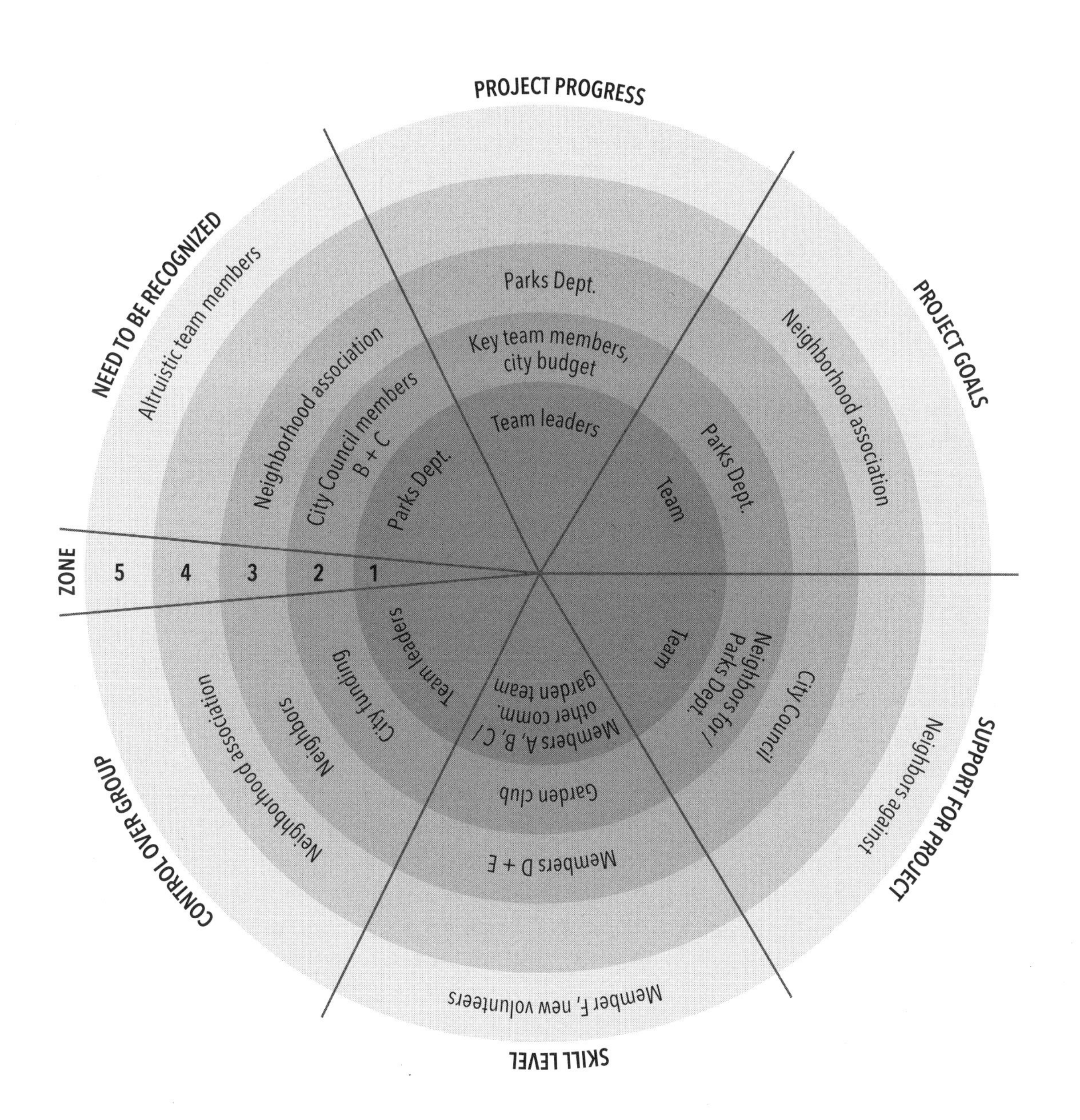


FIGURE 9-3. A human zone and sector analysis for a community garden project, explained in the text.

more diffuse influences such as government or large corporations. This list is just a suggestion, as each person has a different set of influences in their lives, so it's up to you to identify your own personal sectors. Remember, too, that some sectors can overlap with others and that some elements in the design—in this case the people—will be affected by multiple sectors.

Another way to use sectors in the personal realm can help us work with groups of people. What are factors that influence the way that people in groups function? Let's say that a team has formed to plan a community garden in their neighborhood. Their sectors would be the forces that are motivating them and affecting how they work together. One influence is the task itself; that's what is motivating the group as a whole. But within the group are other influences, drives, and needs that will push and pull the members and the whole group in various directions, and those can complicate getting the task done. One of these is each person's relative need for recognition and to be heard. Another is the level of control or influence each person wants to have upon the task and the group. Other sectors might include the ease with which the task is moving forward (the group's success or frustration level), the fear or comfort level each person has with various aspects of the project, the sentiments of the rest of the neighborhood about the project, the level of bureaucratic complexity, and so forth. When a group comes together for a task, it helps immensely to make the human influences explicit, so that the group can talk about them when conflicts or setbacks arise and put mechanisms in place to beneficially use these energies or damp them down when they are causing trouble.

We could easily add zones to our community-garden team map. Zone 1 would include the team members, zone 2 would be the potential users of the garden, zone 3 includes the neighbors affected by the garden, and zone 4 the members of local government and the contractors involved in the project. Now we have a comprehensive overview of the stakeholders in the project and the motivations and forces that influence both them and the progress of the project. I've found these zone and sector maps to be hugely helpful in identifying and working through the challenges and dynamics of group projects.

## PEOPLE NEEDS, PEOPLE RESOURCES

Using the design method of needs to resources for individuals and groups follows the same pattern as using zones and sectors. What are your own personal needs—physical and psychological? Have you designed your life in ways that these are being met? What elements (people and goods) and processes ought to be added or rearranged to satisfy unmet needs? Which needs are causing unnecessary work? Then, what are your yields and behaviors, and how can these be connected beneficially to your own needs and those of others? What is in surplus that could be converted or exchanged to meet a need? Which yields are not serving you or are building up to cause "pollution" in a manner analogous to how physical excess can cause problems?

By now it should be obvious that groups, too—families, businesses, and other organizations of people—have needs and yields that can be charted and linked inside and outside the group. One useful task is to determine which needs are best met from outside the group and which are optimally met from within the group.

A sector analysis of human influences can also help identify needs and yields because the influences that people most notice and respond to often indicate their needs. Some of the sector influences on individuals and groups that we've already identified are direct indicators of needs: people's desire to be heard and their need to see progress toward the task. A needs-and-resources analysis can help us identify and meet these needs by finding processes and resources that provide them.

## ➡ BUILDING THE ⬅ VILLAGE IN THE CITY

Let's move to the physical, built part of the human environment and how it affects how people behave together. Although as a general rule cities today are vibrant and increasing in size, in the late 1960s many cities in the developed world, especially in the United States, had begun to decline, as urban decay and its partner, the flight to the suburbs, shattered and deadened community life in metropolises large and small. How had this happened, and what occurred to reverse it?

As industrialism spread in the nineteenth century, the patterns of growth in cities changed dramatically. Prior to that, most cities had been organically created by the people who lived in them. With the rise of the factory town, urban development was increasingly dictated by top-down planning that viewed the seeming disorder and unplanned accretions of big cities not as the heart of dynamic town life but as problems to be solved by wholesale eradication and rebuilding. The details of this rationalized, clean-sweep redevelopment shifted as pet theories and programs came and went, but the trajectory remained the same: less spontaneity, bigger projects, fewer public spaces, and an erosion of community. After World War II, so-called urban renewal began in force. Whole neighborhoods were bulldozed for monolithic housing projects. The consequences of the American love of the automobile also became apparent in these post-war decades as entire city districts were flattened for highways, and those same highways enabled the flight of the affluent to the suburbs.

Epitomizing the vision of the gleaming industrial city was Robert Moses, one of the most powerful figures in twentieth-century US politics. Moses headed several of New York City's housing, bridge, and highway-construction authorities. As an appointee to posts he had largely invented himself, he worked in near-total immunity from public and elected-official oversight, controlled tens of millions of dollars in tolls and other income from his projects, and could issue bonds himself to fund further construction. Moses guided the building of a long list of the city's expressways and most of the famous bridges and tunnels that link New York City's boroughs and tie the city to its surroundings, as well as public venues such as Shea Stadium and Lincoln Center. More than any other person, Moses and his projects are responsible for the suburbanization of America. In the late 1950s he unveiled a plan for the Lower Manhattan Expressway, which was to tear up a 2-mile path through Greenwich Village and what is now SoHo, displacing nearly 2,000 families and over 800 businesses.

Opposition to the expressway crystallized around Jane Jacobs, whom we met in this book's first pages. Jacobs, a journalist and critic of urban planning who had written a pivotal piece for *Fortune* magazine called "Downtown Is for People," mounted a fierce campaign against the project and against Moses. Her house was one of the thousands that would have been razed for the new highway. After a six-year battle, the expressway was rejected, and by 1968 Robert Moses, whose political machinations had been extensively exposed during the fight, was driven into retirement.

The battle spotlighted the decline of large cities as places to live and was a key step in the shift that began in the 1960s away from urban planning dominated by the view of cities as an extension of the industrial environment toward one of cities as homes for people. Jacobs's 1961 book, *The Death and Life of Great American Cities*, is one of the touchstones of what became the urban placemaking movement. It also introduced the term "social capital."

In the book Jacobs argues that the chaos and unplanned nature of cities is precisely what makes them livable and is what offers opportunities to the most people. She points out that the new buildings created by urban renewal projects are occupied almost exclusively by well-established, high-end, standardized businesses such as chain stores and banks and by wealthy people, or they are uniform ghettos of low-income housing. The places that bring cities to life—the neighborhood cafés, bars, bodegas, thrift stores, and bookstores—largely inhabit old buildings. She also claims that the rationalist's division of cities into separate use areas, such as residential and commercial, creates deadening uniformity and empty streets. Mixed use, she says, is key.

To create this Jacobs advocates four "generators of diversity" that will help cities flourish. To anyone familiar with permaculture principles, these will sound familiar, as they work with function, edge, and relationship.

In order to generate the sort of diversity Jacobs lobbies for these:

1. Each district and its subsections "should serve more than one primary function; preferably more than two."
2. Blocks must be short to allow pedestrians to turn corners and enter new streets frequently.
3. Buildings need to vary in age, condition, and economic yield.
4. Density should be enough to meet the needs of the people there but not greater.

Jacobs felt that mixed use, multiple building styles, and other types of diversity were the key to cities' remaining livable and retaining the sense that each neighborhood was a unique place. Diversity remains a foundation principle of the conscious design of community-enhancing urban spaces, which is a core piece of the placemaking movement.

Two other figures who loom large in the early development of placemaking are Kevin Lynch and William Whyte, and it is worth looking at their central teachings.

Lynch was an urban planner who studied under Frank Lloyd Wright and taught at MIT. His most influential work, *The Image of the City*, grew out of field studies of how people sense and use the urban landscape.[1] In one of his most famous projects, he asked volunteers to take walks around a city and then to sketch, from memory, maps of where they had been. He also observed them as they roamed. He found that different districts and even whole cities had varying degrees of what he called legibility: the ability of a place to conjure up a clear image in an observer's mind that allowed him or her to navigate, identify different districts and features, and feel safe and oriented. Some places present little legibility, and others shine at it. For example, in Boston, people's maps often omitted whole sections of some of the city's very long blocks. The length and monotony of these blocks rendered them invisible; nothing about them stood out. In other cases, a confusing layout of meandering streets left the mapmakers baffled. Illegible places often create aversion or fear in new visitors, affect how people navigate in them, and even change the way users give directions.

Lynch demonstrated that the city is not only a physical place but a mental construct that we experience and represent within ourselves in ways that may not correlate closely with the physical geography, so two different people may construct wildly varying mental maps of the same place. He attempted to show how we build our internal picture of a city and concluded that people perceive five types of *image elements* to construct their mental models:

1. **Paths,** the channels that the subject moves along. Though these may include streets and sidewalks, they don't always correspond to a traditional street network. Paths dominate a person's image of a city because it is our access to places that allows us to see them and build mental pictures of them.
2. **Edges,** which create the boundaries that delineate each region, form the seams that stitch regions together, and make the barriers that wall off one place from another. They can be physical edges or simply perceived disjunctions that separate places in a person's mind.
3. **Districts,** "medium to large sections of the city" defined by common properties distinct enough to notice their boundaries. Most people, Lynch says, use districts to build a mental map of the overall structure of a city.
4. **Nodes,** the junctions or focal points, such as key intersections, hangout spots, plazas, or the center of a district.
5. **Landmarks,** which differ from nodes in that while people enter nodes, landmarks remain external to the observer and serve mostly as reference points.

If any of these elements is indistinct or lacking, our mental image of the city can't be vivid. In a time of infatuation with towering skyscrapers and imposing edifices, Lynch recognized that the user experience, not how much money was poured into the buildings, was the ultimate arbiter of whether a city would flourish or fade. So Lynch's work gives us three important pieces:

1. We construct our personal map and experience of a place with the image elements of paths, edges, districts, nodes, and landmarks.
2. These image elements each need to be based on places and experience that are distinctive and clear enough to be represented in our minds as singular places and events, not as a blur of vague or homogeneous occurrences. That clarity creates legibility, an image we can retain and navigate within.
3. A city's legibility shapes our ability to be comfortable and oriented in a place.

We can use these understandings to create human-centered cities and to diagnose why some places feel unpleasant or undistinguished to us. So here we are gathering our tools for making places instead of spaces, and these real places are where community grows.

A third pioneer in the movement toward livable cities, Whyte was a planner in New York in the 1960s who noticed that plenty of urban planning had been done and projects built but almost no research had been performed to see if the projects were working for the people who used them. One instance that intrigued him was the failure of so many public plazas to attract users. After the success of the plaza fronting Mies van der Rohe's Seagram Building, completed in 1958, New York City zoning was changed to allow greater office-building density in exchange for attached public plazas. The Seagram's plaza became a living heart in the city, a place for people to gather, eat lunch, enjoy the sun with friends, or simply snatch a few minutes' respite from the office. But most of the new parks spawned by the new zoning were dismal, sterile, and unused. With cameras in hand, Whyte and his helpers sought to find out why.

Their findings and suggestions became the basis for a book and film, both titled *The Social Life of Small Urban Spaces*, and found their way into the policies and zoning of cities around the world.[2] Especially important, Whyte found, were pleasant places to sit that were accessible from the street. Using time-lapse photography, Whyte tracked the path of the sun across city parks and found one near-perfect correlation: Seats remained empty when in shadow and filled as soon as the sun struck them. He also noted that because women tend to be choosier than men about the type of public space they will use, a clear sign that the space wasn't working well was if the ratio of women to men using a public space was lower than the average for the locale (in Manhattan men outnumber women 60/40, but in many parks that ratio is even more lopsided).

Manhattan's Bryant Park, next to the New York Public Library, is a case study in transformation using Whyte's principles. The 8-acre park was raised above street level to make room for the library stacks beneath it and ringed by a solid hedge. This hid most of the park from the street, and it quickly became the domain of drug dealers and other unsavory types. In the 1980s the park was rebuilt along Whyte's precepts. City workers tore out the hedge, rebuilt the street entrances as gentle slopes, and scattered benches and, eventually, over 1,000 movable chairs around the park. Bryant Park today is a heavily used, life-filled place that has been showcased in countless media pieces as a placemaking success story.

Although Whyte died in 1999, the New York–based nonprofit Project for Public Spaces (PPS)[3] continues his work and was inspired by it to develop a set of eleven "principles for creating great community places" that are detailed in the accompanying sidebar. These principles—which, not surprisingly, align well with many permaculture precepts—aren't just for designing city parks, or else I wouldn't include them in a book that focuses on the personal and neighborhood scale. They are meant for any public place, and that includes the sidewalk and parking strip in front of your home or apartment building, your front yard, businesses and their parking lots and adjacent street space, traffic circles on side streets, community centers, and any other space that is open to public view and access. The PPS principles are guidelines for bringing to life the places in cities that we use and travel through, and often ignore. All places hold opportunities within them and in their interactions with us. Space in cities is too precious and too full of potential not to use well, by and for the people who live there.

Lynch, Whyte, and Jacobs were in the vanguard of an expanding cadre of advocates for livable cities. Over the following decades, this group of visionaries has expanded to include Christopher Alexander, author of the seminal *A Pattern Language*, which gathers a collection of elements that bring human environments to life and connects them via a grammar of nested relationships; architects such as Elizabeth Plater-Zyberk, Andrés Duany, and other advocates of

## The Project for Public Spaces' Eleven Principles for Creating Great Community Places

1. "The community is the expert." Find the people in the community who know the history: what has worked and what hasn't and how the neighborhood functions. Learn people's stories. This—in essence, the observation step—gets people involved and creates community ownership.
2. "Create a place, not a design." A design is just the beginning. The physical parts—the seating, plantings, pathways, and art—must themselves be welcoming, comfortable, and at human scale. The place itself needs to connect to surrounding retail, residences, and the activities near it. The place should present its own clear, enjoyable image in the minds of those who experience it. Ideally, the location, arrangement, features, and uses all add up to more than the sum of those parts.
3. "Look for partners." Build connections with local institutions and leaders, both in the planning stages for initial funding, vision, and buy-in and for the future, when the project needs ongoing support and activities.
4. "You can learn a lot just by observing." How have other similar projects been done, and what have been their successes and failures? What's missing? Also, after the project is up and running, what's working, where are the bottlenecks and glitches, what needs extra care? Ongoing evaluation is important.
5. "Have a vision." Each community will have its own character to lend to the goals of the place. Some elements need to be present in any vision: What types of activities will happen in the space, what will make it comfortable, what will make it "legible" to the community in Lynch's sense? What will people do there? Why would the place be used?
6. "Start with the petunias: lighter, quicker, cheaper." Here's another permaculture principle in action. Start small, and know that not everything will be perfect at first, so make the elements easy to change and improve. Features such as art, plantings, food trucks and eating spots, seating and conversation nooks, and painting of crosswalks can easily be installed or changed as needed. Concrete hardscapes are permanent and may best be done later, when success has built momentum and the flows and patterns of users have become obvious.
7. "Triangulate." This is Whyte's term for building connections among the elements inside and outside the space that bring people together. If a sculpture, bench, and sandbox are isolated from each other, none is liable to be used much. Tie them together (sandbox beside the bench and sculpture in front of it), and the grouping will attract people. Add a food source—say, a coffee cart—and you have a gathering spot.

8. "They always say, 'It can't be done.'" In making a public place you will nearly always encounter obstacles, so know this pattern and prepare for the likely impediments. Few cities have employees whose job it is to do or approve placemaking, and many placemaking projects don't fit neatly into a zoning-code checkbox. Thus the project should be framed so that it will fit the categories allowed in your city. This is pattern literacy. Also, if you can demonstrate that a small pilot project or a similar site has already built community, saved the city money, or otherwise manifested a real benefit, your path through the approval process will be much simpler.
9. "Form supports function." The planned use for the place should drive the choice and placement of structures in it. Does the choice and arrangement of plantings, seating, paths, and open space support the vision and the planned activities? If it doesn't have a function that is in line with the place's vision, it may not be needed. The community and partners, as well as an understanding of how other places function and the obstacles encountered, can all guide the conception of the place's form.
10. "Money is not the issue." Funding is much easier to get once the community, partners, local businesses, and public officials see the benefits. Get that support first, and the money will follow. This also generates enough enthusiasm so that the costs will be seen as only a small part of the project. Also, once the initial infrastructure is in place, the other elements that bring it to life—seating, food vendors, plants, art—can be brought in at very little cost.
11. "You are never finished." Managing the space after it is built—and having a good plan for that—is the key to success and nearly always represents more work than building the project itself. Places change in how they are used, from weekday to weekend, over the seasons, over the years, and a good management plan and team will accommodate that.

the New Urbanism, which promotes walkable cities and public space; and social critics such as Ray Oldenburg, who argues that the time we spend in our "first and second places" of home and workplace, respectively, needs to be balanced by spending time with other people in accessible, neutral "third places," such as pubs, coffeehouses, barbershops, and cafés, where conversation is the main focus and cost of admission is low.

These professionals, planners, and theorists elaborated a much-needed vision of cities for people, but we are still living in a prebuilt environment that reflects the industrial-city mindset of previous planning and is car-centric. Rather than waiting for architects and planners to build livable cities and towns for us, a new generation of urban activists and grassroots organizers are putting the "public" back into "public space" and simply working with what we have. By identifying underused city- and privately owned land such as streets, office parks, and schoolyards and installing people-scaled, vernacular elements that foster community and a sense of place, emerging teams of community-building activists are reinvigorating neighborhoods and neglected public places. Let's look at some of their work.

## UNMAKING THE GRID

One of the teachings of the livable-cities movement is that our experiences and interactions with each other are shaped by the design of our built environment; design can foster or hinder community. The new generations of urban activists rarely work at the scale of the grand visions of a Robert Moses—where large successes were often offset by even greater catastrophes stemming from failing to envision the social consequences. These designers are not sweeping away the structure of the city but instead are retrofitting and tweaking what currently exists. One of the visionaries spearheading this movement to rework rather than replace is Mark Lakeman, cofounder of the nonprofit placemaking group City Repair. Both the work of City Repair and its evolution as an organization are worth looking at, as they illustrate how building physical environments for people to meaningfully gather is followed very quickly by a need to provide social tools for those people to interact wisely and productively when they are together.

"I was marinated in corporate architecture," Lakeman told me as we sat in the office of Communitecture, his small design firm in Portland, Oregon. "And I had the heroic perspective that I could change the world through sculptural buildings that would inspire people."

Instead, Lakeman found a corporate culture whose structures were monuments to their builders, designed with little regard for their surroundings or inhabitants. An office tower might be a showcase for grand architectural gestures, but, Mark points out, "The people inside are essentially slaves. They get to come out at lunchtime and sit in a wonderful courtyard that I designed, but I don't see them dancing or taking off their clothes."

Discouraged after three years in conventional design firms, Lakeman quit to travel around the world, eager to see societies whose dwellings and cultures still allowed people to connect with each other and the places they lived. He wanted to understand why American buildings and public spaces are so sterile compared to, say, the vibrancy of an Italian piazza. Journeying through Greece, Italy, North Africa, New Zealand, and Central America, he stayed in villages and small towns. "I learned what they don't teach in schools but should: that you can't just design a shell. You're creating a setting for human relationships." Lakeman began to hunger for designs that brought people together.

During his lengthy stay in the rain forests of Central America, a Lacandon Maya elder told him, "Never ask permission to create a public place. Just build it." Lakeman took this wisdom back to Portland, where he helped found City Repair, a collaboration of designers, artists, builders, and urban activists working to remodel the city along the lines of a village and to transform private and public (or, more accurately, government-owned) spaces into communal gathering places. The group's projects have garnered awards from the American Institute of Architects, Portland's mayor, the governor's office, and a host of other organizations.

City Repair's first project was a community café in the Sellwood district of southeast Portland. In 1995, after the exhilaration of his world travels, Lakeman was living in a garage and having a difficult reentry into American life. "I'd been staying in places where people held land in common; where their language, agriculture, beliefs, vision, and philosophy reflected themselves in participation with the earth," Mark says. "I couldn't reconcile that with the violence I was seeing here."

In Lakeman's garage was a pile of windows and doors he had collected over the years. One day, deep in his frustration, he had a vision of a shimmering glass teahouse made from those windows and doors, a place where neighbors could gather and find common ground.

He carried the garage's contents into the yard and arranged the doors and windows in circles around a cluster of trees. Some salvaged timber completed the café, which Mark named the T-Hows. At first a few neighbors dropped by to sit on pillows, chat, and sip free tea. "Then neighbors heard from other neighbors that there was a neat thing happening," Lakeman recalled. "People would just sit, be in communication with each other, and feel wonderful." Word spread, and soon thousands of visitors had experienced the spontaneous community of the T-Hows.

City officials got wind of the structure, which was unpermitted, and demanded it be destroyed. Lakeman cadged a six-month delay. Meanwhile his neighbors decided they wanted to prolong the villagelike atmosphere the T-Hows had created. Lakeman, local designers, and the neighbors began to envision ways to overcome the effects of the urban grid by transforming the nearest intersection into a public place, a project they called Intersection Repair. To slow traffic and identify the new village, they would paint the intersection and streets in bright colors. Every village, the neighbors agreed, needs a heart, with a meeting place, a market, a place for children, and a café. They decided to build simple versions of these essentials on each of the four corners.

The group asked the Portland Department of Transportation (PDOT) for permission to build their design. PDOT said that since no one had done this before, they couldn't do it, and besides, they said, "The street is public property, so you can't use it." However, a sympathetic official in the PDOT office suggested that they apply for a block-party permit and close the intersection for the weekend. "Then if something happened to the intersection over the weekend ..." he said, and let his voice trail off, implying that in this case it might be better to ask for forgiveness after the event rather than permission beforehand.

On the appointed day, the group closed the roads and painted the intersection with brick patterns and rainbows, forming a mock traffic circle. Emanating from the new piazza were broad white stripes down each street. On the corners they built a community notice board, a giveaway box for castoff goods, a tea station, and a library. The weekend ended in a huge celebration.

The following week city officials accused them of vandalism and threatened enormous fines. Delaying once again, Lakeman and his companions went first to the city council and then to the mayor, Vera Katz. Katz took one look and understood. Though unorthodox, the project was meeting the city's goals of enhancing neighborhoods, slowing traffic, creating community, and reducing crime. Katz noted with astonishment and approval that this was all at no cost to the government. The mayor turned to her ombudsman and told him that all of their resources were at the group's disposal.

"The commissioner of transportation has had a complete awakening," Lakeman exulted. "He now says that in Portland the public right-of-ways belong to the people!"

Intersection Repair has spread to other neighborhoods, and as of this writing over thirty-five piazzas now exist. And City Repair's projects are multiplying. One is a mobile reincarnation of the T-Hows called the T-Horse, a portable café that blossoms from a pickup truck. It looks like a giant butterfly, its gauzy wings sheltering tea-sipping visitors from Portland's frequent rain. Another is Dignity Village, an intentional community of homeless people, designed by its residents, who are creating a permanent home on City of Portland land. The self-governed tent city features organic gardens and the beginnings of water harvesting and graywater treatment, as well as solar power generation.

"Our challenge," Lakeman says, "is to give people models of what to do—not just to protest about cutting down trees but to build beautiful places without cutting trees. We need ecological prototypes that are socially inspiring."

In this vein City Repair hosts a ten-day Village Building Convergence every spring, in which hundreds of participants create examples of sustainable building on Portland street corners, storefronts, and front yards: cob archways, walls, monuments, memorial sculptures, a poetry-writing station, a straw-bale studio, earthen benches for passersby. Countless passersby—from ice-cream vendors to cops and little old ladies—see the exuberance of the builders and stop to ask questions or lend a hand. The event exposes a huge population to the benefits of natural building and the community that it engenders, and the participants have tremendous fun.

In his public presentations Lakeman asks people what they want from a city. "The answers are always the same," he says: "Feeling safe, having a voice, connecting with other people." But most people don't get that from their cities. "City Repair says that the way Portland operates now doesn't meet those goals. But in the village model, where those goals are being met, we have a pattern that works every single time. To me the question isn't knowing what to do: We know what to do. It's whether all of us working together can have enough of an effect in time."

One of the major efforts of City Repair and similar organizations is to undo the effects of the urban grid: cities carved into rectilinear blocks, blocks partitioned into fenced-off yards. The urban grid imposes a pattern that is designed in part to provide access to all areas as well as to be easy to comprehend and navigate within, but one consequence—not so unintentionally—is to isolate. "The grid was devised by Rome and other early empires to easily control subjugated people," Lakeman explained. "In any organic village, wherever roads or paths meet there is a gathering place. But where people gather, revolutions can brew." Conquering armies tore out winding village streets and public squares and imposed a grid that eliminated meeting spots. One sentry at a corner could monitor all activity down the length of two ruler-straight streets.

In most later US cities and towns a street grid supplanted the organic village patterns that give character to older cities such as Boston and Philadelphia. By design and neglect, American cities were robbed of communal places. "There's lots of common ground in our cities," Lakeman admits. "But it's all traffic corridors. It's for cars, not people. So City Repair is about restoring the village, taking back the common ground, reconnecting the city's fragments."

City Repair has sponsored the Village Building Convergence every year since 2002 and now has chapters and projects in several cities. Their success stems in large part from working within the existing urban infrastructure rather than replacing it. By starting with the rectilinear, hard elements of the modern city and softening them, curving and rounding their brittle edges, they create places that welcome people and encourage them to linger rather than being funneled through them to work or home. The materials of natural building that City Repair advocates—earth, unprocessed wood, fabric—are soft and maternal, inviting us to touch them and fit into their living curves. In the places that City Repair and their adherents have designed with the principles of placemaking and built at human scale with welcoming materials, people gather.

But City Repair's expansion has also highlighted the challenges of creating community in a society that extols individualism. I spent several years on City Repair's board, which taught me a lot about the human side of placemaking. As with all the elements of sustainability, executing the technical elements—in this case the architectural design of places, the choice of materials, the techniques to build them—was the easy piece. The bumpy part of the ride was the human side: making decisions and implementing them, getting consensus, addressing fears, overcoming inertia, defusing conflict, getting follow-through. City Repair has struggled—mostly successfully—to overcome these common problems.

These same challenges have faced every human group, but they became particularly apparent in the 1960s and 1970s as intentional communities of excited, like-minded young people gathered with visions of building a more cooperative society. We all know what happened. Most of these communities collapsed within months, and only a tiny percentage survived more than a couple of years. The many postmortems of the failures and the analyses of the few successes revealed several common threads: Decisions in groups generated acrimonious, often scarring argument and took forever to arrive at and even longer to implement (if they were implemented at all); power tended to concentrate in the hands of a few, usually the founders, those with money, or the most vocal; assumptions about a supposedly common purpose proved unfounded when it came time to work toward specific goals; and the small incidents and disagreements that all humans encounter daily piled up over time in the pressure cooker of community into longstanding resentments, fragmented factions, and conflicts that had no way of being resolved.

The upside is that the postmortems did reveal these tendencies. Because the moments of true community that had been glimpsed in these experiments seemed real and promising, over the subsequent decades activists, sociologists, and community-builders of all kinds have developed many tools specifically to resolve the issues of power, decision-making, goal setting, and conflict resolution that have dogged human groups since before we were human. The principal issue threatening community now is not the lack of tools but the will to stay in the game. In our culture, unlike tribal and village societies, when the going gets tough it's just easier to walk out and find an alternative place to live rather than work through the challenges. The problem with walking out is that we usually take our problems with us. Then we wonder why each new group we encounter seems to have the same negative qualities as the last one. The common factor, of course, is ourselves and our baggage. As the proverb says, it is easier to put on slippers than to carpet the world. A more reasonable leverage point is to change ourselves rather than everyone else. Hence the recently developed tools for community-building—for changing the old patterns of group dynamics—are necessary keys to making the transition from impersonal big cities to urbanized clusters of villages where deep connections exist within and between each microcosm. Though many of these techniques originated in intentional communities, they apply to any group of people working on a placemaking or community-building project together—and, really, to any group of people working with each other. These tools have been critical in the survival and growth of City Repair and countless other organizations and projects. Just as we have many techniques for building soil successfully, we are blessed with myriad tools for building community, working together, and designing healthy social environments. Let's look at some of those tools, as they apply to everyone who has ever worked in a group—and that's everyone.

The following list of tools is arranged according to the type of task it accomplishes, and each of those tasks is a critical element in creating a successful group of people. The key tasks in this list are a compilation of the wisdom of several of the leaders of the intentional-communities movement, including Bruce Davidson and Linda Reimer, who founded the Sirius Community in Massachusetts in 1978; Lois Arkin, founder of the Los Angeles Eco-Village; Diana Leafe Christian, former editor of *Communities* magazine and author of *Creating a Life Together*; and Brock Dolman and Adam Wolpert of Occidental Arts and Ecology Center, which they helped found in 1994. These elements are also based on my own experience of days to months living in ten or more intentional communities while teaching permaculture, consulting for several fledgling communities, and sitting, like most of us, in meetings of varying degrees of effectiveness, noticing why some worked and some didn't. Groups of all kinds have labored on methods to move coherently toward their goals, but intentional communities played a special role as crucibles in which, driven by dire need, people trying to live together were forced to forge tools for working in groups. Their discoveries and the sophisticated tools developed by their many successors are worth using any time two or more people start a project or discussion. These tools drastically increase the chance of success of any collaboration. The key elements for which every group needs to have clear policies are as follows (we'll get into details directly after):

1. The vision and purpose of the group
2. Making decisions and implementing them
3. Identifying and distributing power and authority
4. Resolving and transforming conflict
5. Agreements on finances and ownership
6. Communicating ideas and concerns among the group
7. Joining and leaving the group

## The Vision and Purpose of the Group

Put this in writing. The very process of drafting the statement brings clarity and focus to a group. Working out the vision together will also reveal how the group members communicate and interact, and it can be a beginning toward working out communication and decision-making styles for the group. The result, in addition to being a crash course in group process, should be a short, clear, statement defining as specifically as possible what the group hopes to be. Ideally, it will be between fifteen and forty words long, and it should offer specific goals while avoiding vague statements full of buzzwords. The temptation in any group—and especially in the nonprofit world—is to default to the most innocuous common denominator, resulting in a vision statement such as (and I'm citing a true example), "We are a diverse group of individuals sharing resources to create a better world." This is so vague that it could apply to every organization from the Audubon Society to the KKK. Make your vision statement specific enough to attract to your goals the people you want and to filter out those with other agendas and interests. That will save a lot of time and angst in the long run.

Here are examples of clear vision statements made by successful organizations:

"Amnesty International's vision is of a world in which every person enjoys all of the human rights enshrined in the Universal Declaration of Human Rights and other international human rights instruments."

"American Red Cross prevents and alleviates human suffering in the face of emergencies by mobilizing the power of volunteers and the generosity of donors."

And you can't beat the Walt Disney Corporation for directness and brevity: "To make people happy."

## Making Decisions and Implementing Them

In permaculture courses, when I ask people to name different decision-making tools, the first word spoken is inevitably "consensus." In my circles consensus—wherein a proposal for action is modified by the whole group until every single person can support it—is often the default. But every toolkit needs more than one tool in it, no matter how good that tool. For some proposals, majority vote, a recommendation from a committee, a coin toss, or any of many other methods may be the best way to decide. Full-on consensus or its equivalent may need to be cranked up for far-reaching issues that deeply affect everyone, but whether to plant curly Scotch or red Russian kale could be handled by a quick show of hands among the garden committee.

Remember, too, that the goal of decision-making is not just to make a decision but to get enough buy-in to *implement* the result with the least delay, conflict, and resistance. The experienced group facilitator Tree Bressen points out that making a major political decision in a dictatorship can be instantaneous—the despot says, "We will sign the treaty," and the decision is made. To decide via majority vote can be fairly fast—draw up the proposal and hold an election—while to decide using consensus may seem to take forever. But how about *implementing* the decision? Our tyrant will have had to imprison some resistance fighters and hang some spies along the way and battle saboteurs and laggards afterward. The voters will debate the treaty, filibuster, and make deal-altering amendments before the decision and perhaps hold a recall vote after it. But the consensus group avoids all that violence and subterfuge, just talking and compiling their knowledge until all can support the decision; then they will quickly get the work done together. The time to implementation may be about the same. But again, simple decisions may not need the complex machinery of full consensus.

If a group wants to use consensus, it's critical that everyone be taught how by a skilled trainer. Most of the problems with consensus that I have heard of are due to its being used incompletely. Also, consensus works only if everyone holds a sincere desire to achieve the group's goals. If there are members who don't share those goals, consensus can turn into a tool to immobilize the group or trumpet the egos of the dissenters. This is where a clear vision statement can help align everyone with the goals.

A new technique called dynamic governance was created in part to overcome some of the perceived shortcomings of consensus. While consensus attempts to flatten hierarchies and place everyone on equal footing in one giant decision-making circle, dynamic governance, also called sociocracy, recognizes that hierarchies exist in most organizations, and it attempts to deal with the resulting challenges. To do that, dynamic governance creates a structure of semi-autonomous *circles*, each a grouping of people with its own aims, functions, and mechanisms for feedback and evaluation of its progress. Each circle is *double-linked* to another "higher" circle—that is, one with more executive power, such as a board or management group—via two members who are full members of both groups. In addition to circles and this double-linking, the third defining element of dynamic governance is *consent*, meaning that decisions can be made only if no one has a "reasoned, paramount objection to it."[4] This is a more stringent criterion for objection than in consensus, because dissenters must give a coherent argument instead of simply announcing that they block the proposal.

Dynamic governance was developed by an engineer, Gerard Endenburg, who was trained in whole-systems thinking and was inspired by pioneers in systems science such as Norbert Wiener and Ilya Prigogine. This whole-systems orientation makes dynamic governance appealing to permaculturists, who have used it in several permaculture organizations and communities with great success. More information can be found at www.governancealive.com.

In my experience, virtually any clear decision-making process is better than an ill-defined one. I've also observed that a key moment in the evolution of a successful organization is when people choose not to be involved in every decision. A feature of healthy maturity is when people are willing to trust their colleagues to make good decisions. Delegating authority spreads both the work and the power around more evenly. And that leads us to the next important element.

## Identifying and Distributing Power and Authority

Power in groups is a hot-button topic. I'll begin with a simple definition: Power is the ability to do work, to make things happen. The economist and systems thinker Kenneth Boulding refines this definition in social situations to "the ability to change the future."[5] In social groups, power takes two principal forms, as power-over and power-with. Power-over is the kind most of us are familiar with, often in a negative sense. It's the ability to direct another person's actions regardless of that person's desires or interests. This is hierarchical power, flowing from greater to lesser. It is usually zero-sum; one person's gain in power-over is another person's loss.

Power-with is holarchical power, flowing horizontally as interconnections among equals, and as with any dynamic linkage, it can generate more than the sum of its parts. New behaviors and possibilities emerge from smart use of power-with, something that rarely happens in power-over because there can be no collective synthesis of wisdom in the one-way flow of power-over. Power-with is sometimes subdivided into several different types: straight power-with, which is mutual reinforcement among equals; power-to, which is the ability of a person or group to shape their experience;

and power-within, an internal sense of self-worth and self-knowledge that serves as the basis for action.

Most of us dislike power-over because of the pathologies that our culture has built around it, but hierarchies do exist and are often useful. There are times when power-over, properly wielded, is appropriate. You don't want your surgical team to break for a round of consensus when your chest has been cut open; the chief surgeon has a justified license to direct the actions of the others. The key in power-over (and often for power-with) is a revocable license to use it. Power can be used beneficially when the authority to use it comes with clearly defined rules for granting and revoking that authority, made by the group affected by that power. In a healthy setting, when the surgical team goes out for dinner to celebrate a successful triple bypass, the chief surgeon no longer has the authority to boss around the others.

Authority is simply the license to wield power. By making that license easily revocable under a clear set of circumstances, we can ensure that power is used effectively and positively. Authority in that case becomes a hat we wear for a purpose, then set down. Pathology enters when the hat becomes a crown that never comes off; when the wearer thinks that the power stems from him or her personally and not from what has been granted by others. Thus key elements in working positively with power in groups are to identify what forms of power are present and what they are to be used for and to develop rules for their use, particularly in limits to the use of power-over.

Like many of the subjects I've merely touched on in this book, social power is a rich field of study that is worth pursuing more deeply. Some of my own core references for understanding the social use of power are included in the endnotes.[6] These will get you started on a rewarding study of this key element of human relations.

## Resolving and Transforming Conflict

Over the past few decades negotiators of all stripes have come to understand that a "How do I win?" stance simply breeds at best unhappy losers and usually more conflict. More recent work has arrived at tools that optimize everyone's outcomes. These tools usually have a common format: a method for communicating each party's views in neutral language, a disinterested party to mediate (in simple conflicts neutral and agreed-upon language may fill this role), and a set of steps to generate options, explore them, and negotiate a solution. There is no single "right" tool for conflict resolution, and having *some* method is better than having none.

One of the breakthroughs in thinking about conflict came in 1981 with the publication of *Getting to Yes* by Roger Fisher and William L. Ury.[7] Their key insights were that almost all parties in a conflict or bargaining process have negotiable interests, that is, some wiggle room in the amount of benefit to their welfare they will be satisfied with, and those interests are where the negotiation should focus. The conflict will quickly become intractable if the discussion strays to negotiating rights, power, or values. For example, in a shared-living arrangement, when the residents decide whether or not they will let each other know when visitors will be stopping by, they are negotiating their interests as opposed to arguing whether they have a right to know whenever one of their housemates has a visitor.

Some negotiators prefer the idea of conflict *transformation* rather than resolution. One proponent of this idea is John Paul Ledarach, who, while doing human-rights work in Central America in the 1980s, often used the term "conflict resolution."[8] But many of his Latin American colleagues worried that this quick-fix approach—resolving and moving past the problem—didn't address the changes needed at the root of the conflict. "Conflicts happen for a reason," they told him, and it was those deeper causes that needed to be worked with. The goal of conflict transformation is to build constructive change out of the energy created by conflict. Conflict helps us spot where potentially creative energy is bottled up and gives the participants a chance to change the underlying social structures and the patterns that shape relationships. Like trifocal glasses, conflict transformation holds three lenses within a single frame to view conflict: a close-up view of the immediate problem, a midfocus look at what patterns of human relationships are creating the context, and a long-range view of how to reshape the context and the structure of those relationships so that the energy driving them can be used beneficially.

This means that we can think of conflict as a sector, much like wind or sun, that is a result of energy being developed and radiated through a system. We can harness and use this energy as a productive rather than destructive force by running it through a defined pattern (a set of steps in a method of conflict transformation) in a designed framework.

## Agreements on Finances and Ownership

These are the contractual agreements around who pays for what and who owns what. When all transactions are done with money, these policies can be straightforward, but in many nonprofits and shared-living arrangements, members may contribute varying amounts of money and sweat equity, or labor. In our culture there is often a perceived or real power imbalance between those who pay money for land or membership and those who buy in via their time. Once again, clear, written agreements can adjust these imbalances and increase the odds of harmonious relations around money and work for the group.

## Communicating Ideas and Concerns among the Group

Having a common language is the basis for smooth-flowing relationships. Again, my experience has been that simply agreeing upon a communication format of almost *any* type is better than not having one. If you're forming a group that plans to work together for a long time or on a large project, it's worth bringing in a facilitation or group process trainer. This will help the group achieve its goals with minimal discord. Some people are task-oriented—"Let's decide this and get it done!"—while others are process-oriented—"We need to hear everyone's views and address their concerns before we make a proposal"—and an agreed-upon format for airing and discussing ideas, issues, concerns, and proposals will go a long way toward defusing the friction that can easily arise when both types are in a room together. Some of my favorite resources for developing effective group communication are Tree Bressen's group facilitation website, www.treegroup.info, which offers a wealth of exercises and handouts; *Facilitator's Guide to Participatory Decision-Making*, by Sam Kaner; *The Skilled Facilitator*, by Roger Schwartz; and the four-player model of group dynamics developed by David Kantor and explained in his book, *Reading the Room*, as well as on his website, www.kantorinstitute.com.[9]

### Joining and Leaving the Group

Joining and being a harmonious part of any group is made smoother by having clear criteria for what the group is looking for in its members. That's another reason for a well-thought-out goals and vision statement, as it helps all parties decide if there is a good fit and a suitable niche. It's important, too, to have a written description of the rights and responsibilities that go with membership and a clear description of a title and compensation package if appropriate.

We've all heard about the importance of exit strategies, and this applies to groups as well. Inevitably, someone will need to leave a group, or the group itself may decide to disband. How will assets and shares be paid for and distributed when this happens? How will tasks be reassigned? (And this last suggests that each person's tasks or role should be defined well enough for others to understand what they are and how to do them; an operating manual is always a good idea.) Endings are never pleasant to face, but they are much better addressed before they occur than during the crisis of a surprise departure.

## BUILDING COMMUNITIES FOR OUR TIME

Humans are social animals, and the community is the basic unit of social cohesion. For most of human history, communities arose organically and formed via natural affinities: the family, clan, tribe, village, spiritual connection, or those within a watershed or other geographic delineator. In the past few centuries, powerful forces have cleaved our natural communities, and the fracturing has accelerated in the past few decades. Many organic communities have shattered or are hanging by threads, often because of the extreme mobility and other rupturing forces brought on by the redistribution and concentration of power in many forms as a result of uncorking and burning the energy of many millions of years of stored sunlight in a couple of centuries. Renewing the bonds of organic communities is thus essential for several reasons. Their loss is an impoverishment of the human spirit; the impersonally administered, trickle-down "community" of welfare programs, subsidies, and employer benefits disempowers us by robbing us of the chance to care for ourselves; and the decline of the fossil-fuel age is virtually certain to mean the end of the energy spree that built the complex bureaucracies that today deliver the benefits that organic communities once provided.

We need to relearn community-building. Fortunately, many excellent models still exist or have arisen to fill the gaps. Many of the services once provided by family, friends, and the village are now supplied by a rich network of nonprofits and similar organizations. Although many of these institutions are grounded on what I view as an unsustainable or at least unhealthy relationship—a dependency on the largesse, continuing prosperity, and enthusiasm of wealthy donors—I'm seeing a timely evolution of the nonprofit model. Many are moving from grant-based support to community-based support, where their funding and workforce derive from a large, natural network of supporters who in turn benefit from the organization, rather than from a few rich funders who are not dependent on a return from their generosity.

Another renewal of community resources is the resurgence of voluntary associations such as the Grange, Rotary, Elks, and similar groups. Although they flourished for many years in the nineteenth and early twentieth centuries, in recent decades these associations have foundered, and some were tainted with often justified accusations of racism and sexism. For their members, however, they wove a social safety net that did not exist until the New Deal entitlement programs and employer-benefit mandates came along that made the associations' chief function redundant. Most of these associations had a similar structure: Members paid dues that built a fund that was invested carefully, and that fund then supported members in need. Peak-oil blogger John Michael Greer writes how the children of his great-grandfather, a policeman who was killed in the line of duty, were supported until they were eighteen by their father's Odd Fellows lodge, and his widow received an ample stipend from the lodge until her death.[10] In 1920 roughly half the US population—of both sexes and many ethnicities—belonged to one of the more than 3,500 voluntary associations then extant.

Many of America's hospitals, schools, libraries, churches, and local highways were built by short- or long-lived voluntary associations that were chartered to meet a community need, and millions of disabled and impoverished people were supported by these groups. It seems likely that the federal government won't be able in the near future to fund all of its welfare and entitlement programs, and employers are struggling to pay for their own benefit programs. Voluntary associations that have local chapters as well as other community and grassroots organizations let us fill vital functions by returning support and governance of the people in the community to that community. They can help build resilience in uncertain times. The benefits of local control and support of a community's welfare are obvious and many.

This brief overview can offer only a glimpse of the well-stocked toolkit that has been developed over the last few decades for creating community, restoring a sense of place to our surroundings, and working together in groups. We have so many more techniques, in fact, that the challenge once again is to decide which methods out of the collection are best for the given circumstances. Permaculture's protocols offer an abundance of flexible decision-making methods to help us identify, select from, and use the most effective community tools.

# CHAPTER 10

# Tools for Designing Resilient Cities

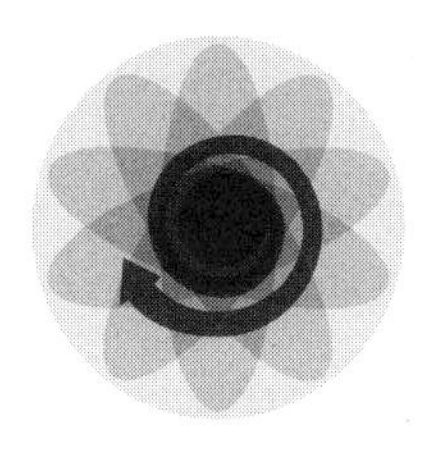

Although I've emphasized in this book that permaculture is about much more than growing food, for many the garden gate is an entry to ecological living and thinking and one of the most common paths for experiencing whole systems. Food brings us together for sharing meals, seeds, and garden play but also through the diverse interests it can unite. Food can make common cause among all of us who care about nutrition and the well-being of our children and what they eat at home, in school, and around town. Food opens our eyes to the health of urban and rural soil, water, and air; it connects urban eaters to farmers, soil scientists, land-use planners, water stewards, and polluters. Seeing the horizon-spanning monocultures that commodity food rips out of prairie and forest forces us to ask where all that biodiversity went. Gorging on the abundance of a garden when so many are hungry raises awareness of food justice, which in turn brings attention to communities where finding healthy food isn't even a dream—and that makes it hard to ignore the correlation between those neighborhoods and the disproportionate number of their members unemployed, at risk, and imprisoned. Food connects us to policymakers, whether through zoning for urban agriculture, plots for community gardens, labeling standards, or cottage-industry laws to encourage small food-based businesses at low entry cost.

Food is a welcoming, universal entry point into the rest of the web of life outside our yard, including the human web. It's the easiest place for many of us to see our direct connection to nature and to each other. But the other petals of the permaculture flower offer connections and lessons every bit as rich. This is why a major focus of this book has been on applying permaculture's four major design methods—needs-and-resources analysis, zones, sectors, and highest use—across many examples in food, water, energy, livelihood, and community. The methods identify leverage points, suggest strategies, reveal patterns, and show where relationships can be built or strengthened, and they do this for every need or discipline that I've seen them applied in. It's a powerful toolkit.

Just as growing food ecologically builds diverse links to people, policies, and nature, solving our need for water places us within an equally large network that we can strengthen and play a wiser role in by using permaculture's tools. Caring about water attunes us to the management of farms, forests, and fisheries in our watershed and bioregion. Catching some of our own water transforms our relationship with rain and makes us acutely aware of its presence and absence. Reusing graywater reminds us of how much water we use, what we put in it, and what it can do for us even when we think we're done using it. At the community and neighborhood scale, curb cuts, bioswales, and rain gardens beautify the streetside, cut sewer bills, and weave a common landscape theme through the blocks lucky enough to have them. We can work with water at many scales, from disconnecting downspouts so they won't tax storm drains but will build biodiverse yards instead to supporting policies for stream daylighting to advocating dam removal to running for a seat on the local water board. Again, permaculture's methods and principles give us formal tools for seeing where those leverage points are and suggesting strategies for working with them.

Moving to another petal on the permaculture flower, mindfulness about our energy use and the buildings and cars that consume energy interlinks us with our communities, with the region, and to nature as well, nudging us back toward the solar budget that every other species adheres to. It reminds us of the mixed but many blessings of this era of cheap, energy-dense fuel and prepares us for the near certainty of that era's end. The tools for thinking about energy—emergy, transformity, life cycles—gently prompt us without coercion to avoid waste and consider the impact of consumption.

These petals of the permaculture flower, the visible structures of human life, tell us the *what* that we need to care for. We have a wide array of ways to meet those needs, and how we do so matters. Solving for the invisible needs on the flower—of community, livelihood, spirit, health, and justice—is the how. It's obvious how livelihood ties to the visible needs; that's the exchange we make to satisfy them. The other petals are part of the how as well. You can grow all your own food, but it's less work, more resilient, and probably healthier to meet your food needs via a community of food suppliers at multiple scales. In meeting our physical needs, permaculture's tools can guide us toward doing it in ways that build rather than degrade health, spirit, community, and justice.

Just as the visible needs are linked—we can grow food in ways that improve or deplete our sources of water, waste, and energy—the invisible structures are tied to the other petals. We can gain livelihood in ways that damage our community, spirit, and planet or that strengthen them. How justice is meted out determines how wealth is distributed and affects the welfare of our spirit and community. Thus we need intelligent, even wise tools for working with all the petals of the flower and deciding how to structure the ways we meet our needs.

In this book I've argued and tried to demonstrate that permaculture design can provide those tools. Permaculture's ethics start us on the path to ecologically and socially sound decisions, ruling out the acts and ideas that obviously harm people and planet. The principles provide a more sophisticated layer of filters, keeping us in tune with the design wisdom accumulated in life's

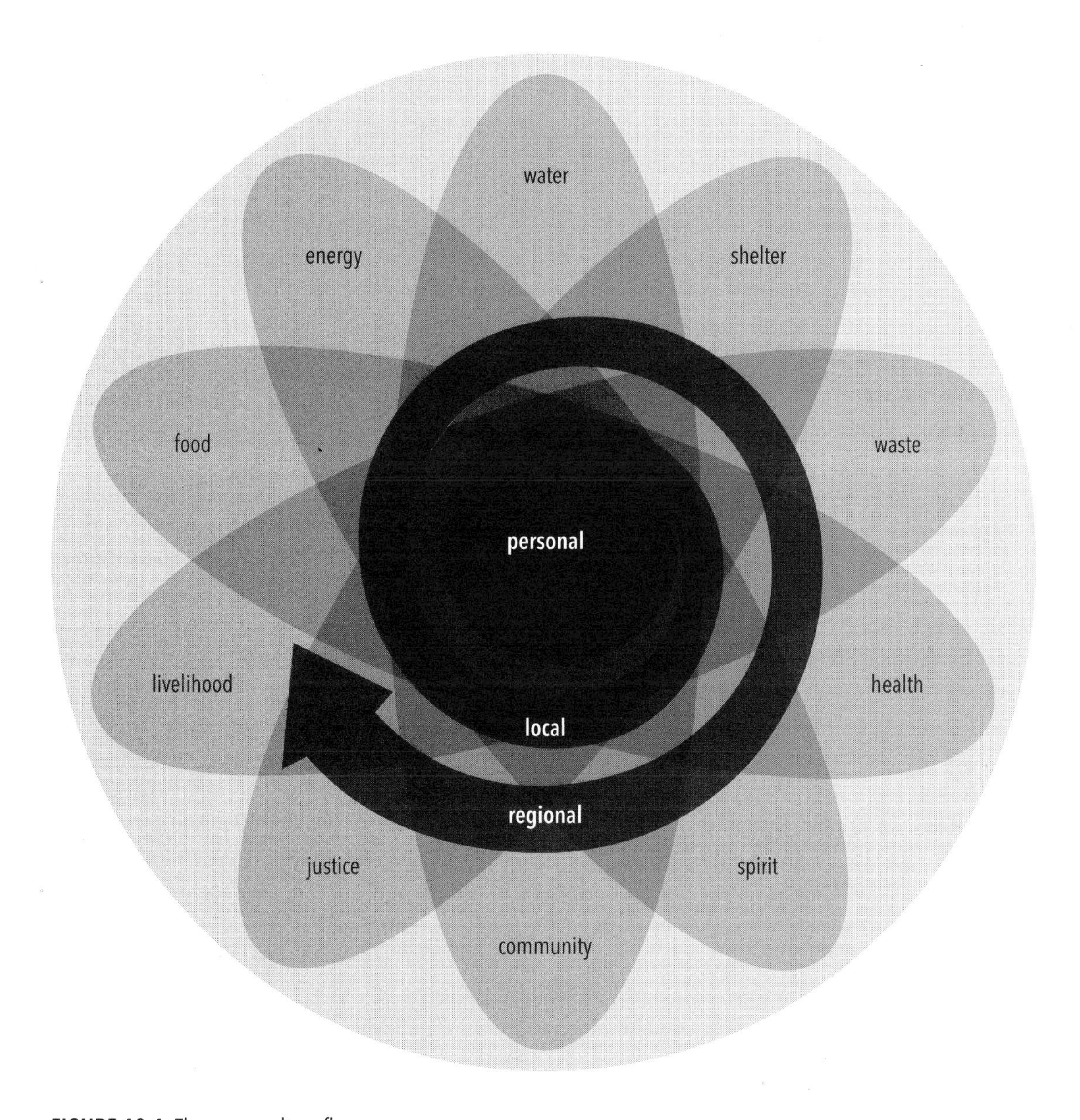


**FIGURE 10-1.** The permaculture flower.

four-billion-year research and development program. The principles do more than just rule out foolishness, though; they encourage us to think like an ecosystem, helping us spot solutions that enhance diversity, resilience, productivity, interconnectedness, and elegance and encourage lean use of resources.

Permaculture's methods are the heavy-lifting equipment in the regenerative toolbox. An ecologically designed yard, a healthy community, successful business, friendly neighborhood, livable metropolis, dynamic city region—all are complex adaptive systems, and none can function or even exist without mutually beneficial relationships among its parts. Suggesting, creating, and enhancing those connections is what permaculture's design tools do better than any other system I know of. When we need them, we can choose from the palette of conventional design tools, such as data overlays and bubble diagrams, but they aren't specifically designed to tie functions and design elements together in dynamic, supportive relationships. That is permaculture design's strong suit.

I've shown how thinking permaculturally can identify potent leverage points and critical paths and help spot important resource flows and remove their bottlenecks, connect them to needs, and help us work with them in ways that improve rather than deplete them. When Mark Lakeman realized that the largest area of public space in any city is the street and that making intersections act as villages was a way to create community, he was thinking permaculturally. When Ianto Evans designed the rocket mass heater to burn clean and trap heat in thermal mass, when Jenny Pell's class envisioned a public food forest that would meet multiple needs in a struggling urban community, they were thinking permaculturally. It's no coincidence that each of these visionaries has been formally trained in permaculture design.

The four principal permaculture design methods work in concert to build relationships on multiple levels, constructing a tightly linked network of functions, flows, yields, and support wherever they are applied. I've offered examples of ways that these methods work not just in the landscape but in the human context of cities, towns, and suburbs and all that we do there. To briefly recap the big four:

**Highest use** is the ordering of tasks, activities, and processes in time or space so that the first activity in the sequence leaves as many possible yields or subsequent uses as possible. It creates relationships among design elements or processes in time.

**Needs-and-resources analysis** identifies the necessary inputs and the outputs of many kinds (products, behaviors, characteristics) of each element in a system (including people) and connects the needs of one element with the outputs of another. It builds relationships among the design elements themselves.

**Zones** organize design elements according to how often they are used or need attention. The most used elements are placed nearest (in space or time) to the user or center of activity. Zones build and organize the relationships between an element and the user or center of use.

**Sectors** are influences from outside; that is, influences that can't be altered from within the system in question. They can be physical influences, such as wind or fire, or intangible influences, such as zoning, customs, personal power, or emotional triggers. We place design elements in relation to sectors to harvest or block sector energies or let them pass through with no effect. Sectors create relationships between the parts of a system and the forces or influences coming from outside the system.

Those abstract terms "element" and "system" are just thinking tools that remind us how many different contexts permaculture design can be applied in. Once we're actually designing or working in a specific place, those elements and systems become very real. In a yard an element pops out of the abstract and into form as a comfrey plant, downspout, or pathway. When the system we're working with is a business or community, most of the elements are people, acting in the roles they play there.

One of the beauties of permaculture's design approach, besides the network of mutually reinforcing connections that it builds, is that when we apply these methods—especially if we've got the time and energy to use each of the big four—we can be just about certain that we're putting things in an optimal, multifunctional place. If highest use, needs-and-resources analysis, zones, and sectors all tell us that our chicken coop goes next to the fence or our social media director should work on *that* project, we can sink into a sense of confidence that we've made a smart decision.

At its heart, permaculture is a tool for developing pattern languages to work intelligently with the basic human needs. Just as Christopher Alexander, looking for the reasons that some structures are alive and others feel dead, created a pattern language of potent, reusable, identifiable solutions in architecture, permaculture can help us find and work with critical patterns in any field it is applied to. What are the common patterns that always emerge in community, energy use, water conservation, power structures, the workplace, any system? Identifying those and developing regenerative tools for working with them is the task that permaculturists have before them. Permaculture designers and others are already working on pattern languages for food production, small ecosystem design, community living, decision-making, and in many other fields. In each of the chapters of this book, I've attempted to show how permaculture's tools can identify a few of the key patterns for working with human needs, to spot the leverage points in fields as diverse as microclimate management, energy use, water harvesting, community decision-making, and livelihood security (to pick out just a few of the places for which the design methods can develop strategies).

## SPIRALING TOWARD RESILIENCE AND REGENERATION

Good design, and especially permaculture design, as I've emphasized, is based on building relationships. Connecting the pieces of a landscape, both living and nonliving, is the basis of the backyard ecosystem. In the same way, it is the relationships among people, local commerce, and place that build resilient, prosperous, and dynamic neighborhoods, communities, and towns. This is the key process in urban permaculture: designing and living within a web of mutually beneficial relationships that link the elements of the cultural and social world as well as those of the world of nature.

Meeting our needs is grounded in creating healthy relationships with the resources that satisfy those needs, which means learning to develop ways to provide for each need that are not exploitive or depleting and that won't impoverish future generations.

At the center of the permaculture flower, a spiral uncoils, which symbolizes the ways we can travel around the circle of needs that the flower represents. In one sense the spiral reminds us that we are constantly cycling through our list of needs. That's because in a dynamic world we can't just set up a food or energy system, then walk away from it. The resource flows and our relationship to them are constantly changing, so we continually reevaluate and reassess: How are we doing? Does our design for the place and system that provides for us need reworking?

Also, each time we transit around that spiral to reconnect with these resources, we're in a slightly different place. The Heraclitean river of resource flows is constantly changing, and we can never step into it in the same way twice. Each time we return to a need, we've learned more, made new friends, found better tools.

The spiral also tells us that meeting our needs doesn't stop at the individual. In many cases it makes sense to focus first on solving for our needs at the personal level, for ourselves and our immediate family. That's the self-reliant homesteader dream: Go around the petals of the flower and solve for food, water, energy, and the rest of the basic needs. There's a reason that flight attendants tell you to strap on your own oxygen mask first before trying to help others. This makes sense as a starting point and also as a way to gain valuable skills. But in that journey around the flower, especially when we come to livelihood, unless we're utterly determined to make absolutely everything ourselves and do it all ourselves (fuel? a bicycle? health care?) we are inevitably pushed to move beyond the personal to the community or local realm. Not all of our challenges can be solved at the personal level. The cliché about its taking a village to raise a child—and to do much else—is true.

The second turn of the spiral around the flower lifts us to the local level. One way to think of this is as an invitation to take a literal path of multiple circuits around the spiral, solving for our personal needs on the first pass before moving outward to meet the needs of our community. We can also take this inclusion of multiple levels as telling us that everyone can plug their skills into the needs list somewhere. There is a role not only for those growing their own food but for those working on, say, local justice or energy issues or regional water policy. The multiple turns of the spiral around the needs list also suggest that solving for needs only at the personal level isn't sufficient. It's not enough—nor is it a secure or even an ethical existence—to be an island of abundance surrounded by a sea of want and poverty.

"Local" covers a lot of ground. It includes the whole region that begins outside your skin or at your property line and extends some ill-defined distance away: A day's walk? The hundred-mile dinner? Your watershed? The county line? I often think of "local" as the region inside which my voice can be heard, where the workers in my favorite shops recognize me, where I can get a few minutes in front of the neighborhood association or city council; in short, where I am a person rather than a demographic. For just that reason, local is increasingly being understood as a critical leverage point. At the local level, one person or one small organization can make a tremendous difference. For every Gandhi or Darwin who has changed how a whole culture thinks, there are millions of people who have rewritten a city ordinance, cleaned a polluted river, brought down a corrupt mayor, or in some way bettered the lives of their neighbors in ways that mattered. Thirty years ago my next-door neighbor in Seattle organized a tree-planting party in our barren neighborhood, persuading the city to donate 200 street trees. Today a half dozen once-sterile blocks around my old home are lined with thick-trunked, spreading street trees, and the neighborhood is shaded and cool, musical with the sounds of songbirds, and often filled with strolling residents, thanks to that one person. Multiply that tiny act by millions, and you have a livable culture.

At the local level, we're all activists. Where you buy your food determines which local businesses thrive or die, whether nearby farms succeed, what methods are used to grow that food, how much oil is burned to ship it, how many trees are felled to package it, what the health of your watershed and soil is, whether your income enriches the community or some far-off CEO. Whether you walk, bike, bus, or drive to work shapes the transportation network and policies of your city and cleans or dirties the air you breathe. It all matters and has tangible effects on your life and that of all the creatures around you. With every act, whether at home or out in the community, we have the choice between making the world a little better or a little nastier.

Urban life is impoverished without a strong connection to the local. Rural homesteaders might be able to feel that they are going it alone, living in the wild and not bound to society (although, as I've said, much of that is illusory: Who made their ax and their stove?), but urbanites, whether hipsters or homesteaders, know that it is the ties to others that enrich their life. Thus part of urban permaculture—and what permaculture's design methods are superb at doing—is identifying, preserving, and building those mutually supportive relationships, and not someday but *right now*. Many of us hold the dream that one day we'll live in a community, or at least a town, where we'll have just the right friends and neighbors. I would love it if that could come true for all who dream of it. But meanwhile, we have the lives we live now. We do live in a community, all of us, just not one we designed, and it's doubtful that many of us will have the chance to design our own.

One of permaculture's credos is, we must start where we are. Right now, that dream community isn't where we are—it isn't local; it's off in the future. So let's look at what is around us right now and solve for basic needs based on that. The questions we need to ask aren't very hard; in fact, they are boringly banal: Who will take care of you if you're sick or hurt? Who will feed your cat when your family is on vacation? Who will, spur of the moment, lend you some eggs? Who would check on you if the power goes out? Are there people in your life right now who can help you with those kinds of basics? I'm not talking about finding people who are in perfect alignment with your politics and dietary preferences. I'm just talking about someone who can feed your cat. If you go around your own version of the permaculture flower, who in your neighborhood or circle of friends—your personal zones 2 and 3—can help you meet those basic needs, and whom would *you* want to help in meeting theirs? Most of a good life boils down to that kind of everyday collaboration and kindness. We don't need to wait for the perfect intentional community to solve for basic needs. Mostly, we just need good relationships with the people around us. That's the power of local. It's where we have the ability to get things done and make a difference, immediately and with the least effort.

Local is also the place where a great deal of power is concentrated in individuals and small organizations. In traditional cultures every village or neighborhood has the equivalent of a ward boss or godfather type, known variously as a padrone, alcalde, cacique, or a host of other terms in hundreds of languages. This is the person the villagers see to make something happen. When it's time to enlarge a house, tap into the irrigation canal, or change jobs, it's the padrone who smooths the way or throws up walls. In industrial cultures these positions are often held by zoning committees and permitting officials, but their members are still people who have personal interests; remember the story of City Repair and the DOT official who went out of his way to offer them a permit for a block party.

These power-holders can emerge in idiosyncratic ways. In older towns I've seen remarkable and quirky influence wielded by the historic-building committee. What seems like a casual drop-in for tea by a few members of this group may instead dictate whether you are allowed to replace a porch railing or paint your house a slightly brighter color. This is how social systems work. Whatever the local culture, a key to implementing innovative and regenerative projects is identifying the people in these positions and building alliances with them.

Local power is also the realm of the small nonprofit, church, and civic association. A handful of people, properly organized, can drive enormous changes in a city's dynamics. I'll offer yet another example from Portland, Oregon. A group of water-conservation enthusiasts, frustrated at the illegal status of graywater reuse in the city and state, formed an organization called Recode. Although many in the group were young, among them they had built solid relationships with a number of local officials, business leaders, and other key people in the politics of the area. Recode pooled their respective connections to gather together relevant stakeholders, such as health officials, state legislature staff, the plumbing board, and developers. To the surprise of all, everyone at the meeting supported graywater use. So, everyone wondered, what was up? A state legislature staffer in attendance zeroed in on the main obstacle: There was no provision in the state codes for graywater. Legally, all of Oregon's water fell into one of two categories, potable water or sewage. Since graywater was not potable, it had to be considered sewage. The staffer told them, "So, all we need to do is create a third water category, graywater." They drafted a resolution doing that, got it to their state representative, and it passed at the next legislative session. After three subsequent years of bureaucratic wrangling and gentle pressure from Recode, graywater use became legal in Oregon. Recode then tackled urban composting toilets as their next target for legalization.

Local policy shaping need not be that formal. A community organizer I knew in Seattle lived on a houseboat, and across the canal from her sat a small shipyard that left its floodlights on all night, eliminating the night sky as well as sound sleep for a score of houseboat families. She called the shipyard about it and was told, "You're the only one who has complained, so unless we hear from a lot more people about it, we don't feel the need to do anything." To a community organizer like she was, this was an irresistible siren song. She left the shipyard's number with a dozen or so of her neighbors, and within the week the shipyard went dark every day at the close of work.

Local successes like these, using the ecological principle of "growing by chunking," can provide models for regional, national, and even global change. The Transition movement, for example, was birthed during a permaculture design course as a class project when a design team drew up a theoretical energy descent plan for the town of Totnes, England. Much like the Beacon Hill Food Forest mentioned in chapter five, the possibilities suggested by this classroom exercise inspired some of the students to take the next step toward making it a reality. From this humble and very local beginning in 2004, the Transition Network had spread by 2013 to over 1,100 communities in 43 countries.[1] This is a familiar pattern for many innovative organizations: Build a successful project in one location, garner regional attention from the success and impact, and be replicated, either by franchise or imitator, in first a few, then dozens or hundreds of other communities. This model of starting small also spares the originators the anxiety and potential for grand failure that taking on a giant project might generate. I'm sure that many founders of now-global organizations would have been petrified by the notion of beginning at the national or world scale.

## TYING CITIES TO THEIR REGIONS

The next turn of the spiral expands to the regional scale. Again, "region" can mean many things, but since we are using our permaculture lenses, we're usually speaking of the bioregion, an area that in the United States ranges in size from a portion of a state to several states. Unless you live in a small country, your bioregion is smaller than a nation-state, but it's usually bigger than what you'd call local. My bioregion, according to the official list for California, is the Bay/Delta region, ranging from about 80 miles north of San Francisco to the same distance south.[2] That's larger than local. The average person will be interested in and affected by what goes on in his bioregion, but unless he wields unusual clout, he won't have a lot of influence at the bioregional level. However, people and groups can band together to shape bioregional policies. Alliances are powerful tools for generating leverage.

In the technical definition of a bioregion, the plants and animals have qualities that distinguish them from nearby bioregions. I've noticed, though, that the people do, too. Living an hour north of San Francisco, I identify more with the grape growers in Napa and the gritty urbanites in Oakland than I do with the industrial food-producing communities in the Central Valley, good folks that they are, and I'm sure the feeling is mutual. Hence the bioregion is a natural social as well as ecological unit.

You'll notice that the spiral doesn't extend to the national or global scale. My own sense of this is that if enough of us get our acts together on the personal, local, and regional levels—meeting our needs at each scale while preserving ecosystem and social health—the national and global scales will take care of themselves.

But there's another reason that it doesn't make sense, from the point of view of this book, for us as permaculture designers to solve for needs at the national or global scale, and it takes us back to the importance of cities. A city and its region bond into a natural economic and social unit, one that is more resilient, older, and likely to last longer than the nation-state. Various nation-states that claimed Rome, Istanbul, and Mexico City have come and gone many times, while the cities themselves remain almost timeless.

The relationship—that word again—between a city and its region is what shapes the well-being of the people living in both, and it determines how the inhabitants treat the land as well. To unpack that, we return to the first chapter's heroine, Jane Jacobs. Two decades after *The Death and Life of Great American Cities*, she wrote *Cities and the Wealth of Nations*, where she developed the concept of the city region. Jacobs makes a strong claim: the source of economic, social, and cultural vigor is a dynamic city, not the national economy or abundant natural resources. This also cycles us to our opening theme of cities as centers of innovation and vitality.

Jacobs overturns yet another economic myth (it seems that we are fed quite a few of those, including the myths of barter and asset-backed currencies, as we've seen). We're told that most cities develop because the underlying farmland is rich enough to spawn a large population. Not true, Jacobs says: For at least the last few millennia, it's the metropolis that develops the farms. A look at the cities built over the last 2,500 years or so shows that this is usually the case. The land surrounding Rome, London, Paris, Nairobi, Rio de Janeiro, Tokyo, New York, and many other metropolises isn't particularly fertile. The biggest cities don't correlate with the best farmland. And the innovations that boosted farm productivity—the mechanized reaper, the cotton gin, the tractor, the Haber-Bosch process, and most of the rest—originated in cities. The prosperity (and, I will add, the ecological health) of what Jacobs calls the hinterlands, the rural region around a city, depends on the prosperity of that city.

An energetic city builds a city region via a dynamic give-and-take. As a city's economy and culture flourish, that wealth spreads to the region around it. Often cities are portrayed as sucking the fertility, products, and people out of the land around them, but Jacobs shows that those are the effects of a moribund or otherwise corrupted city. A healthy one does just the opposite. Farmers thrive from demand for their produce and can purchase productivity-increasing equipment from the city. Farmworkers made jobless by those machines find better-paying work, as well as education and opportunity they'd never get on the farm, in the vigorous nearby urban economy. The area's small towns swell with affluent commuters and the local businesses they support; capital and tax revenue fund infrastructure, public services, and cultural centers both in town and out; trade, tourism, and weekend getaways from the city bring more income to the increasingly attractive hinterlands.

That's all fine, you might say; it's just a novel perspective on the classic development pattern. But if it's true, we haven't gained anything unless we know what makes a city thrive. Jacobs says that the initial engine of urban prosperity is *import replacement*. As one of several examples of this, she describes Japan's emergence as an industrial power. At the dawn of the twentieth century, the Japanese found that bicycles were ideal transportation; no feed nor fuel needed, no manure, no noise, easy to keep running, and no infrastructure revision required to support their use. Tokyo imported them by the thousands. Bicycle repair shops soon sprang up, their parts supplied by cannibalizing broken bikes. Since the bottleneck here was the availability of parts, local artisans began fabricating the most needed ones. This earned them the skills to build more complex bicycle hardware, which they did, and soon they were making entire bicycles at lower cost than the imports. They prospered, and the bike riders got a good product cheaply. When the Tokyo market was saturated, the makers began exporting bikes to other cities, which started the cycle anew in those towns.

In this way meeting a need—and eventually a set of needs—locally, via import replacement and all the skills it fosters, can build a healthy city region. We're back, then, to the permaculture method of matching needs to resources and how it creates resilient designs. There are important feedback loops in the process as well—in this case the supply of parts and the rise of local skills—that drive the upward dynamic. This cycle creates not only jobs but also innovation and expertise as the residents learn to find solutions and become skilled. From that experience they see how to develop strategies for solving similar problems. If you can manufacture bicycles, you can manufacture cars and farm equipment, and, as the Japanese quickly learned, it's not a long leap to fabricating electronics. Another result of this process is the creation of real wealth, because money and raw materials that once drained to a far-off exporter now are caught and stored in the local economy.

Although the industrial model that I'm using to illustrate the idea of city regions isn't a sustainable one, the principles and logic behind Jacobs's argument hold true for economies in general. Nations that lack healthy cities, no matter how rich their resource base, tend to be poor (and their resources are often exploited by others). Although globalization—which may be a brief bubble inflated by cheap fuel—has linked producing areas to far-off markets, the prosperity and tenor of rural regions still tend to mirror those of the nearest major city. The collapse of Detroit took a good bit of Michigan down with it. The vitality of Portland and San Francisco spreads far into its exurbs and surrounding farms. The nascent return of Pittsburgh and Philadelphia at this writing is lifting Pennsylvania and parts of surrounding states.

Readers of this book can draw any number of lessons from Jacobs's notion of the city region. For one, initiatives that beef up the connections and flows between a city and its region will benefit both. An example is farm-to-table programs of all sorts, especially the kind that don't just pull food in from farms but also draw urbanites out to the farms for recreation, education, and time with nature. Many permaculture centers and institutes as well as schools and nonprofits have both urban and rural bases. A useful leverage point for rural as well as urban prosperity is to create flows of people, skills, goods, and other matches between real urban and rural needs and yields.

The idea of the city region also highlights the value of local currencies. When Houston is booming while the dollar is weak, that city's economy is dragged down because workers are being paid in undervalued currency. If Buffalo is in the doldrums when the United States is prosperous, the city gets precisely the wrong feedback for its economy, because imports are cheap, hurting the region's industry. Local currencies are linked more accurately to regional conditions, giving smarter feedback.

In addition, healthy city regions rarely suffer because of overspecialization. Meeting the many needs of its residents fosters a diverse economy. When Seattle crashed in the 1970s during a rough period for Boeing or when Detroit tanked, it wasn't because the region's people stopped needing goods and income. These were one-industry towns that had neglected to link their prosperity to their regions. A city that ignores its connection to its region, to the land and people around it, will be unbalanced and unable to meet the needs of its people during busts and become rapidly gentrified and unaffordable during booms.

Some of the needs on the permaculture flower are people based and others are land based. Connecting people resources to land resources is another way to damp volatility. The flows of resources and skills to and from a city and its region are the alternating current that powers the prosperity of both.

In the course of writing this book, I've gotten the sense that the city with its region is a natural political, economic, and cultural unit. Most nation-states are Johnny-come-latelies, grown up from city-states or carved off from arbitrary colonies. Nation-states didn't become widespread until the fossil-fuel era enabled rapid communication and mass-scale movement of people and goods over great distances. The founding of the nations of Germany, Italy, and many others; the consolidation of the continental United States; and the overall hardening of national borders didn't occur until the nineteenth century.[3] The city-state has been the predominant political unit for at least 6,000 years, from early Thebes, Babylon, Athens, Tikal, and Rome to nineteenth-century Bangkok and the free city of Frankfurt. Nationalism didn't emerge as a popular rallying cry until two centuries ago. I suspect that unless we find a cheap replacement for fossil energy, the large nation-states will fragment, and the city-state will return.

## THE ROAD TO A REGENERATIVE CULTURE

Vibrant cities are not designed. As we've seen from Jacobs and the other urban thinkers I've cited, top-down planning is not the route to healthy cities. This not-surprising fact is good news for permaculturists, because it means that we don't have to intervene at the whole-city level—we don't have to become the benign equivalent of Robert Moses or Le Corbusier—to have a positive effect on our hometowns. Cities self-organize, and much of that organization happens at the personal, neighborhood, and community levels. As the examples throughout this book illustrate, creating the conditions for community, prosperity, and resilience to grow can start with something as simple as a neighborhood yard sale, potluck, or seed exchange. In fact, it's smarter to start with something small. The founding members of Portland's Recode had each mastered much less daunting projects, from social work to organizing river cleanup days, which let them gain experience, connections, and momentum before scaling up to drafting legislative bills and expanding sanitation codes.

Many of us live in a love-hate relationship with cities. They are often the places where the worst aspects of human nature manifest, in crime, corruption, and pollution or as simple unkindness and apathy. But they are also where we are the most creative, collaborative, visionary, artistic, and productive. Cities are a leverage point, the place where 50 percent of humanity now lives, and that number is rising. Cities aren't going away any time soon. If we include the suburbs and towns where many of the basic patterns of urban life still apply—traffic, commerce, public space, workplaces—the vast majority of people live there. It's where the work of sustainability, of moving human beings onto the path of a regenerative culture, needs to be done. The demands and depletion emanating from cities lie at the heart of the planet's serious ecological crises. There's plenty to be done to rebalance the relationship between cities and the wild and once-wild places that provide for them, these places that keep Gaia alive.

Wild nature could take care of itself without our help; the mostly urbanized human population is the problem. We can plant trees and fund restoration projects all we want, but if the ethics and actions of those who live in our cities and towns are not grounded in a regenerative paradigm, it won't matter. But our mostly urbanized humanity is also the solution. We have the tools, we have shown we have the spirit, and now we're seeing that we have the need.

One book can offer only a glimpse of what truly resilient cities, people, and cultures can look like. I leave it to you to add to the knowledge, the work, and the stories of a few of the many people who are finding their way to that new paradigm, of meeting human needs while preserving ecological and social health.

# Acknowledgments

Every book, even one with a single name on its spine, is a collaboration. I am indebted to my many mentors, colleagues, friends, students, and a large group of helpful strangers for providing me with ideas, critical thoughts, stories, and materials. I am sure that I am forgetting an embarrassingly large number of those who have aided me in this project, but any list of those to be thanked must begin with Bill Mollison and David Holmgren, the originators of the permaculture concept, for their writings and for helpful conversation. Larry Santoyo, Kevin Bayuk, and Michael Becker have provided me with key concepts and have immensely broadened my understanding of urban permaculture and invisible structures, and I apologize to them all for writing this book before they got around to finishing each of theirs. I'm also grateful to Paul Stamets for many walks and talks in the woods; the sheer presence of that kind of genius repeatedly inspired me to be smarter than I am.

A special thanks goes to Erik Ohlsen for bringing me to Sonoma County and for all the support in so many ways that he has given, and immense gratitude to Patricia Flora and Amazing Grace for a stimulating and magical setting in which to write, and to Redheart for financial support and many key insights. Brock Dolman, Kendall Dunnigan, Vanessa Carter, Adam Wolpert and the rest of the OAEC crew, Penny Livingston, James Stark, Pandora Thomas, Peter Bane, Keith Johnson, Jenny Pell, Mark Lakeman, Trathen Heckman, Bill Wilson, Robert Waldrop, and other permaculturists far and wide have shared their ideas and wisdom with me. To all those who provided me with photos, I am deeply grateful. And I am keenly aware of the unpayable debt I owe to the many urban theorists, planners, and activists whose writings and work gave me broad shoulders on which to stand.

One of the joys of where I live is the incredible pool of talented people in the Bay Area, and not least are the members of the Wednesday Back Alley Group, whose conversations and readings on complex adaptive systems, emergence, and similar themes have enriched my life and, I hope, this book.

My thanks to Dr. Earl Herr, Rosemary Devitt, Dr. Phil Harriman, Paul Nicholson, Ken Ackerman, and Dr. Richard Ely for their first-rate intellectual engagement and friendship.

To my editor, Makenna Goodman, as well as Patricia Stone, Alice Colwell, and all the staff at Chelsea Green, I thank you not just for the labor and talent you have put into this book but for your trust and belief in my work and for reducing the number of hoops I needed to jump through in the process. The deep experience and solid presence of my literary agent, Natasha Kern, was invaluable in keeping me calm and confident through the uncertainties of the writing and publishing process. And my unending gratitude as always, to my muse and soulmate, Kiel, for her love, patience, reassurance, and steadfastness once again during my disappearances into and unpredictable reemergences from this book.

# Notes


## Introduction: Looking at Cities through a Permaculture Lens

1. Jane Jacobs, *The Death and Life of Great American Cities* (New York: Random House, 1961).
2. "Permaculture Flower," http://permacultureprinciples.com/flower/, accessed March 14, 2014.

## Chapter 1: The Surprisingly Green City

1. David Owen, "Green Manhattan," *New Yorker*, October 18, 2004, 111–123.
2. United Nations Population Fund, *State of World Population 2007: Unleashing the Potential of Urban Growth*, http://www.unfpa.org/swp/2007/english/introduction.html, accessed June 30, 2014.
3. Gwendolyn Leick, *Mesopotamia: The Invention of the City* (New York: Penguin, 2002).
4. Edward L. Glaeser, "Are Cities Dying?" *Journal of Economic Perspectives* 12 (1998): 139–160.
5. Fernand Braudel, *Civilization and Capitalism, 15th–18th Century*, vol. 1, *The Structures of Everyday Life* (New York: Harper & Row, 1979); Lewis Mumford, *The City in History* (San Diego, Calif.: Harcourt, 1961).
6. Kathleen M. Kenyon, *Digging Up Jericho: The Results of the Jericho Excavations, 1952–1956* (New York: Praeger, 1957).
7. J. Wolf, M. S. van Wijk, et al., "Urban and Peri-Urban Agricultural Production in Beijing Municipality and Its Impact on Water Quality," *Environment and Urbanization* 15 (2003): 141.
8. K. Bettencourt, J. Lobo, D. Helbing, C. Kühnert, and G. B. West, "Growth, Innovation, Scaling, and the Pace of Life in Cities," *Proceedings of the National Academy of Sciences* 104 (2007): 7301.
9. J. Portugali, J. Meyer, H. Stolk, and E. Tan, eds., *Complexity Theories of Cities Have Come of Age* (Berlin: Springer, 2012).
10. David Harvey, *The Condition of Postmodernity: An Enquiry into the Origins of Social Change* (Oxford: Blackwell, 1989).
11. Le Corbusier, *The Radiant City: Elements of a Doctrine of Urbanism to Be Used as the Basis of Our Machine-Age Civilization*, trans. Pamela Knight (New York: Orion Press, 1964).
12. Madhu Sarin, "Chandigarh as a Place to Live," in *The Open Hand: Essays on Le Corbusier*, ed. Russell Walden (Cambridge, Mass.: MIT Press, 1977).

13. Christopher Alexander, "A City Is Not a Tree," *Architectural Forum* 122, no. 1 and 2 (1965): 58–61 (pt. 1) and 58–62 (pt. 2).
14. Christopher Alexander, Sara Ishikawa, and Murray Silverstein, *A Pattern Language* (New York: Oxford University Press, 1977).
15. Stefan Lämmer and Dirk Helbing, "Self-Stabilizing Decentralized Signal Control of Realistic, Saturated Network Traffic," Santa Fe Institute Working Paper #10-09-019, 2010, http://www.santafe.edu/media/workingpapers/10-09-019.pdf.

### Chapter 2: Permaculture Design with an Urban Twist

1. Richard Polenberg, ed., *In the Matter of J. Robert Oppenheimer: The Security Clearance Hearing* (Ithaca, N.Y.: Cornell University Press, 2001).
2. Henry Mintzberg, "Patterns in Strategy Formation," *Management Science* 24 (1978): 934–948.
3. Richard Rumelt, *Good Strategy, Bad Strategy: The Difference and Why It Matters* (New York: Random House, 2011).

### Chapter 3: Designing the Urban Home Garden

1. Elena Lazos Chavero and Maria Alvarez-Buylla Roces, "Ethnobotany in a Tropical-Humid Region: The Home Gardens of Balzapote, Veracruz, Mexico," *Journal of Ethnobiology* 8, 1 (1988): 45–79.
2. William Hawk, "Expenditures of Urban and Rural Households in 2011," U. S. Bureau of Labor and Statistics *Beyond the Numbers* 2, 5 (2013). http://www.bls.gov/opub/btn/volume-2/expenditures-of-urban-and-rural-households-in-2011.htm#_edn2.

### Chapter 4: Techniques for the Urban Home Garden

1. Centers for Disease Control and Prevention, "Low Level Lead Exposure Harms Children: A Renewed Call for Primary Prevention," 2012, http://www.cdc.gov/nceh/lead/ACCLPP/Final_Document_010412.pdf.
2. Carl J. Rosen, "Lead in the Home Garden and Urban Soil Environment," University of Minnesota Extension, 2010, http://www.extension.umn.edu/distribution/horticulture/DG2543.html.
3. Susan Trulove, "Iron-Rich Soil Can Help Remove Lead; Manganese Also Important," *Science from Virginia Tech*, 2002, http://www.research.vt.edu/resmag/sciencecol/iron.html.
4. H. Heinonen-Tanski, S. K. Pradhan, and P. Karinen, "Sustainable Sanitation—A Cost-Effective Tool to Improve Plant Yields and the Environment," *Sustainability* 2 (2010): 341–353.
5. Carol Deppe, *The Resilient Gardener: Food Production and Self-Reliance in Uncertain Times* (White River Junction, Vt: Chelsea Green, 2010).
6. Audrey Ensminger et al., eds., *Foods and Nutrition Encyclopedia*, vol. 1, 2nd ed. (Boca Raton, Fla.: CRC Press, 1993).
7. Larry M. Geno and Barbara J. Geno, *Polyculture Production: Principles, Benefits and Risks of Multiple Cropping Land Management Systems for Australia* (Barton, Australia: Rural Industries Research and Development Corporation, 2001).
8. Josep A. Gari, "Biodiversity and Indigenous Agroecology in Amazonia," *Etnoecológica* 5 (2001): 21–37; M.

A. Altieri, *Agroecology: The Science of Sustainable Agriculture*, 2nd ed. (Boulder, Colo.: Westview Press, 1995).

9. V. D. P. Risso, "Diversity, Productivity, and Stability in Perennial Polycultures Used for Grain, Forage, and Biomass Production," PhD diss., University of Iowa, 2008.
10. Charles Darwin, *The Formation of Vegetable Mould through the Action of Worms with Observations on their Habits*. (London: John Murray, 1881).
11. Clive A. Edwards and P. J. Bohlen, *Biology and Ecology of Earthworms*, 3rd ed. (London: Chapman & Hall, 1996).
12. Diane Thomson, "Competitive Interactions between the Invasive European Honey Bee and Native Bumble Bees," *Ecology* 85 (2004): 458–470.

## Chapter 5: Strategies for Gardening in Community

1. Christian Peters et al., "Mapping Potential Foodsheds in New York State," *Renewable Agriculture and Food Systems* 21 (2009): 72–84.
2. Christopher Weber and H. Scott Matthews, "Food Miles and the Relative Climate Impacts of Food Choices in the United States," *Environmental Science and Technology* 42 (2008): 3508–3513.
3. USDA Agricultural Marketing Service, "Farmers Markets and Local Food Marketing," http://www.ams.usda.gov/AMSv1.0/ams.fetchTemplateData.do?template=TemplateS&leftNav=WholesaleandFarmersMarkets&page=WFMFarmersMarketGrowth&description=Farmers+Market+Growth, accessed June 12, 2014.
4. Jake Claro, *Vermont Farmers' Markets and Grocery Stores: A Price Comparison*, Northeast Organic Farming Association of Vermont, 2011, http://nofavt.org/sites/default/files/NOFA%20Price%20Study.pdf, accessed June 12, 2014; Rich Pirog and Nick McCann, *Is Local Food More Expensive? A Consumer Price Perspective on Local and Non-local Foods Purchased in Iowa*, Leopold Center for Sustainable Agriculture, 2009, http://www.leopold.iastate.edu/pubs-and-papers/2009-12-local-food-more-expensive, accessed June 12, 2014.
5. Steven McFadden, "The History of Community Supported Agriculture," pt. 1: "Community Farms in the 21st Century: Poised for Another Wave of Growth?" Rodale Institute, 2004, http://www.newfarm.org/features/0104/csa-history/part1.shtml.
6. Timothy Woods et al., "Survey of Community Supported Agriculture Producers," Cooperative Extension Service, College of Agriculture, University of Kentucky, 2009, http://www.uky.edu/Ag/NewCrops/csareport.pdf.
7. US Department of Agriculture, *2007 Census of Agriculture*, 606, table 44.
8. C. Smith-Spangler, "Are Organic Foods Safer or Healthier Than Conventional Alternatives? A Systematic Review," *Annals of Internal Medicine* 157, 5 (2012): 348–366.
9. Philip H. Howard, "Consolidation in the North American Organic Food Processing Sector, 1997 to 2007," *International Journal of Sociology of Agriculture and Food* 16, 1 (2009): 13–30, https://www.msu.edu/~howardp//organicindustry.html.
10. Kari Hamerschlag, *Meat Eater's Guide to Climate Change and Health*, Environmental Working Group, 2011, http://ewg.org/meateatersguide .

11. Comments on Robert Mellinger, "Nation's Largest Public Food Forest Takes Root on Beacon Hill," *Crosscut.com*, February 16, 2012, http://crosscut.com/2012/02/16/agriculture/21892/Nations-largest-public-Food-Forest-takes-root-on-B/?pagejump=1.

## Chapter 6: Water Wisdom, Metropolitan Style

1. Los Angeles Department of Water and Power website, https://www.ladwp.com, accessed November 8, 2013.
2. Glenn A. Richard, "New York City Water Supply," Stonybrook Earth Science Educational Resource Center, http://www.eserc.stonybrook.edu/cen514/info/nyc/watersupply.html, accessed November 9, 2013.
3. Jessica Pupovac, "Bursting Chicago's Water Bubble," *Great Lakes Echo*, April 21, 2010; City of Chicago Department of Water Management, "What We Do: Water Supply," http://www.cityofchicago.org/city/en/depts/water/provdrs/supply.html, accessed November 8, 2013.
4. US Geological Survey, *Ground Water Atlas of the United States : Alabama, Florida, Georgia, South Carolina*, publication HA 730-G, http://pubs.usgs.gov/ha/ha730/ch_g/G-text4.html.
5. Michael M. O'Shaughnessy, *Hetch Hetchy Water Supply*, Bureau of Engineering of the Department of Public Works, City and County of San Francisco, California, 1925, http://archive.org/stream/hetchhetchywater00osha/hetchhetchywater00osha_djvu.txt.
6. Brad Lancaster, *Rainwater Harvesting for Drylands and Beyond*, vol. 1 (Tucson, Ariz.: Rainsource Press, 2006).
7. Benjamin D. Inskeep and Shahzeen Z. Attari, "The Water Short List: The Most Effective Actions U.S. Households Can Take to Curb Water Use," *Environment: Science and Policy for Sustainable Development*, July–August 2014, http://www.environmentmagazine.org/Archives/Back%20Issues/2014/July-August%202014/water_full.html.
8. Center for Rainwater Harvesting, "Selection and Consequences of Roofing and Gutter Materials for Rainwater Harvesting," 2006, http://www.thecenterforrainwaterharvesting.org/2_roof_gutters2.htm.
9. Peter J. Coombes et al., "Rainwater Quality from Roofs, Tanks and Hot Water Systems at Figtree Place," 3rd International Hydrology and Water Resources Symposium of the Institution of Engineers, Perth, Australia, Hydro 2000 Proceedings, vol. 1.

## Chapter 7: Energy Solutions for Homes and Communities

1. Reiner Kümmel, *The Second Law of Economics: Energy, Entropy, and the Origins of Wealth* (New York: Springer, 2011).
2. Luis De Sousa, "What Is a Human Being Worth (in Terms of Energy)?" *The Oil Drum* (blog), July 20, 2008, http://www.theoildrum.com/node/4315, accessed January 2, 2014.
3. Roger Andrews, "How Cheap Is 'Cheap' Oil?" *Energy Matters* (blog), August 20, 2914, http://euanmearns.com/how-cheap-is-cheap-oil/, accessed August 30, 2014.
4. K. Klein Goldewijk and G. van Drecht, "History Database of the Global Environment," in A. F. Bouwman, T. Kram,

and K. Klein Goldewijk, eds., *Integrated Modeling of Global Environmental Change: An Overview of IMAGE 2.4* (Bilthoven: Netherlands Environmental Assessment Agency, 2006).

5. Vaclav Smil, *Enriching the Earth: Fritz Haber, Carl Bosch, and the Transformation of World Food Production* (Cambridge, Mass.: MIT Press, 2004).
6. Paul Ehrlich, *The Population Bomb* (Cutchogue, N.Y.: Buccaneer Books, 1968).
7. G. T. Miller, *Living in the Environment*, 12th ed. (Belmont, Calif.: Wadsworth Thomson Learning, 2002).
8. Amory Lovins, "Energy Strategy: The Road Not Taken?" *Foreign Affairs*, October 1, 1976, http://www.foreignaffairs.com/articles/26604/amory-b-lovins/energy-strategy-the-road-not-taken.
9. Vaclav Smil, *Energy in Nature and Society: General Energetics of Complex Systems* (Cambridge, Mass.: MIT Press, 2008).
10. H. T. Odum, M. T. Brown, and S. Brandt-Williams, "Introduction and Global Budget," *Handbook of Emergy Evaluation*, folio 1 (Gainesville: Center for Environmental Policy, Environmental Engineering Sciences, University of Florida, 2000), http://www.epa.gov/aed/html/collaboration/emergycourse/presentations/Folio1.pdf.
11. Kazuhisa Miyamoto, ed., "Renewable Biological Systems for Alternative Sustainable Energy Production," Food and Agriculture Organization of the United Nations Agricultural Services Bulletin 128, 1997, http://www.fao.org/docrep/w7241e/w7241e00.htm.
12. Christopher Weber and H. Scott Matthews, "Food Miles and the Relative Climate Impacts of Food Choices in the United States," *Environmental Science and Technology* 42 (2008): 3508–3513.
13. G. Tyler Miller, *Energetics, Kinetics, and Life: An Ecological Approach* (Belmont, Calif.: Wadsworth, 1971), 293.
14. Howard T. Odum, *Environmental Accounting: Emergy and Environmental Decision Making* (New York: John Wiley and Sons, 1996).
15. International Society for the Advancement of Emergy Research, "Transformities List," http://emergydatabase.org/transformities-view/all, accessed January 30, 2014.
16. Mason Inman, "The True Cost of Fossil Fuels," *Scientific American* 308, 4 (2013): 59–61.
17. Pedro Prieto and Charles A. S. Hall, *Spain's Photovoltaic Revolution: The Energy Return on Investment* (New York: Springer, 2013).
18. E. Mearns, *The Global Energy Crisis and Its Role in the Pending Collapse of the Global Economy* (Aberdeen, Scotland: Royal Society of Chemists, 2008).
19. US Energy Information Administration, *Residential Energy Consumption Survey 2009*, http://www.eia.gov/consumption/residential/, accessed January 30, 2014.
20. Bill McNary and Chip Berry, "How Americans Are Using Energy in Homes Today," *ACEEE Summer Study on Energy Efficiency in Buildings*, 2012, http://www.aceee.org/files/proceedings/2012/data/papers/0193-000024.pdf, accessed January 30, 2014.
21. The search phrase "passive solar roof overhang design" will bring up several calculator tools on the Internet. As of February 2015, one can be found at http://www.borstengineeringconstruction.com/

Passive_Solar_Roof_Overhang_Design_Calculator.html.
22. Two excellent resources for energy-saving tips are the US Department of Energy handbook, *Energy Savers*, available at http://energy.gov/sites/prod/files/2013/06/f2/energy_savers.pdf, and a book by Kelly Coyne and Erik Knutzen, *The Urban Homestead: Your Guide to Self-Sufficient Living in the Heart of the City*, rev. ed. (Port Townsend, Wash.: Process, 2010).

## Chapter 8: Livelihood, Real Wealth, and Becoming Valuable

1. B. Malinowski, "Kula: The Circulating Exchange of Valuables in the Archipelagoes of Eastern New Guinea," *Man* 20 (1920): 97–105; Marcel Mauss, *The Gift: Forms and Functions of Exchange in Archaic Societies*, trans. Ian Cunnison (London: Cohen & West, 1966).
2. Peter Bernholz, *Monetary Regimes and Inflation: History, Economic and Political Relationships* (Cheltenham, UK: Edward Elgar Publishing, 2003).
3. Board of Governors of the Federal Reserve System, "Reserve Requirements," http://www.federalreserve.gov/monetarypolicy/reservereq.htm, accessed February 6, 2014.
4. M. McLeay, A. Radia, and R. Thomas, "Money Creation in the Modern Economy," *Bank of England Quarterly Bulletin* 54 (2014): 14–27, http://www.bankofengland.co.uk/publications/Documents/quarterlybulletin/2014/qb14q1prereleasemoneycreation.pdf.
5. "List of Community Currencies in the United States," *Wikipedia*, http://en.wikipedia.org/wiki/List_of_community_currencies_in_the_United_States, accessed April 5, 2014.
6. Joseph A. Tainter, *The Collapse of Complex Societies* (Cambridge: Cambridge University Press, 1988).
7. Diana Leafe Christian, *Creating a Life Together: Practical Tools to Grow Ecovillages and Intentional Communities* (Gabriola Island, BC: New Society, 2003).
8. Donella Meadows, *Thinking in Systems: A Primer* (White River Junction, Vt: Chelsea Green, 2008).
9. Larry Santoyo, personal communication.
10. Richard Heinberg, "Fifty Million Farmers," *Energy Bulletin*, November 17, 2006, http://www.resilience.org/stories/2006-11-17/fifty-million-farmers, accessed April 17, 2014.
11. Don Lancaster, *The Incredible Secret Money Machine* (Carmel, Ind.: Howard W. Sams, 1978).
12. Aspen Institute, *At the Five-Year Mark: Outcomes Reported by U.S. Microenterprise Clients*, March 2010, http://fieldus.org/Publications/MultiyearDataRpt09.pdf.

## Chapter 9: Placemaking and the Empowered Community

1. Kevin Lynch, *The Image of the City* (Cambridge, Mass.: Joint Center for Urban Studies of MIT and Harvard University, 1960).
2. William H. Whyte, *The Social Life of Small Urban Spaces* (New York: Project for Public Spaces, 1980).
3. http://www.pps.org.
4. John A. Buck and Gerard Endenburg, *The Creative Forces of Self-Organization* (Rotterdam, Netherlands: Sociocratic Center, 2012), http://www.governance

alive.com/wp-content/uploads/2009/12/CreativeForces_9-2012_web.pdf, accessed June 7, 2014.

5. Kenneth Boulding, *Three Faces of Power* (Newbury Park, Calif.: Sage, 1989).
6. J. R. P. French and B. Raven, "The Bases of Social Power," in D. Cartwright and A. Zander, *Group Dynamics* (New York: Harper & Row, 1989); James Hillman, *Kinds of Power* (New York: Doubleday, 1995); Steven Lukes, *Power: A Radical View*, 2nd ed. (New York: Palgrave Macmillan, 2004), and www.powercube.net.
7. Roger Fisher and William L. Ury, *Getting to Yes*, 3rd ed. (New York: Penguin, 2011).
8. John Paul Lederach, *The Little Book of Conflict Transformation* (Intercourse, Pa.: Good Books, 2003). An abridged version of the book can be found at http://www.beyondintractability.org/essay/transformation.
9. Sam Kaner et al., *Facilitator's Guide to Participatory Decision-Making* (San Francisco: Jossey-Bass, 2007); Roger Schwarz, *The Skilled Facilitator* (San Francisco: Jossey-Bass, 2002); David Kantor, *Reading the Room* (San Francisco: Jossey-Bass, 2012).
10. John Michael Greer, "In a Time of Limits," *The Archdruid Report* (blog), February 20, 2013, http://thearchdruidreport.blogspot.com/2013/02/in-time-of-limits.html, accessed June 14, 2014.

## Chapter 10: Tools for Designing Resilient Cities

1. "Transition Initiatives Map," https://www.transitionnetwork.org/initiatives/map, accessed February 9, 2015.
2. http://ceres.ca.gov/geo_area/bioregions/Bay_Delta/about.html, accessed February 23, 2015.
3. E. J. Hobsbawn, *Nations and Nationalism since 1780: Programme, Myth, Reality* (Cambridge: Cambridge University Press, 2012).

# Index

Note: ci refers to color insert pages

# About the Author

PHOTOGRAPH BY ART VARGAS

Toby Hemenway is the author of the first major North American book on permaculture, *Gaia's Garden: A Guide to Home-Scale Permaculture*. After obtaining a degree in biology from Tufts University, Hemenway worked for many years as a researcher in genetics and immunology, first in academic laboratories at Harvard and the University of Washington in Seattle, and then at Immunex, a major medical biotech company. At about the time he was growing dissatisfied with the direction biotechnology was taking, he discovered permaculture, a design approach based on ecological principles that creates sustainable landscapes, homes, and workplaces. A career change followed, and Hemenway and his wife spent ten years creating a rural permaculture site in southern Oregon. He was the editor of *Permaculture Activist*, a journal of ecological design and sustainable culture, from 1999 to 2004. He teaches permaculture and consults and lectures on ecological design throughout the country. His writing has appeared in magazines such as *Whole Earth Review*, *Natural Home*, and *American Gardener*. He lives in Sebastopol, California. Visit his web site at www.patternliteracy.com.